Computer Vision in Smart Agriculture and Crop Management

Scrivener Publishing
100 Cummings Center, Suite 541J
Beverly, MA 01915-6106

Publishers at Scrivener
Martin Scrivener (martin@scrivenerpublishing.com)
Phillip Carmical (pcarmical@scrivenerpublishing.com)

Computer Vision in Smart Agriculture and Crop Management

Edited by

Rajesh Kumar Dhanaraj

Balamurugan Balusamy

Prithi Samuel

Malathy Sathyamoorthy

and

Ali Kashif Bashir

Scrivener Publishing

WILEY

This edition first published 2025 by John Wiley & Sons, Inc., 111 River Street, Hoboken, NJ 07030, USA and Scrivener Publishing LLC, 100 Cummings Center, Suite 541J, Beverly, MA 01915, USA

For more information about Scrivener publications please visit www.scrivenerpublishing.com.

Wiley Global Headquarters
111 River Street, Hoboken, NJ 07030, USA

For details of our global editorial offices, customer services, and more information about Wiley products visit us at www.wiley.com.

Library of Congress Cataloging-in-Publication Data

ISBN 978-1-394-18629-7

Front cover images supplied by Adobe Firefly
Cover design by Russell Richardson

Set in size of 11pt and Minion Pro by Manila Typesetting Company, Makati, Philippines

Printed in the USA

10 9 8 7 6 5 4 3 2 1

Contents

Preface

As editors, we feel privileged to have been asked to edit the 1st edition of *Computer Vision in Smart Agriculture and Crop Management*. The need for a sufficient food supply and societal demands has driven the evolution of smart agriculture. Smart agriculture leverages information and technology to assist farmers, particularly in remote locations. A key component of advancing agricultural automation systems is Computer Vision (CV) technology. By integrating computer vision with other intelligent systems, such as deep learning, it is possible to manage agricultural productivity using massive datasets, addressing persistent agricultural challenges and enhancing the economic, general, and operational efficiency of agricultural automation tools. Automation in agriculture offers significant benefits, including reduced costs, high performance, and increased accuracy, all contributing to sustained growth. Aerial imaging is commonly used during the growing season to monitor crops. High-throughput phenotyping research is expected to provide substantial insights into yield attributes on a large scale, supporting effective crop management decisions.

Machine learning and computer vision techniques help farmers distinguish between fertile soil, natural remedies, and pest control measures. These technologies can also assess the color, size, thickness, and surface texture of crops to detect impurities in agricultural yields and identify contaminated food products. Computer vision-based Artificial Intelligence (AI) models are highly effective for detecting and monitoring plant health. Furthermore, AI aids in developing more accurate seasonal forecasting techniques, which enhance production and operational efficiency. Innovations in computer vision, intelligent systems, and machine learning will eventually lead to the development of advanced remote sensing technologies for detecting and managing plants, weeds, diseases, and pests, thereby supporting sustainable agriculture and farming. The goal of this specialized area is to utilize computer vision and other emerging technologies to create smart agriculture and improve crop surveillance for societal benefit. This book focuses on using deep learning to extract relevant image content features from digital photogrammetry and applying data mining and machine learning techniques to organize these features into clusters for intelligent applications, ultimately achieving smart agriculture and sustainable farming.

Rajesh Kumar Dhanaraj
Pune, India
Balamurugan Balusamy
New Delhi, India
Prithi Samuel
Chennai, India
Malathy Sathyamoorthy
Coimbatore, India
Ali Kashif Bashir
United Kingdom

1

Computer Vision-Based Innovations for Smart Agriculture and Crop Surveillance: Evolution, Trends, and Future Challenges

M. Nalini* **and B. Yoga Bhuvaneswari**

Department of Electronics and Instrumentation Engineering, Sri Sairam Engineering College, Chennai, India

Abstract

The economic prosperity of a nation relies heavily on its agricultural industry. Due to population expansion, frequent climate change, and material restriction, it has become harder to supply the food needs of the present population. Precision agriculture, also frequently referred to as smart agriculture, is a cutting-edge method for addressing the challenges associated with agricultural sustainability. This cutting-edge technology is powered by a system based on computer vision (CV) and artificial intelligence (AI). These AI and CV techniques have made remarkable contributions to the agricultural industry in the areas of plant health detection and monitoring, planting, weeding, harvesting, and modern weather forecast analysis. So many use cases of smart farming have an influence on the whole food supply chain by offering insightful data on the entire agricultural process, easing operational decision-making in real-time, and improving farming methods by adding smart sensors and equipment to the field. The use of CV in agriculture has recently increased. Computer vision has enormous potential to improve the whole functioning of the agricultural industry, from lowering production costs through intelligent automation to increasing output.

Artificial intelligence includes the field of CV. Because of breakthrough technology, machines can now understand and perceive the visual environment in a manner that is comparable to that of humans. Computer vision techniques combined with image capturing from remote cameras provide agriculture-specific, non-contact, and scalable sensing solutions. The CV–AI models have made several

**Corresponding author*: nalini.ei@sairam.edu.in

Rajesh Kumar Dhanaraj, Balamurugan Balusamy, Prithi Samuel, Malathy Sathyamoorthy and Ali Kashif Bashir (eds.) Computer Vision in Smart Agriculture and Crop Management, (1–24) © 2025 Scrivener Publishing LLC

contributions to the agricultural business in areas including harvesting, enhanced weather analysis, weeding, planting, and plant health detection and monitoring. This chapter covers the evolution, trends, and future challenges of smart farming using CV.

Keywords: Computer vision, smart farming, precision agriculture, artificial intelligence, intelligent farming, internet of things-based smart farming, machine learning, precision farming

1.1 Introduction

Precision agriculture is an agricultural management idea built on surveilling, measuring, and reacting to crop variances. By maximizing input returns while protecting resources, precision agriculture research tries to establish a decision-making system for farm management. This type of agriculture has benefited greatly from machine vision, which makes automated alternatives to chores that are often done by hand. Manual processes are usually laborious and vulnerable to mistakes. Accurate, precise, and effective solutions may be provided by machine vision to help agricultural activities. Additionally, machine learning (ML) techniques make it possible to analyze enormous amounts of data swiftly and reliably, opening the door for the adoption of machine vision applications in agriculture. For agricultural production to meet the issues of productivity, environmental impact, food safety, and sustainability, smart farming is crucial. Due to the continually growing world population, a significant increase in food production must be made while simultaneously maintaining high nutritional quality and availability everywhere and safeguarding the natural ecosystems by employing sustainable agricultural practices. Advanced computer vision (CV) algorithms and ML-based artificial intelligence (AI) can continually analyze the data gathered from multiple sources. Artificial intelligence, deep computing, and computational intelligence are all related technologies created for intelligent systems. A larger definition of AI is the development of intelligent computers that can mimic human thought and behavior. Simply expressed, the basic goal of AI is to develop computer systems that are intelligent enough to resemble human intellect. These AI systems need knowledge engineering to function properly.

Machine learning, on the other hand, is a branch of AI that enables computers to learn from collected data without explicit programming. Machine learning makes accurate predictions on new data by using sophisticated algorithms that repeatedly cycle over the massive data set. One dynamic and fascinating area of AI is CV, which replicates the intricate human visual system. The basic objective of CV is to enable computer systems to detect and locate objects in images and videos in a similar way to how people do. Two examples of cutting-edge methods that have considerably enhanced CV are deep learning and neural networks. Deep learning algorithms like the convolutional neural networks (CNNs) and their derivatives are frequently employed in in-plant phenology investigations. For classifying images, CNNs are frequently employed. The main building blocks of CNNs are convolutional layers, and to create a feature map, a window (filter) is utilized to scan the pictures and seek for certain features. Convolution, pooling, and fully connected layers are utilized to construct an entire CNN, which produces precise feature recognition and accurate picture categorization. With little information loss, pooling layers attempt to minimize the dimensionality of the feature map produced by the convolutional layers [12].

1.2 Artificial Intelligence in Agriculture

Artificial intelligence techniques are frequently employed in agricultural industries to solve a unique set of issues and improve production and operating methods. Agriculture can quickly adapt to using AI and ML when reading each sentence of agricultural products and agricultural techniques in the topic. Because it can comprehend, learn, and react to a one-of-a-kind (mostly based on learning) rising efficiency, of particular importance, cognitive computing is positioned as the catalyst in the next great agricultural revolution. With the use of AI, farmers may gain valuable information about their farm's weather, temperature, water consumption, and soil monitoring, which will help them increase their earnings. By discovering which crops they can grow, creating high-quality hybrid seeds, and maximizing resource efficiency, AI technologies aid farmers. Both the quality of harvested products and the accuracy with which they are harvested may be improved with AI [11].

Precision agriculture uses AI technology to assist in the early diagnosis of diseased plants, insect infestations, and nutritional issues in agricultural areas. Artificial intelligence-powered sensors can quickly identify weeds and advise users to apply pesticides. Utilizing this method can save cultivation expenses since herbicides were made more widely known and sprayed throughout the entire field. Farms of any size across the world are using AI technology to operate more effectively and meet global food demand. With the use of AI, forecasting models may be more accurately and effectively learned. These models can predict weather accurately months in advance. The most efficient method for assisting small farmers is seasonal forecasting. Applications of AI are shown in Figure 1.1.

Artificial intelligence is positively transforming Indian agriculture in several ways. The value of AI applications in agriculture is predicted to increase from $852.2 million in 2019 to $238.38 billion in 2030, or a growth of about 25%. Access to markets, inputs, loans, and crop insurance are all improved by technology. A supply chain that is driven by demand may be built with the use of appropriate timing and precise data. Many AI models become inexpensive and accessible by utilizing sensors, phone photos, drones, agricultural weather data, and data on the condition of the soil. Various challenges, such as climate change, population increase, and job issues, are being addressed by AI, which is functioning as a catalyst for improved yield. Nowadays, relatively few individuals are interested in farming, which causes a labor shortage on many farms. In traditional farming, numerous laborers are needed to seed the crops and make the fields profitable. The use of AI farm bots is a way to deal with the labor scarcity. The AI bots are used in a variety of ways to supplement human

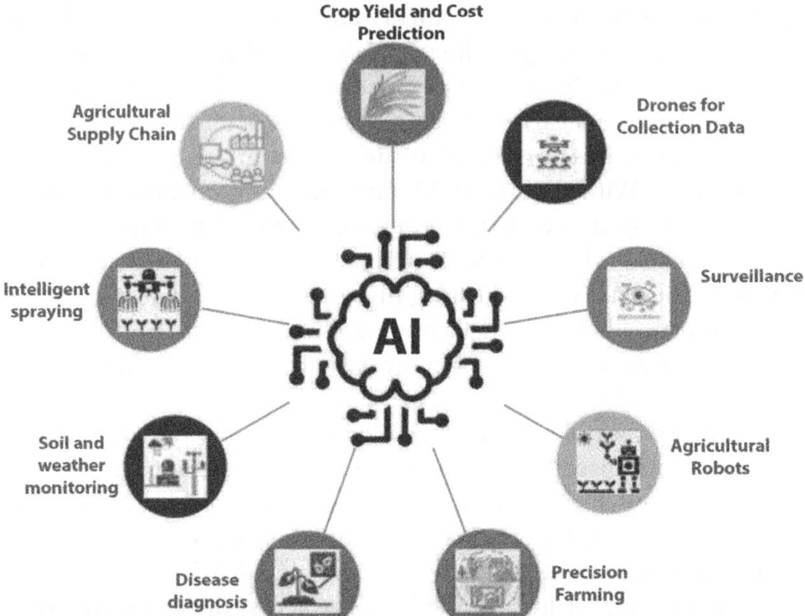

Figure 1.1 Applications of AI in the agricultural sector.

labor. Bots can harvest crops more quickly than human laborers can, and they can readily spot and get rid of weeds, which can lower expenses [10].

Thanks to a multitude of technologies, such as wireless, wired, and cellular connections (5G or beyond) for the Internet of Things (IoT) and robotics (agricultural drones and agribots), smart agriculture (SA) has started to be deployed. Specialized hardware and expertise in wireless communication are required for AI in 5G and beyond. Deep learning (DL) and machine learning (ML) are frequently employed for communications that require a massive number of inputs, outputs, and beamforming to support 5G and beyond. The DL, reinforcement learning, and CNN are utilized in channel coding for effectively using the air interface for 5G communications. Long short-term memory-type algorithms are used to foresee resource requirements for 5G slicing, enabling the operator to deliver diverse services over a single infrastructure.

1.3 Evolution of Smart Agriculture

Around the turn of the 20th century, the first real-world uses of so-called smart agricultural technology have been concentrated on the potential of new automated data management methods to improve the efficiency of machines used in sowing or harvesting operations. The ability to connect the quantities of input/output material flow to an exact position of the equipment in the field at that moment resulted in a paradigm shift in how agricultural operations were performed (through suitable geo-spatial references). This entailed overcoming the restrictions of geographical variability to get a more thorough and logical understanding of production locations, while introducing new issues connected to the vast amount of data that needed to be managed. These characteristics are also evident in several more names that precision agriculture is still known by today, including location farming systems, target farming, and prescription farming. These terms frequently emphasize the necessity of viewing precision agriculture as a collection of technical possibilities designed to provide an accurate (i.e., geographical) understanding of field activities to produce a parallel timely management of the same. Yet, in this sense, the use of "precision" in precision farming (PS) seems to be limited to a straightforward "spatial" idea, which also raises questions about the necessity for "accuracy." Without a doubt, the advancement of the positioning system technology has been a major factor in the spread of PF approaches [9].

Precision farming, according to an exhaustive description provided in 1997 by the U.S. Nuclear Regulatory Commission (NRC), is a "management technique that employs ICT in order to collect data gathered from different sources in view of their eventual application in decisions about production operations." At first, this idea was restricted to systems of arable farming, primarily involving cereal crops. Later, it was expanded to integrate additional farming techniques, including orchards, vineyards, and animals. Notwithstanding the continued applicability of the initial concept, the primary purpose of PF has historically been to enable the introduction of sophisticated automated applications into the agricultural industry. Potential customers were left feeling very confused and frequently let down by their entirely unmet expectations. Several farmers and industry researchers have just lately begun to refer to the practice as "smart agriculture" (SA) [5–8]. This phrase is sometimes replaced by "Farm 4.0" (Agri-4.0) by analogy with Industry 4.0 (Industry-4.0), sometimes more to represent a vogue than to assert a new technical idea. The phrase "Industry 4.0" is warranted since the industrial sector has been experiencing what could be considered its fourth industrial revolution for the past 10 years. A comparable strategy can only be proven for agriculture if it shares a historical evolutionary vision like the one suggested in Table 1.1, and Figure 1.2 summarizes a visual comparison of the technical advancements that have taken place in the industrial sector [3].

We must acknowledge that a step has been skipped in the technical advancement of agriculture based on the examination of both Table 1.1, as well as Figure 1.1. It is the one that relates to the comparison to the third industrial phase (Industry 3.0, which roughly spanned the years 1970 to 2010), which was characterized by the introduction of information and communication technology (ICT) into manufacturing processes and the slow but steady spread of enterprise resource planning (ERP) systems in business management. In other words, there has not really been a revolution within agriculture pushed on by information technology (IT). On the contrary, during the Industry 3.0 phase of Table 1.1, electronic improvements were distributed, and during the Industry 4.0 phase, an additional integrated approach using the previously discussed PF logics was made [4].

Although the lack of expertise and IT tradition in the administration of agricultural operations continues to be a hindrance for the quick adoption of technological advancements in the industry, SA may also be considered from this perspective as an advanced step of Agri-4.0. Of course, there are some deviations, and they are all application domains that most closely reflect the logic of the business sectors regarding both their administrative and structural traits. Examples of these domains include vegetable nurseries, dairy farms, and the cutting-edge technology of the so-called vertical farming.

Table 1.1 Suggested potential classification of the developmental stages of agricultural technology advancements.

Agriculture	Period	Name	Description
Industry 0.0	Until 1920s	Traditional farming and ruralism	The utilization of manual labor and animal traction is common; in the latter, mechanized traction is gradually included.
Industry 1.0	1920s–1960s	Motorization	The introduction and adoption of advancements in the tractor industry, such as diesel-powered engines, pneumatic circuits, three-point linkages, and the "inventory" of tires; continuing manual effort in the fields; due to fascism's autarkic agrarian policies, this era has continued in Italy; acceleration of change after World War II, thanks to the electrification of rural areas.
Industry 2.0	1960–1980	Mechanization	Improvements in tractors, such as an ongoing increase in their nominal strength and efficiency, the arrival of powerful operating machines in all agriculture-based manufacturing regions, the rapid substitute of manual laborers and the emigration of rural residents, and a sharp rise in primary yields due to developments in the chemical and genetics industries.

(Continued)

Table 1.1 Suggested potential classification of the developmental stages of agricultural technology advancements. (*Continued*)

Agriculture	Period	Name	Description
Industry 3.0	1980s–2000s	Humanism and electronics	Improvements to mechanized systems that pay more attention to the human–machine interface (HMI), the use of computerized control system on board tractor models and initial fixed point process automation approaches; the initial efforts to digitize farm management were never completely and widely strengthened, except in particular sectors like livestock husbandry; and finally, the use of electronic control systems on board vehicles and other portable machinery (especially in dairy farms).

(*Continued*)

Table 1.1 Suggested potential classification of the developmental stages of agricultural technology advancements. (*Continued*)

Agriculture	Period	Name	Description
Industry 4.0	After 2000	Precision agriculture, traceability, and cyber physical systems (smart farming)	Combination of gadgets and automation in all agricultural regions, with a focus on automating mobile-point operations (site-specific control); adoption of sensors for monitoring purposes and on-board tractor positioning system; protocols for communication between the device (CAN bus, ISOBUS, wireless networks, Wi-Fi, Bluetooth, etc.); encounters with computerization and the use of integrated information systems, particularly in large farms; proprietary farm information system proposals from major farm machinery makers; process connection and machine-to-machine connections; IoT, cloud, and fog computing; and processes and products certification geared toward traceability systems.

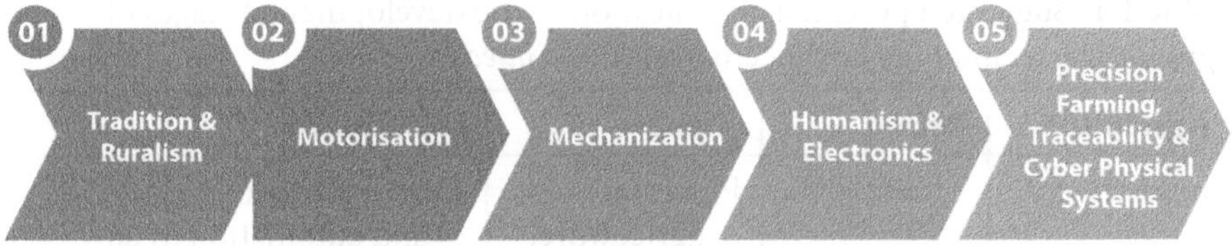

Figure 1.2 Evolution of precision or smart farming.

1.4 AI Technology Trends in Computer Vision

There are three fundamental steps in CV: (1) acquiring the image/video using a camera, (2) processing the acquired image, and (3) understanding the processed image.

1.5 Benefits of Artificial Intelligence in Agriculture

Artificial intelligence approaches are widely utilized to solve a unique set of problems and to improve the production and operating processes in the industries of agriculture, food, and bio-device engineering. The use of DL, ML, and AI, as well as the production of agricultural goods, is accelerating in the agricultural sector. The most revolutionary generation in agricultural services will likely be AI, which can perceive, learn, and react to different situations (totally based on learning) to boost efficiency [2]. The efficiency of current procedures has been considerably increased by cloud computing, big data, and the IoT. The IoT is expected to transform into the "Internet of Action" by the 2030s, with sensors and devices equipped with built-in AI and data analytics capabilities that enable self-optimization and autonomous activity initiation. This is how it could be beneficial:

1.5.1 Improving the Whole Supply Chain

The use of AI technology enhances the whole food supply chain by helping with the production of healthy crops, management of pests, soil surveillance, environment growth, data organization, manual work reduction, and enhancement of several agricultural tasks. The opportunity for AI in Indian agriculture is huge due to the large amount of data that Indian farmers can supply to develop AI solutions for the nation.

1.5.2 Agricultural Robotics

For farmers, a primary issue is weed control, which is difficult to accomplish because pesticide resistance is on the rise. Using agricultural robots, farmers may more effectively develop weed-control strategies for their crops. After sending visual data to a computer or a robot's system, the robot sprays on the weeds in cotton plantations to stop or reduce the growth of weeds.

Spraying accurately can stop herbicide resistance.

1.5.3 Policy, Governance and Market Access

A broad variety of big data is becoming exponentially more accessible because of farmers' extensive adoption of digital technology, which can help with improved policymaking and the transformation of the agricultural industry. The ICT-powered innovative media platforms fill the gap between agricultural extension agents and researchers on the one end and farmers on the other. It is a less expensive way to improve smallholders' familiarity with the most recent agricultural techniques and markets. Market information services facilitate farmers' access to neighboring marketplaces and increase their knowledge of current consumer wants by transmitting information from dealers with affordable mobile phones, Internet, and other information-dissemination services, as well as expanding rural farmers' access and understanding of climate-smart technologies.

1.5.4 Early Warning System

Governments and communities can get up-to-date information on disaster management and prevention through early warning systems. By giving individuals prompt advice on risk reduction techniques, they also facilitate more effective communication and increase the effectiveness of an emergency response.

1.5.5 Food Safety and Traceability

Mobile phones, software programs, RFID tags, data entry websites, and GPS-enabled sensors are just a few examples of the simple and complex technologies that manufacturers use to collect and analyze trustworthy data while still abiding by global traceability rules.

1.5.6 Financial Inclusion and Risk Management

It facilitates small-scale farmers' and rural farmers' access to financial services, helps them locate cost-effective insurance plans and risk management instruments, and informs them about the financial services that are accessible to them.

1.5.7 Capacity Building and Empowerment

It serves as a crucial educational instrument for the development of regional community. By expanding the reach of women, young people, and also other beneficiaries, they open the door to more recent business prospects that can enhance livelihood and earnings.

1.5.8 Growth Driven by IoT

Every day, enormous amounts of unstructured and independent information are produced. In agriculture, they can be about soil reports, fresh research, rainfall, insect infestation, pictures taken by drones and cameras, among others. Cognitive systems for the IoT can detect all these and offer reliable insights to boost crop output. Artificial intelligence may be used to develop intelligent systems that can be integrated into equipment that can work faster and more accurately than people.

1.5.9 Image-Dependent Insight Generation

An important area of farming that is currently most frequently discussed is PF. Drones generally help with intense topic evaluation, crop monitoring, field scanning, and other tasks. Drone data, IoT, and computer-generated vision may all be used to ensure quick and more efficient production with the help of farmers. Real-time signals can be generated by feeds from drone photo records to increase PF. Disease diagnosis, crop readiness identification, and field management are among the areas where computer-generated vision may be used.

1.5.10 Identification of Optimal Mix for Agronomic Products

Cognitive solutions give farmers recommendations about the many ways healthy crop production and yield can be achieved with hybrid seeds and plants based on a few elements like soil quality, climatic prediction, a particular epidemic, and a certain kind of seed. Suggested actions can also be tailored depending on the needs of the farm, the environment, and information about successful agricultural practices from the past. Farmers will be able to make an informed choice by taking into account outside factors including market trends, costs, and client preferences [1].

1.5.11 Monitoring of Crops and Soil Health

Building agricultural metrics across hundreds of acres requires remote sensing techniques along with hyperspectral imagery and 3D laser scanning. It may signal a gradual shift in terms of how farmers manage farms from both a labor and a time standpoint. In case of abnormalities, this generation may be utilized to exhibit vegetation together with their entire lifespan. One of the largest areas of agriculture that continues to provide drone-technology based remedies in cooperation with AI and web imaginative and prescient technology is crop monitoring and health evaluation. Drones equipped with high-precision cameras gather precise pictures that may be sent through a CNN to locate areas where weeds are present, which plants need water, and the amount or degree of stress some plants in the mid-bloom stage are. By analyzing plant life in both near-infrared and RGB light, drone technology can easily provide multispectral photos of ill and affected plants. This makes it much simpler to identify which plants in a specific region are affected so that remedies may be used immediately. Hyperspectral images and 3D scanning techniques are used in multispectral images to create a spatial data device that may be used to delineate acres of land.

1.5.12 Automation Techniques in Irrigation and Enabling Farmers

Irrigation is an agricultural practice that is human labor intensive. Machines knowledgeable on past climatic trends may improve soil, choose the right plants to grow, automate watering, and increase overall productivity. Automation in irrigation can help farmers better regulate use of water because irrigation consumes close to 70% of global freshwater.

1.5.13 Drones: The New Buzz in AI-Driven Agriculture

The production of agricultural drones has increased over the past few years, and farmers are now more aware of the usage of drones in agriculture.

Drone applications in agriculture range from crop dusting and spraying to surveying and mapping. When dealing with bad weather, productivity enhancements, precision agriculture, and yield management, drone-based solutions in agriculture are very important. Before a crop cycle, a 3D field map with precise geography, drainage, soil viability, and irrigation might be provided by the drone. Drone technology may also be used for nitrogen level regulation. The earth is sprayed from above with pods containing seeds and plant vitamins, providing plants with essential dietary nutrients. In addition, depending on the topography, drones may be programmed to springle by adjusting their distance from the ground. Crop health monitoring at mid-season is the most common use of drones in farming with the use of near-infrared region (NIR) or normalized difference vegetation index (NDVI) sensors to scan crop growth from a height of roughly 100 meters. Traditionally, this duty was carried out by humans who entered the fields with a notebook. A modern drone enables the capture of data that cannot be seen by the normal human eyes, as well as cover larger surface areas in a shorter amount of time (like the NDVI or near-infrared). Also, it greatly reduces the possibility of human mistakes in traditional inventory work, while it is still advised to physically verify a potential problem location after seeing the images [14].

1.6 Precision Farming

"Right place, right time, right product" is a catchphrase used to describe PF. The repetitious and labor-intensive portion of farming is replaced with a more accurate and controlled method. Additionally, it provides guidance on the best times for planting and harvesting, crop rotation, water and fertilizer management, insect attacks, and more. The goals of precision agriculture are strategic selection of crops and markets: This is necessary for profitabilityand estimating ROI using cost and margin data. Efficiency: A precise algorithm enables better, quicker, and more affordable agricultural options. Overall accuracy and resource efficiency are enhanced by doing this. Sustainability: Better social, environmental, and economic performance guarantees that all performance metrics gradually advance each season.

Precision farming can significantly improve agricultural practices to boost productivity and lessen the environmental effect of pesticide waste. It is likely that adopting technology with the necessary capabilities to offer fully supported PF will be an evolutionary process, with farms steadily enhancing their IT capabilities. Majority of the necessary components for a full PF system have had significant advances, enough for the system to be used in full by farmers in the near future. The software that assists farmers in choosing appropriate treatments and treatment doses to be spatially administered is the area that is expected to require ongoing improvement for the foreseeable future. As more businesses enter the market, the cost of IT components is anticipated to decrease. This new strategy is not only rational, but it is also technically possible, as well as being environmentally and economically justified [15].

1.7 Future Challenges

One of the crucial industries that contributes to a nation's security and stability is agriculture, which is also most impacted by climatic and environmental changes. Throughout their growth and development, plants are subjected to a variety of biotic and abiotic stressors. Some of the important challenges are shown in Figure 1.3.

Issue with a single standard solution: To increase agricultural productivity, several technologies have been developed. Due to the agriculture industry's continually changing environment, there are still no reliable predictions. The performance of suggested models is negatively influenced by the elements that have an impact on prediction accuracy, including soil quality, downpours, weather, and pests. As a result, the models'

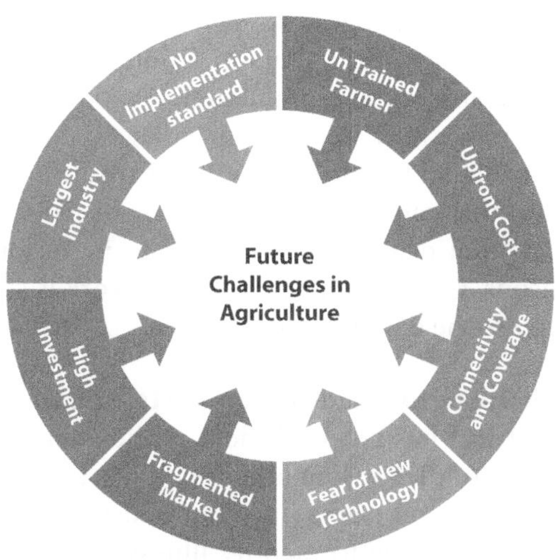

Figure 1.3 Future challenges in agriculture.

performance in one field differs from their performance in the other, making it impossible to reproduce efficiency.

A shortage of interactions among scientists and farmers: The breakthrough technical advancements that could facilitate decision-making are not yet adopted by farmers due to their lack of access to such advancements. Some researchers are also not aware of any production-related challenges that developing technology may present to farmers. The study found that the best method to close this gap is to bring together researchers, agricultural experts, and farmers in a single group.

Data security and privacy: As smart agricultural technology develops, one of the main concerns among farmers is the issue of data privacy. Huge and varied data from numerous farmlands are needed to create an effective ML model. Yet, it is challenging to keep track of agricultural data. Transfer learning is a novel ML method that was created to overcome issues with data privacy. By allowing several data providers to simultaneously train and utilize a single model, it is important to protect the privacy of local training data. Ubiquitous ML allows training across dispersed machines utilizing local data samples without information transmission. Federated ML techniques may be used in the future to identify crops using satellite data [13].

Large public image data collections are scarce: Large public picture data sets are hard to come by, which makes creating expert systems and cutting-edge CV for smart farming a huge task. Nevertheless, several attempts have been conducted during the past 5 years to produce new collection of data for the efficient development and evaluation of models based on DL. The main online`benchmark now is the plant village data set. Furthermore, gathering data from actual farmlands is still a costly, time-consuming, and error-prone process that calls for subject-matter specialists. There are now just a few data sets that deal with various crops in various geographical locations and seasons. Consequently, significant work is still needed to fill this void and increase the data sets for SA.

Recognition of diseases in various crop parts: The illness in the upper section of the plant leaf has received most of the attention in studies up to this time. In terms of diagnosing other plant diseases, other body parts including flowers, fruit, and stems have not had much attention. Only a few attempts have been made to identify illnesses in various plant components. Therefore, technologies must undergo the laborious process of picture annotation and data set labeling for various plant sections. Finally, a technology that can effectively identify agricultural diseases needs to be developed.

New learning model development: Several models constructed using ML have been reported in the literature and have demonstrated promising outcomes. Numerous DL models are based on Alex net, Resnet, Visual Geometry Group, Inspectionv3, and GoogLENet's five CNN designs. The accuracy and speed of calculation are frequently improved using CNN architecture. The problem of determining the number of layers and the type of layers required for an optimal solution remains unsolved. To adapt the improved CNN design, a hit and trial method is used. Additionally, these models may fall victim to the overfitting problem since they are unable to handle a variety of data sets. New learning models must be created to address the issue.

Hardware and software IoT shortcomings: In the previous two decades, the usage of smart devices in the agriculture sector has significantly increased awareness on the IoT. However, there are three significant obstacles to real-time implementation of IoT-based systems. First, the lack of expert sensors is the main impediment to the development of sophisticated farms, especially in plant phenotyping. Second, communication shortages in networks, particularly among rural locations, lead to erratic data transfer, which could have an impact on the overall system. Third, ongoing research and development are yet to provide a dependable and strong system that transfers data across wired and wireless media, especially in rural areas. Furthermore, because of the diverse nature of agricultural applications, a single approach is not feasible due to the environment's variable behavior. Therefore, a mechanism for evaluating how well a communication medium performs in various contexts needs to be developed.

Functionality and interoperability challenges: Diverse business and technical models cannot be supported by a single organizational structure. In addition, during data transfer, several security risks, including wiretapping, replay attacks, and data tampering, may be tried, endangering security. Thus, it is necessary to guarantee data integrity, confidentiality, and authenticity during data collection and transfer. To prevent unwanted access to data, various route authentication policies should be used. Another issue that could impede the expansion of a smart farming business is capital investment in agriculture. Additionally, it is difficult to get affordable IoT equipment for small-scale farmers.

Managing the big computing data: Several IoT gadgets generate a large collection of data. Intelligent infrastructure is needed to manage, process, and archive information for the automated execution of various activities. Infrastructure as a service solution is also quickly gaining favor because of they are inexpensive. There are many IoT systems available, including Salesforce IoT Cloud, Cisco IoT Cloud Connect, Amazon Web Services, and ThingWorx. These platforms can aid farmers in the future by making economic transactions easier. The challenges of technology still include data collection, hardware and software compatibility, decentralization, remote places with poor connectivity, and an inadequate number of techniques for making decisions.

Technology hurdles: To increase the use of agricultural robots, a number of technological obstacles related to robotics and autonomous systems must be overcome. The first difficulty is finding the hardware necessary to create agricultural robots. Second, for communication between humans and robots, there must be reliable wireless connectivity. Wireless devices can be used to track the whereabouts of autonomous systems, particularly when there is no visual contact. Third, creating software for agriculture robots is difficult. This entails creating software that can effectively manage various kinds of sensor, steadiness on uneven ground, and the capacity to dodge obstacles. Fourth, effective robot operation requires robot supervision to assure effective human–robot interaction.

Organizational challenges: For robotic and automated technology to be used in SA, farmer engagement is essential. Most farmers struggle to use technology correctly, which keeps them in the dark about important issues. To overcome this, there is a need for knowledgeable individuals who can help farmers use robots. Additionally, a government policy framework will help in a large-scale distribution of robots for agriculture to small farmers.

Financial difficulty: The cost of automated agriculture for autonomous and robotic systems is one of the major problems. Now, pushing robotics is seen as a way to increase agricultural productivity. An infrastructure made up of hardware and software is required to achieve this. However, the cost of the infrastructure's technological installation is prohibitive. Additionally, many farmers might not be able to afford it. This could make using agricultural robots unfeasible in the first place. Therefore, to meet the impending issues of increasing food consumption, rising wages, a declining workforce, and environmental concerns, considerable long-term evolutionary adaptations are needed. The abilities needed to manage a farm may also change because of these modifications.

Managing mobility and electricity challenges: One of the most crucial elements in the realm of smart farming is the issues associated with mobility and power management. For real-time analysis, smart farms require an elevated degree of mobility assistance because they frequently gather fresh data from the agricultural field. If the farms are solely connected via the cloud, certain features are unavailable. To collect data from farms, fog computing has the capacity to do so. Furthermore, to do this operation, a consistent fast speed is needed. Better power management is needed to address this issue. Since a variety of devices, including mobile phones, are used to collect data, sensors and actuators are employed, and delivering data to an edge node demands a lot of power, which is deficient in some circumstances. It is possible to increase the lifespan of sensor nodes by using sources of clean energy like solar and wind power.

High hardware costs and trouble connecting to the Internet: Since edge devices transmitted real time data constantly to cloud servers. Smart infrastructure is required when handling, analyzing, and storing data to automate numerous activities. This task necessitates the use of numerous hardware components, each of which consumes more energy. Therefore, to reduce the hardware cost issues in SA, effective cost management is necessary. Additionally, bad Internet connectivity is another typical issue, particularly in rural locations. An Internet connection with a fast speed is necessary to handle the massive amount of data. However, when connected to the cloud, inadequate Internet connectivity in actual farmlands suffers a number of difficulties, including computational delays, delayed replies, data loss, and slow data uploading speeds.

Environmental issues in cyber-physical systems (CPS): As IoT devices are used in agricultural fields, CPS have attracted a lot of interest. However, there are a number of significant obstacles to real-time CPS adoption in smart agriculture for a variety of reasons. The first is the lack of fresh drinkable water, which is necessary to boost agricultural productivity. Second, dwindling agricultural land due to growing urbanization. The potential for soil erosion, which has a negative impact on productivity, is ranked third. To address these difficulties, a sustainable SA system with highly adaptive technologies is required. Furthermore, because the environment behaves differently depending on the application, a single solution cannot be used for all agricultural applications. Despite this, the advancement of modern technology makes use of CPS, one of the finest areas for application. However, further research is necessary to determine whether CPS will be successful in smart farming given the issues raised above.

Challenges in real time implementation: Using CPS to build SA applications in the real world is fraught with difficulty. There is not a single structure that can satisfy the requirements of every gadget. The various behaviors of weather conditions present a number of functional issues. Long-term preparation of animals and crops in accordance with consumer expectations is also necessary, in addition to sophisticated techniques to assess the condition of the soil and weather forecasts. Additionally, the sensors that monitor temperature, humidity, and the quantity of nutrients in the soil are being placed in actual fields to perform a constant quality check. Another difficulty is developing a productive management system for intelligent farming that combines real-time monitoring and a cooperative agent-oriented approach. The cost of implementation, however, impedes the development of the smart agricultural sector.

Challenges in design decisions: Design decisions have a variety of effects on present and forthcoming block chains, including OriginTrail, AgriLedger, FarmShare, and AgriDigital. For instance, authorization is required (i.e., partners are trusted) for an organizational structure that is free, and someone can join or leave. Depending on how the current permission-free block chains are set up, transactions may take anywhere from a few minutes to an hour to complete before everyone updates their financial information. Because of these architectural choices, block chain systems are rigid. Therefore, under certain situations, some blockchain-based solutions might not be as useful as traditional and centralized ones.

Usefulness of smart surveillance systems: The agricultural industry has placed a lot of pressure on technological advancements in smart farming. One of the biggest problems in the agricultural area is the smart monitoring of crop health on large fields. Swarm intelligence systems are necessary to make agricultural systems intelligent. The monitoring of several agricultural applications, such as water and soil, irrigation systems, surveillance of assets, greenhouse expansion, remote controlling, and diagnostics, are covered by swarm intelligence. To be able to monitor agricultural diseases, agricultural robots can be used in swarms. For farmers to take prompt action, it is be possible to develop a system for delivering real time information.

Agriculture autonomous system: Over the past 10 years, several autonomous systems have been created to boost production and dependability, and they have partially replaced humans. Agricultural operations are becoming more intelligent with the aid of modern robotics and autonomous systems. These tools are used in a variety of SA production methods, including plant factories, 3D printing of food, aerial pesticide application, farming that is biodiverse and self-sufficient, and automated husbandry. However, there are still a lot of issues that need to be resolved, such as exact identification, autonomous route navigation, and operational efficacy. Some cognitive systems are still required for smart farming, such as exact task distribution, item translation, and human–robot interaction.

Generating fresh data sets: The main challenge is the creation of new data sets because massive data sets are necessary to train models of DL or ML. Generative adversarial networks (GANs) have been used to combat this in several fields, and they are currently being used in AgroVision. Data wrapping and GANs are two recent examples of data augmentation techniques used to train models rather than creating synthetic images that are representative of actual farming conditions on a big scale.

Model choice: Difficulties in image improvement, crop identification, early-stage virus identification, plant-disease identification, and salt tolerances are some of those addressed by using GANs in SA. To examine diverse applications, a variety of customizable GAN models have recently been introduced. OneGAN, DoubleGAN, CGAN, DICNN and ARGAN, Auxiliary Classifier—GAN, Leaf GAN, Style GAN, Cy-cle Consistent-GAN, Wasserstein GAN-GP, and Enhanced Super-Resolution GAN are some of the GAN models that have been presented to the agriculture industry. For scientists and academics, choosing the best GAN model for a specific crop is a difficult issue.

1.8 Conclusion

The establishment of a strong and resilient agricultural system is essential for the future of agriculture. Currently, SA is structured. Likewise, the development of new automation technologies in agriculture may be hampered by a lack of awareness about technical breakthroughs. As a result, this chapter provides a thorough analysis of the most cutting-edge technologies used in smart farming. One of the main obstacles to the forecast of issues related to smart farming, according to the study, is the collection, initial processing, and formatting of the data sets. It should also be noted that a number of problems have been effectively solved using DL models. However, acquiring a sizable amount of labeled data—which can occasionally be difficult in real-world situations—is required for training a model using DL. Agro-informatics, AI, and nanotechnology are just a few of the numerous innovations in the agricultural industry that define our modern period. All these technologies seek to increase both the quantity and quality of crop output, tolerate a variety of biotic and abiotic challenges, and attempt to identify agricultural techniques that can provide sustainable production that satisfies the demands of the anticipated population rise in 2050. Hence, to assure the growth of sustainable agriculture, researchers must unite their efforts to apply these technologies in an integrated manner. Finally, the fact that we live in a time of challenges necessitates that we prioritize based on the local conditions that are currently in place before integrating these scientific developments in terms of methods and technologies to accomplish multiple goals including various perspectives during the stages of growth and development of various species, which can to faster and sustainable development.

References

1. Williamson, H. F., & Leonelli, S. (2023). "*Towards Responsible Plant Data Linkage: Data Challenges for Agricultural Research and Development* (p. 317). Springer Nature."
2. Pinstrup-Andersen, P. (2030). "Challenges to agricultural production in Asia in the 21st Century. *Water Agric., 116*."
3. Wakchaure, M., Patle, B. K., & Mahindrakar, A. K. (2023). "Application of AI Techniques and Robotics in Agriculture: A Review. *Artif. Intell. Life Sci.*, 100057".
4. Demestichas, K., Peppes, N., & Alexakis, T. (2020). "Survey on security threats in agricultural IoT and smart farming. *Sensors, (2022)*, 6458."
5. Bacco, M., Berton, A., Ferro, E., Gennaro, C., Gotta, A., Matteoli, S., ... & Zanella, A. (2018). "Smart farming: Opportunities, challenges and technology enablers. *2018 IoT Vertical and Topical Summit on Agriculture-Tuscany (IOT Tuscany)*, 1-6."
6. Purohit, S., & Purohit, S. (2021). "Digitization in Indian Agriculture: Evolution from Simple to Smart Farming. *Indian J. Econ. Dev.*, *9*(1), 1-6."

7. Charania, I., & Li, X. (2020). " Smart farming: Agriculture's shift from a labor intensive to technology native industry. *Internet Things, 9*, 100142."
8. Lytos, A., Lagkas, T., Sarigiannidis, P., Zervakis, M., & Livanos, G. (2020). "Towards smart farming: Systems, frameworks and exploitation of multiple sources. *Comput. Netw., 172*, 107147."
9. Sharma, V., Tripathi, A. K., & Mittal, H. (2022). "Technological revolutions in smart farming: Current trends, challenges & future directions. *Comput. Electron. Agric.* 107217."
10. Debauche, O., Trani, J. P., Mahmoudi, S., Manneback, P., Bindelle, J., Mahmoudi, S. A., ... & Lebeau, F. (2021). "Data management and internet of things: A methodological review in smart farming. *Internet Things, 14*, 100378."
11. Pang, H., Zheng, Z., Zhen, T., & Sharma, A. (2021). "Smart farming: An approach for disease detection implementing IoT and image processing. *Int. J. Agric. Environ. Inf. Syst., 12*(1), 55-67."
12. Dahane, A., Benameur, R., Kechar, B., & Benyamina, A. (2020, October). "An IoT based smart farming system using machine learning. In *2020 International symposium on networks, computers and communications (ISNCC)* (pp. 1-6). IEEE."
13. Sørensen, C. A. G., Kateris, D., & Bochtis, D. (2019). "ICT innovations and smart farming. In *Information and Communication Technologies in Modern Agricultural Development: 8th International Conference, HAICTA 2017*, Chania, Crete, Greece, September 21–24, 2017, Revised Selected Papers 8 (pp. 1-19)." Springer International Publishing.
14. Sheela, K. S., & Chakravarthi, A. D. (2019). "Smart farming with e: Technology. *IJECS, 1*(1), 32-35."
15. Glaroudis, D., Iossifides, A., & Chatzimisios, P. (2020). "Survey, comparison and research challenges of IoT application protocols for smart farming. *Comput. Netw. 168*, 107037."

2

Cyber Biosecurity Solutions for Protecting Smart Agriculture and Precision Farming

Balakesava Reddy Parvathala[1]* and Srinivas Kolli[2]

[1]IT, Vallurupalli Nageswara Rao Vignana Jyothi Institute of Engineering & Technology, Hyderabad, Telangana, India
[2]Department of CSE, Koneru Lakshmaiah Education Foundation, Bowrampet Campus, Hyderabad, India

Abstract
As the demand for food increases, agriculture implements smart technologies to meet. Cyberattacks on the economy are more likely as a result of these automations and the corresponding rise in connectivity. Security in the food supply chain is necessary, as evidenced by the rising frequency of cybersecurity breaches in various industrial sectors. Modern life has undergone a complete transformation thanks to the Internet of Things (IoT) and clever computing technology. The IoT and the data-driven services it enabled were unimaginable just 10 years ago. They are currently pervasive and have an impact on many industries, including healthcare, smart homes, and driving. These kinds of technical inputs are especially welcome in the agricultural sector. Smart technology is used by everyone, from farmers to business owners. In recent years, the agricultural sector and the research community have been interested in smart agriculture (SF) and precision agriculture (PA). The combination of SA and PA enables farmers to reduce inputs (such as fertilizers and pesticides) more successfully but does so at the expense of increased safety risks that may ultimately make this aim more difficult to achieve. Being aware and taking the proper precautions.

***Keywords*:** Cyber attacks, IoT, agriculture, precision, pesticide, bio security, farming, technology

**Corresponding author*: balakesavareddyp@gmail.com

Rajesh Kumar Dhanaraj, Balamurugan Balusamy, Prithi Samuel, Malathy Sathyamoorthy and Ali Kashif Bashir (eds.) Computer Vision in Smart Agriculture and Crop Management, (25–56) © 2025 Scrivener Publishing LLC

2.1 Introduction

By 2050, the world's food output must rise by 70% to feed 10 billion people, according to the Food and Agricultural Organization (FAO) of the United Nations [2, 3]. To serve the food sector and industrial agriculture, various cutting-edge Internet of Things (IoT) solutions have been developed [4, 5]. Yet, several safety issues are brought up by these industrial advances and applications [6–8]. Furthermore, because of population growth restrictions on available land and water, agricultural practices aimed at better utilizing resources and boosting crop yields are encouraged. A workable way to address issues in modern agriculture and boost productivity is through IoT technology. We want smart farms (SF) and precision agriculture (PA) to become more commonplace because of this PA. Project categories are used to group the security components of SF and PA. The cyberattacks that might transgress these features are listed below.

Thus, based on how they relate to various stages of the cyber disruption chain, we classify cyberthreats into SF and PA. An advanced persistent threat (APT) will narrow your search for online dangers. Lastly, we look at risk mitigation options and related actions and create a future research roadmap. They are currently pervasive and have an impact on many industries, including healthcare, smart homes, and driving. People in all walks of life, from farmers to business owners, frequently utilize smart devices. Yet, there is a high risk of cyberattacks due to the widespread use of IoT and smart communication technologies. Environmental vulnerabilities in smart farming the economies of nations that significantly rely on agriculture could be affected by such cyberattacks. The "bioeconomy" is seeing a rapid evolution of national infrastructure and data security challenges. Meanwhile, the U.S. food and agriculture system's discussion of cybersecurity (also known as cyberbiosecurity) is disjointed and lacking. The U.S. economy (6.7 tons) and 15% of jobs in the country are both accounted for by food production and agriculture (43.3 million jobs). The U.S. economy (6.7 tons) and 15% of jobs in the country are both accounted for by food production and agriculture (43.3 million jobs). Due to the industry's high level of diversity, it necessitates cutting-edge computing, big data, cloud storage, and internet access technologies and capabilities.

There is evidence from numerous studies that the improper use of these technologies can have a negative impact on food production systems. So, to alleviate climate-related issues, it is crucial to investigate, create, and put into practice technical solutions specific to each region's features.

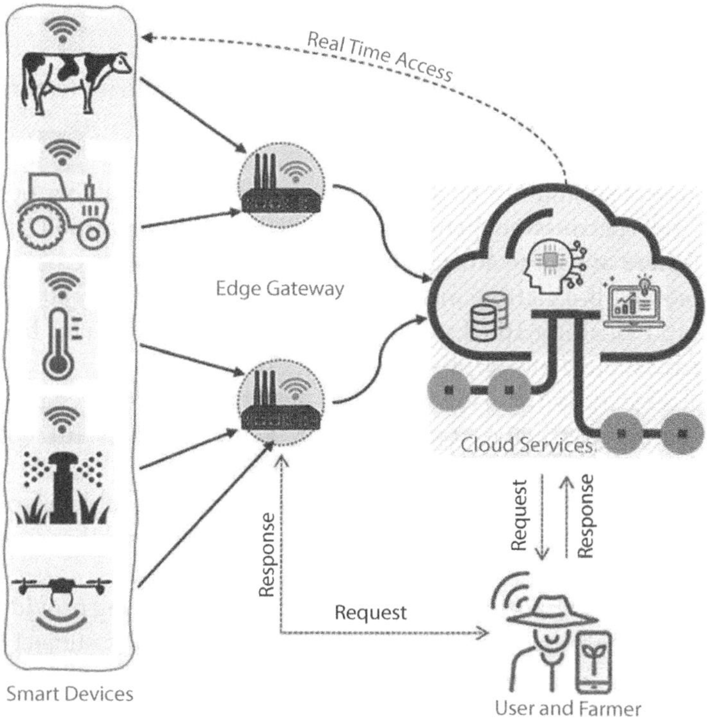

Figure 2.1 End-to-end interaction model for various smart agriculture stakeholders.

The ecosystem of smart agriculture's many players is depicted in Figure 2.1 in terms of their overall interactions. Using a user application, physical and animal sensors on the ground produce data and take motion commands. In order to interface with farm equipment, filter sensor data, and perform real-time agricultural analytics, these farm equipment work with gateways that support end nodes. In addition, monitoring data, such as sensor and smart machine status, power management, and environmental information (such as soil moisture and fertility status), are all stored in enormous quantities in cloud data lakes. Please be aware that, from a security and data protection perspective, modifying and disclosing such data, in addition to hacking into actual hardware and software systems, might have serious repercussions. Using communication networks to develop "smart farming," or interconnected agricultural technologies, is a significant answer. The food production industry can greatly benefit from research and development (R&D) in smart agriculture, which is supported by many developed nations.

2.2 Cyber-Attacks on SF and PA

The following might be used to summarize the key security features of SF and PA suggested in recent studies.

- Users should not be able to access other users' data without their permission, hence data protection should prohibit this. There are some attacks that can result in the release of personal information, including physical attacks, replay attacks, and cloaking attacks. The qualities of SF and PA have been the subject of several study studies [6].
- Integrity is the assurance that data has not been changed while it is being stored or sent. Several studies [6] have examined the consistency between SF and PA.
- Privacy: Due to the way that SF and PA safeguard data from illegal access, some academics are interested in these two areas.
- Existence: Existence ensures that the services being provided will continue. Smart farms and PA have been the subject of several recent research projects [4].
- Non-repudiation: Users are not allowed to backtrack on actions they have taken within the system thanks to non-repudiation. SF and AP has been mentioned by certain academics [5].
- Trust: Prevent people from using false identities. Many studies on SF and PA creativity have been published in the literature [5].

Because of the above-mentioned lack of proper security rules, SF and PA will utilize a variety of attacks to harm, steal from, modify without authorization, or destroy these environments and the accompanying intelligent IT systems. Figure 2.2 depicts the classification of attacks on SF and PA.

Target components are then used to categorize SF and PA assaults.

Hardware attacks can be carried out by knowledgeable attackers using specialized tools [19] to take advantage of IoT and physical electrical equipment's known and undisclosed weaknesses (and other hardware). targeting cyber-physical and IoT systems out of design considerations for stability [23].

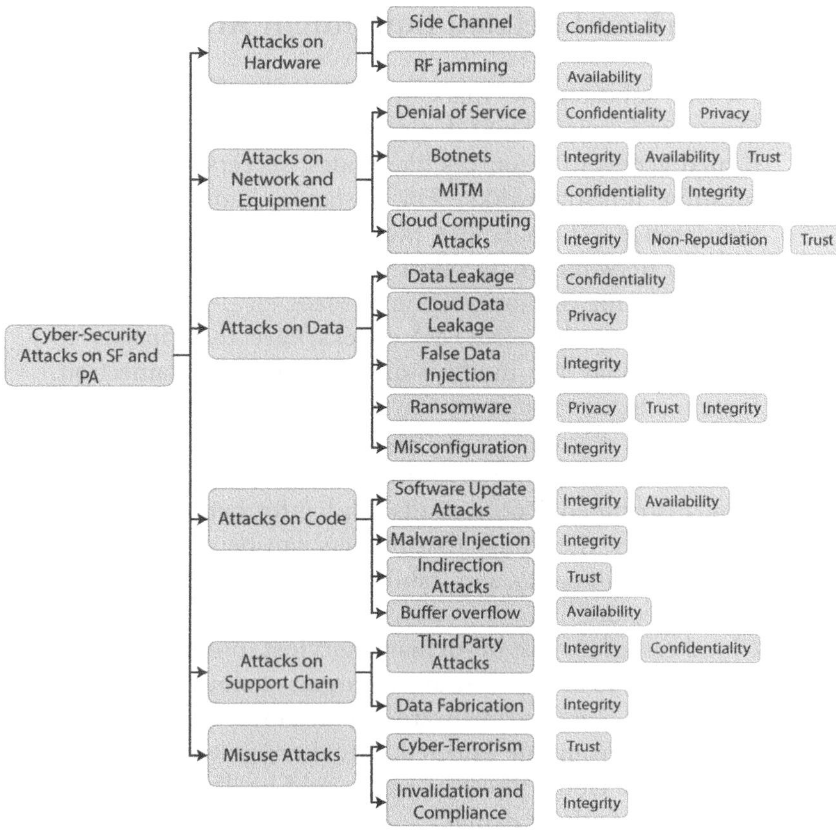

Figure 2.2 Classification of attacks on SF and PA.

- Subchannel attacks [2, 9] keep track of physical variables like currents and voltages to collect unauthorized information about the specifics of system implementation. The system's secrecy is put at risk by this sort of attack.
- Radiofrequency interference [2, 8]. In order to open wireless channels and withstand interference, wireless network technologies are designed to be vulnerable to attacks. System availability in SF and PA locales, including greenhouses, can be compromised by this kind of attack.

2.3 Network and Related Equipment Attacks

There are several subcategories that can be applied to this attack type:

- Denial of Service (DoS) denies a person or device authenticated access to resources like nodes, servers, or communication links/networks. For instance, radio frequency interference [2] obliterates the radio frequency spectrum used by networks in an effort to prevent connected nodes from receiving communication services.
- Man-in-the-middle (MITM) attacks have the capacity to covertly replay (or modify) data sent over a connection. System integrity or privacy may be the target of these attacks.
- A robot is a group of networked devices (sensors) that are each running one or more robots [7]. Spam distribution, distributed DoS assaults, and information theft are just a few of the things that botnets can be employed for. Availability, integrity, and stability could all be hampered by its architecture.
- Cloud resources like on-demand services, proprietary provisioning, and autoscaling are all targets of cloud computing assaults [2]. Infected malware, for instance, can be swiftly transferred from one infected virtual computer to another through the cloud. These assaults could be directed against dishonesty, faith, or denial.

2.3.1 Attacks on Data

When the computer system stores, sends, or processes data, it pushes data. The following subgroups of this attack category are listed [5].

- A data breach [2] is the unlawful movement of data from approved internal sources (people or equipment) to outside targets. These attacks go against the privacy of data.
- Ransomware [12] guards private keys and encrypts whole files, partitions, or volumes in an effort to get the owner [7] to cough up a ransom. Privacy, trust, and integrity are all violated by these attacks.
- Forgery [9] describes attacks that introduce harmful data or command and control instructions into a system. The integrity of data is the target of this attack.

- Misconfiguration [2], the act of setting up SF or PA reporting systems to display false information about running a farm, can result in expensive and perplexing choices and actions for farmers. Integrity may be compromised through attacks on misconfigurations.

2.3.2 Attacks on Code (Applications)

- Attacks on installed software's update procedure [2] disrupt it, endangering the availability and integrity of the system.
- Malware-infected nodes and devices are the target of assaults known as malware injection [2]. This assault is an integrity violation.
- Buffer overflows [10] are software flaws or vulnerabilities that can be used by hackers to break into company systems and acquire confidential information. Usability is put at risk by this attack.
- A database server is tricked into running malicious a structured query language (SQL) code that has been inserted into a database input field by an indirect attack (SQL injection) [2] using code injection techniques. Indirect attacks destroy trust.

2.3.3 Attacks on Support Chain

- Particularly made to sustain the various links in the chain.
- Third-party attacks take place when an adversary gains access to a system through a partner or service provider that is a third party but has access to the system and data [2, 8]. System confidentiality or integrity may be jeopardized by outside threats.
- Data generation [8] describes the creation of harmful data or procedures that make use of particular access rights for other objectives. An integrity breach of the system might happen as a result of this.

2.3.4 Misuse Attacks

Attacks that take advantage of SF and AP defense capabilities to target specific people, companies, and other targets.

- The IoT platforms and cyber-physical tools can be used by cyberterrorism [2] to launch remote attacks against targets. Systems for SF and PA may no longer be trusted as a result of this.
- Disabling and/or compatibility [2, 3] indicates that bogus data manipulation has compromised the authentication process. Attaining system integrity is the goal of this attack.

2.4 Security Threats to SF and PA Using the Cyber-Kill-Chain (CKC) Taxonomy

A sophisticated attack is carried out in seven steps, according to the Cyber-Kill-Chain (CKC). This analysis enhances an analyst's comprehension of an assault and aids in determining the tactics, strategy, and procedures used by an adversary. Because the CKC attack phases are interconnected, the entire assault scenario fails if the adversary does not complete this step.

The actions listed below make up a typical compound attack as suggested by the CKC model.

1. Reconnaissance: The assailant gathers as much data as possible to identify and assess the target. They may be looking for relevant information, such email addresses, during the search phase. This action is done primarily to highlight the victim's weakness. The victim's defense strategy cannot refute the confession, either. It is possible to think about both passive and aggressive reconnaissance techniques. Passive detection is also known as fingerprinting when it comes to cyberattacks. Sweeping is a common term used to describe active scouting. To find out which IP addresses are connecting to the physical host, pinging every IP address on the target network is a straightforward test. Because of this, questionnaires offer more precise yet extensive information than fingerprints do. A target may be made aware of a potential assault if the scan produces a lot of odd connections that should be ignored.
2. Weaponization: Attackers create and use remote access malware (weapons), including worms, viruses, and backdoors, to match the weaknesses of the target (discovered during the reconnaissance phase).

3. Delivery: The perpetrator transmits the victim remotely executable malware (for example, via a USB drive, email attachment, or website).
4. Deploy remote access malware as part of exploit [4]. Now, the attacker accesses the victim's network remotely using malware to take advantage of the weakness.
5. Installation: The attacker sets up an appropriate command and control (C2) server to gain ongoing access to the victim.
6. C2: To control the victim, the assailant establishes a connection with the C2 server.
7. Attacker Activity on Target: The attacker carries out the attack script in order to get the victim to the intended target. The procedures above are shown in Figure 2.3.

Remember that not all attacks follow each of the phases exactly and explicitly. Also, it is possible that some acts will not be verified, followed up on, or reported.

Figure 2.3 Typical cyberattack stages as described by the CKC (Courtesy [8, 9]).

2.5 The Taxonomy

Known online dangers to CKC divisions SF and PA are covered by the taxonomy in this subsection.

2.5.1 Threats Pertaining to the Phase of Reconnaissance

According to our findings, a large portion of contemporary sci-fi cyberattacks concentrate on multilayered structures that acquire data during the intelligence phase [7]. The IoT gadgets are also SF [1, 2] or PA [1, 2] or PA [2, 3] systems. Hence, the two main types of cyberthreats to SF and PA are:

- Threats to IoT devices that include those posed by hardware assaults, such as master password attacks, despite the fact that software programs frequently safeguard IoT devices [2, 7].
- Risks to layered architecture: these threats aim to compromise various layers of the layered architecture, as can be shown in Figure 2.4 below.
- Throughout the phase of reconnaissance, this hierarchy poses a threat of several attacks. Physical sensors and gateway devices are a part of the architecture, which are dispersed

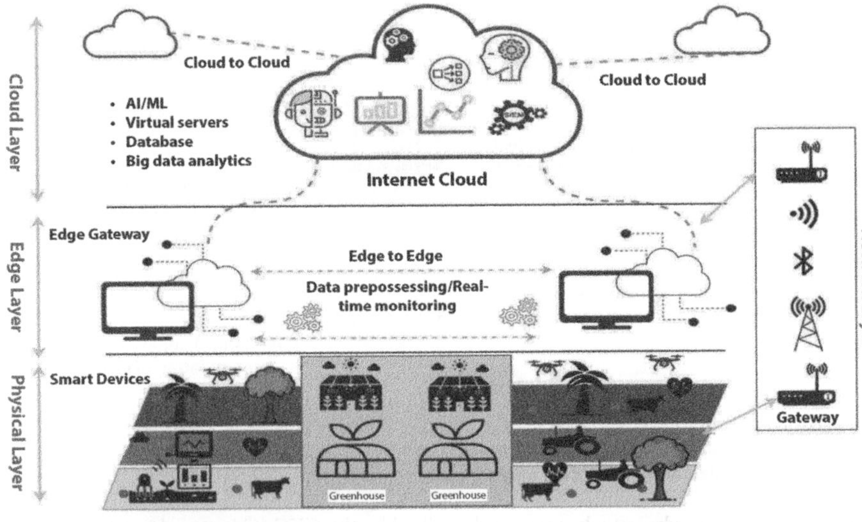

Figure 2.4 A typical SF architecture.

across greenhouses and farms. Public drones, self-driving tractors, sensors built into animals, and centralized devices that connect to a central cloud or other smart things are some examples of these gadgets. The use cases offered by these gadgets are extremely sophisticated because they provide data and work together with other devices based on the information gathered during agriculture. Gather and communicate to Smart or Cloud Wind current data on weather, soil moisture, or cow temps. For instance, data gathered from soil moisture measuring devices can be processed in the field, at the border, or in the cloud to calculate how much water a farm requires, optimize irrigation schedules and provide helpful information for aspiring farmers.

- Edge layer: The SF and PA environments have real-time or almost real-time computation, as well as services, at this level. Attackers may be able to disable these devices, leading to disruptions, delays, complaints from users, and financial issues such as loss [2]. Collection, filtering, encryption, and real-time data stream encryption are all included in data collection services. Machine learning models that have been trained centrally, at the Cloud, and at the Edge are frequently the foundation of predictive services. They can be used to forecast soil erosion or other occurrences related to a certain crop or animal, as well as to estimate crop yields, evaluate plant or animal health, and determine how much fertilizer or water the soil needs to boost harvests. Anomalies can be monitored in real time and classified as malicious or benign using security monitoring and detection techniques that can be applied for prediction and/or classification. Services like equipment failure forecasting and defect detection are included in this.

The network layer manages the hardware heterogeneity between farms and greenhouses. This layer also offers connectivity, enabling use of the gadget in both SF and PA. This level frequently sees DoS assaults and botnets [2].

"Connectivity" is a recurring topic in the majority (but not all) of contemporary technologies. The notion of a network of smart gadgets has materialized as the need for unrestricted internet usage develops. The concept, referred to as the "Internet of Things" (IoT), enables linked objects to monitor, manage, and exchange data with one another. A number of applications can be made from this data after analysis. The network layer in smart agriculture not only makes it easier for the edge layer and physical layer to communicate with one another, but it also acts as an interface for cloud layer interaction. From the change in soil temperature via a P2P sensor communication system to Cloud data storage via a high-speed mobile network like 5G [8] or farm-to-farm quality updates via a customized wireless architecture [9], the network has offered a communication tool layer that connects all other classes. The network layer's primary duties in an intelligent control system are two. A variety of heterogeneous gadgets are firstly present in each stratum of the smart farming system. Wired, wireless, and mobile subnets can all be connected to one another using the network layer's secure and effective network stack. Upkeep and accessibility improvements are the second duty at this level. The fundamental layer of network connection for the entire system, from the big data processing system that analyses the data gathered to the individual sensors that gather data on the ground.

Precision agriculture and Cloud computing models are a step forward in improving the PA relationship. In the data center, the Cloud layer is typically virtualized, and it connects to other layers via the internet. These Cloud computing systems leverage the platform-as-a-service (PaaS) architectural concept, which frees users to concentrate on using apps and importing data. A distributed file system (DFS) is used by PaaS providers to store records that are collected by data agents from data received from the backend layer. Using this saved data, analytics tools can extract information. This data analysis component computes and delivers the data to end users via a client application running on the user's PC. John Deere, Farmers Business Network, etc., are well-known producers of agricultural equipment. The company has developed a number of Cloud-based products that let customers keep an eye on the numerous sensors and equipment used on farms. The majority of the time, these PaaS systems run cloud computing platforms like Amazon Web Services, Google Cloud15, etc. There are a number of security issues that the ecosystem for smart agriculture faces as a result of the way these well-known cloud computing platforms are employed to construct these farm PaaS systems.

2.6 Data Collection

During a two-day virtual workshop in the fall of 2020, the information used in this paper was gathered as a component of a project funded by the USDA African Research Institute (Cybersecurity and Agriculture, Food, and Economic Protection, 2020). Eight sessions with a shifting agenda made up the online workshop that was held through Zoom (see Table 2.1). The protection of digital, physical, biological, and information systems as they relate to the local environment is a topic covered by speakers from throughout the country and the state at this gathering. In addition to the breakout rooms, Zoom meeting trainers and Google presentation coordinators will monitor the session to assist participants in structuring their ideas, contributing in various ways, and involving as many participants as possible. Producers lead discussions in groups about how to be a good leader and offer management and input. Have a look at the Google-based documentation and ask any more questions. Since this is a virtual workshop, the number of participants per session, which organizations are present at each session and how many participants there are are determined

Table 2.1 Lists the topics to be covered during the SAFE virtual workshop (6–7 October 2020), as well as the agenda and types of data analyzed.

Day	Session topic	Data type used
1	Introduction and Keynote – Protecting our Bio Economy – The Importance of Securing	Zoom audio
	Our Agriculture and Food System-Preparing for a Cyber Pandemic	Zoom audio, Session notes
	Cyberbiosecurity: Convergence and Emergence in the Digital, Physical, and Biological Systems	Zoom audio
	Your Opinion What are the weaknesses of the present system?	Zoom audio, Breakout rooms
2	Where do you fit? is the rallying cry. How come? Who? How?	Zoom audio, Session notes
	Crisis Scenarios: How prepared are you?	Zoom audio
	Priorities' following steps: Formal partnerships and collaborations, regulations, science and technology, awareness-raising, and education	Zoom audio, Session notes
	Priorities' next steps include formal partnerships and collaborations, policies, science and technology, and education and awareness.	Zoom audio, Session notes

by the information in this report. Each session only has a single participant registered. What was heard was difficult to pinpoint. Throughout the conference, information was gathered from a variety of sources, including Google-based artefacts, Zoom audio, and Zoom video. The setup, configuration, and transmission of all recorded material via Zoom was done in collaboration with conference organizers. Only the study team is permitted to use the Deep Sessions data, which is also stored in shared storage.

2.6.1 Participants

Food, agribusiness, cybersecurity, and national security are just a few of the professional sectors represented at the conference. In partnership with the Office of Continuing and Professional Education at Virginia Tech, the university manages registration for workshops and conferences. The announcements will be publicized via emails sent out through various local and national listing services, posts on LinkedIn, Facebook, and other social media, website information, and workshops planned within Virginia, to name just a few. The first publication and subsequent local, national, and international media coverage; technological news; and tales. Invitations are spread via word-of-mouth. Additionally, businesses in the private sector that represent state and federal agencies, colleges, law enforcement, commodities producers, food and agricultural consultants, and other businesses in the agriculture sector are also mentioned. In the workshop, 170 people participated and helped in various ways. They were also invited to participate in several conference sessions. Figure 2.5 depicts the total number of participants for various arrangements.

2.7 Vulnerability of the Food and Agricultural System and the Bio Economy

According to Feeding the Economy [15], the Fd+Ag system in the US has a major negative impact on the regional green economy, accounting for 20% (6.7 tons) of it, which puts the world's food security at risk. Around $20 billion is thought to be the agricultural data science market [18]. The Fd+Ag system is made up of several different areas with a wide range of sizes (staff counts) and functions. It is also loosely linked and geographically dispersed. Large multinational firms produce the majority of the

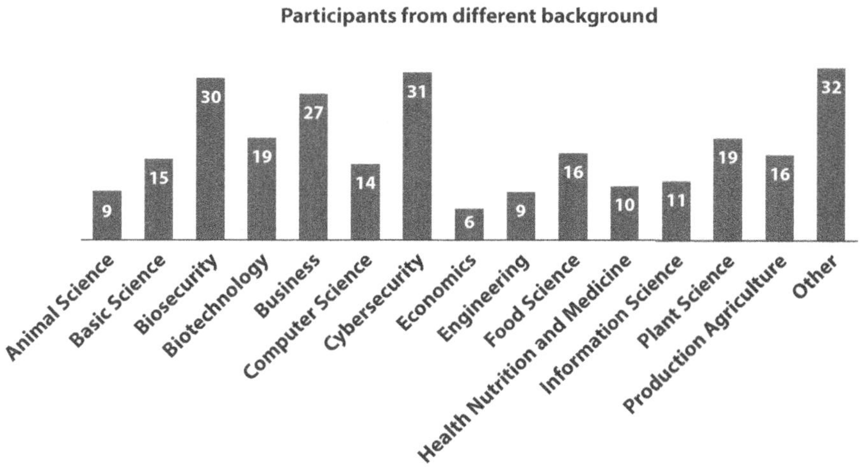

Figure 2.5 Participation in the Securing Agriculture and Food Economy (SAFE) using Cyberbiosecurity virtual workshop, October 6–7, 2020, by participants from many disciplines.

Fd+Ag system's economic value. The economic and social elements of social farms are present on small family farms (family farms). Farmers have a private and protective perspective on data [18]. Small farmers and nutritional businesses typically appreciate the need and value, but they are hesitant to use computer security technologies (selection, configuration, and control). Cyberattacks frequently target enterprises with less than 100 people, putting the economics of medium-sized businesses, including many in the food processing and support sectors, at risk. Military lunch production is one of the food and agriculture (Fd+Ag) systems' offerings. No specific knowledge of the food production process is required of the attacker. Attackers are only knowledgeable about technical uses of tools and procedures, such as lowering the temperature of meat containers before packaging [20]. Unauthorized people are more likely to access data when an enterprise control system is built on proprietary software. Bioinformatics and genetic technologies have also sped up the creation of new goods and crops. Information technology and biology are interdependent, which opens up new prospects for biological dangers that could erode public trust. Openness is essential. A number of broad categories can be used to categorize the effects of cybersecurity failures in Fd+Ag systems 2018 [21].

- Data protection – threats to secrecy
- Breaches of data (personal raw data breaches, small business cybersecurity hacks or potential threat labs).
- Gathering personal information in order to increase profits or revenues as predicted.
- Risks to integrity include security threats, production disruptions, and theft or destruction of intellectual property. Inappropriate use of intellectual property rights (e.g. animal and plant species development and genetics).
- Automated control of crucial processes via computers (heat treatment time and temperature for food safety).
- Autonomous or robot-driven vehicle steering (e.g. performance bugs, microrole switching).
- Risks to supply of agriculture/food include interruptions to those industries' production and supply.
- False information that undermines customer confidence and cooperation inside the Fd+Ag system.
- Addressing anticipated tool, material, or completed item shortages; inability to conduct vulnerability studies and develop emergency action plans (e.g., safeguarding rivers, surface water, and drinking water sources).

The advancement and use of biology, genetics, PA, and informatics technology have reached a turning point in the food and agricultural industries. Generally, we must assess potential hazards and appreciate the limitations of biological and genetic data systems. Figure 2.6 is shown below.

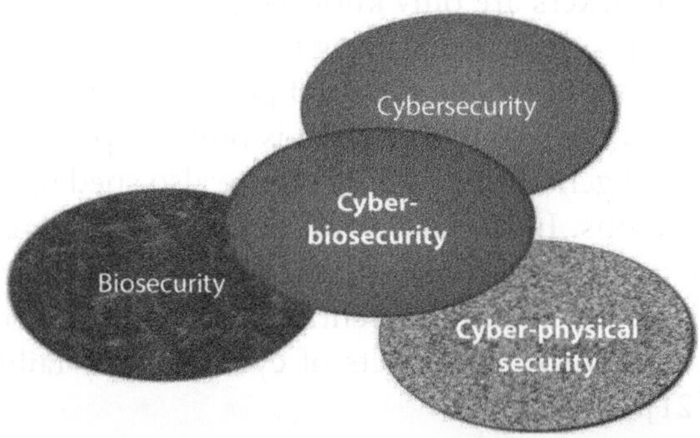

Figure 2.6 Cyber security is an emerging discipline for protecting the data, functions, and processes (or infrastructure) of the life sciences and bioeconomy.

2.7.1 Risk Mitigation Strategies and Countermeasure

The examination of SF and AP attacks and security threats emphasizes the necessity of investigating mitigation and response tactics for various IoT attacks and security risks in SF and AP. The relevant studies are condensed in Tables 2.2 and 2.3. Although agricultural expertise is not required for all farmers, cybersecurity concerns and technologies FarmerScan has certain cybersecurity measures it can do to lower security threats. Data encryption, frequent password changes, two-step user authentication, and necessary software updates are some examples of these [22, 23]. We advise staying away from emails, dubious links, and saving personal and financial data to your browser or autocomplete [21, 22]. Farmers also need to wait until the system has been shown to be secure before making data-driven judgements. The cybersecurity standards of their data ecosystems must be strengthened because businesses frequently share data with reliable advisors, researchers, suppliers, and customers. Less security threats can also be achieved by regulating physical access and other physical processes. For instance, it is important to remember that safety precautions for preventing complications in the workplace and among coworkers could be ineffective [24]. The severity of the repercussions can be lessened if farmers are unable to access stolen data because of data backups. Cameras, external

Table 2.2 Risk mitigation strategies for SF and PA.

Strategy	IoT layer	Related security aspect
User Authentication [11]	Middleware	Privacy, Confidentially, Authenticity
Secure Passwords [12]	Middleware	Privacy, Confidentially, Authenticity
Device-Level Encryption [13]	Edge layer	Privacy, Integrity Middleware
Communication-Level Encryption (SSL/TLS) [14]	Access Gateway	Confidentially, Integrity, Authenticity
Port Hardening [3]	Internet, Cloud	Privacy, Authenticity, Non-Repudiation
Remote Login Deactivation [15]	Internet, Cloud	Privacy, Authenticity, Non-Repudiation

Table 2.3 Attack countermeasures for SF and PA.

Measure	IoT layer	Related security aspect
Password Recovery Made Safe [16]	Middleware	Privacy, Confidentially, Authenticity
Unauthorized Account Deactivation [17]	Edge layer	Privacy, Confidentially, Authenticity
Periodic Device Evaluation [21–25]	Edge layer	Privacy, Availability

caches, and other security measures can help reduce security threats [3]. Store important information, documents, and equipment in a secure location, like a locker. For fault tolerance, access control, and third-party censorship, blockchain technology has been incorporated into a number of architectural designs [3]. It also ensures data privacy and security. The IoT and ICT use in agriculture research has been shown to improve security and data protection. In an environment of intelligent management, risks to security functions and threat layers are addressed by a list of potential responses in Table 2.2. In fact, a good method to lessen security risks in SF and PA situations is to recognize and address this problem. Industrial control systems (ICS) cybersecurity risk evaluation is developing quickly. For evaluating the cybersecurity risks related to ICS, [25] looked at 24 different assessment techniques. The evaluation of security risks to adequately safeguard Fd+Ag systems is a cybersecurity risk assessment technique that takes into account the special security issues at the nexus of biological, physical, and cyberspace is essential. Cybersecurity management in the Fd+Ag domain faces different economic challenges. Due to the network features of the system, the most urgent problems are external impacts and improper incentives on individual agents. Individual company losses may not even come close to comparing to the dangers posed by cybersecurity attacks and societal costs. It is possible that no single movie will make the network's social security system function at its best. Agents are aware that their security depends on investments made by third parties, yet it is still free. Oversupply in the public sector or in the linked private sector will probably yield the quickest profits [26]. In terms of political and editorial editing of the Fd+Ag address system, there are numerous alternatives. When it comes to strict government laws that support conservation efforts in the Fd+Ag sector, cultivated plants may occasionally be irrelevant or relevant. Regardless of the final policy decision, it is critical to strike the correct balance between the expense of system upkeep and the likelihood and severity of harm brought on by cybersecurity threats. A variety of regulatory strategies will determine the best strategy to safeguard the Fd+Ag system. Table 2.3 depicts the attack counter measure for SF and PA.

2.7.2 A Consideration of the Diversity Within and Between the Plant, Animal, and Environmental Sectors of the Food and Agricultural System

The Hazard Analysis and Critical Control Points (HACCP) concept assesses risk and identifies crucial control points particular to a given establishment, however it cannot be effectively applied to all food establishments. As a result, it is challenging to apply this approach to cybersecurity risks, checkpoints, and important checkpoints because of the diversity of businesses in the sector. moreover, within the Fd + Ag system. As an illustration, the Fd+Ag system targets various portions of the commodity segment using an image of a multi-carriage train (Figure 2.7 above). Every time an organization or agency receives funding, the entire system is put

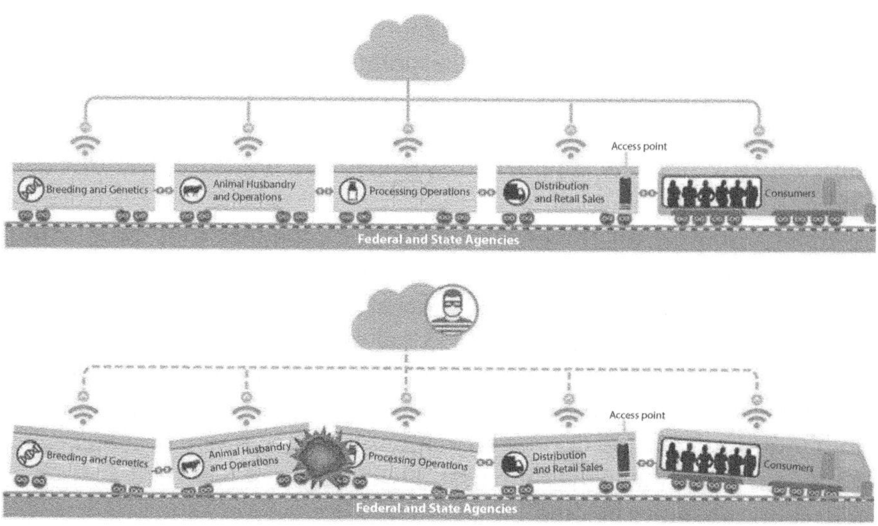

Figure 2.7 (top) The little contact and data sharing across the various stages of the Fd+Ag system for each commodities sector; (see below) An attack on cyber security might have devastating effects on the overall Fd+Ag system.

in danger, posing persistent hazards to the essential economy and the U.S. food supply (Figure 2.7 below). Many solutions to this issue's potential mitigation exist. Protection of physical and digital property (data) associated to Fd+Ag priority requires careful planning and execution of cybersecurity measures. For instance: animal and plant germplasm resources (including economic value preservation and data exchange), such as old-world maize germplasm resources and microbial banks (pathogens, fermentation, microbial communities); networks for diagnosing animal and plant diseases and exchanging information; fermentation process and heat treatment control parameters; supply of fresh water, potable water, and sanitation systems.

A few of the distinctive qualities of the numerous Fd+Ag compounds are also described.

Dairy: In order to breed successfully, genetic information is crucial. When identifying productive animals, milk production statistics are crucial. The majority of American dairy farms are small and medium-sized, and the majority of them sell their milk through cooperative arrangements, despite having extremely large cows (over 2,000). Establish a system for tracking medication use and employee health. Security of data varies and is frequently constrained. Producers of dairy products and fluid milk do not keep thorough records of each cow's production or farming methods, which results in information monitoring gaps and possible data breaches.

Bush crops: The capacity to trace grains is severely constrained because they are frequently combined at the point of sale and gathered during storage and processing [27]. Machine positioning and performance are enhanced in modern farms employing precision farming techniques. Agriculture technology suppliers frequently have influence over this data [28].

Water resources for the environment: The economy, food production, consumer health and safety, and the health and wellbeing of ecosystems all depend on the safety of drinking water. The percentage of people who have use certified and regulated sources of drinking water—which is close to 90%—is rising [29]. An estimate of 2.3 billion people worldwide do, however experience illnesses brought on by contaminated drinking water. Both industrialized and developing nations have had severe occurrences of tainted drinking water over the past 30 years, which have had a negative impact on consumer health.

2.7.3 Technological Solutions for the Intelligent Production of Food

The development of technology for food production has significantly altered the way many people operate, especially farmers, researchers, and everyone else involved in agriculture.

- Using agricultural technology mainly benefits agricultural systems' productivity and reduces pollution.
- Using herbicides, fertilizers, and water efficiently.
- The cost of groceries is going down.
- Playback and real-time observation.
- A smaller environmental impact.

Technological and financial expertise in technologies used on smart farms in Central America and the Caribbean, such as greenhouses, small and big tunnel structures, monitoring, control, and automation systems, smart irrigation systems, and soil and water analysis and management systems are is explained in detail below.

2.7.3.1 Structures of Greenhouses, Micro-Tunnels, and Macro-Tunnels

One of the most widely used strategies to combat climate change is manufacturing in controlled environments. Plants are grown using the controlled environment culture method, which creates ideal climatic conditions for plant growth by using closed structures that are tailored to non-climatic situations [23]. Examples of growing in a controlled environment include Production produced utilizing cutting-edge technology, managing all climatic conditions in a fully enclosed environment inside a building, resulting in high production costs. The purpose of greenhouses is to generate various microclimates that are suitable for plants. They are tall, closed structures that are covered with transparent material (polycarbonate plastic sheets, glass, fiberglass, or cardboard). According to the design of the structure, it can be classified as swept, asymmetrical, liturgical, double liturgical, tunnel, semi-cylindrical, glass or ventro, grating or shaded greenhouses. Temperature, humidity, sunlight, and pests are all controlled in greenhouses to aid in the growth of plants. In order to manage radiation, rain, wind, pests, and diseases during the early phases of plant development, mesh support structures called microtunnels are small, straightforward, inexpensive, and easy to install. Most often, medium-sized constructions, large tunnels are them. Such items as bamboo tables, PVC pipes, or a greenhouse are made of galvanized iron that has been wrapped in plastic, agrotextiles, or insect screens. At different stages of crop development, radiation, rain, wind, pests, and illnesses. Surveys of the wind, the soil, the water quality, and the infrastructure at the site [27]. When building a greenhouse, there are several things to take into account.

- 0°C or more is the minimum temperature that plants need. Analysis of the minimum soil temperature, the highest soil temperature, the daylight hours, and the absolute minimum temperature (0°C), among other variables, is crucial. Relative humidity and temperature should be kept to a minimum. Knowing the environment both inside and outside the building is necessary for the best adaption.
- It is crucial to be aware of the external factors that will facilitate building construction in advance, including height, temperature, precipitation, topography, wind speed, water supply, water quality, and staff availability.

2.7.3.2 *Monitoring, Control, and Automation Systems*

The production of food using old methods in conventional agriculture is characterized by a lack of knowledge of the technological and ecological behavior of these methods, which leads to production systems with low yields and serious land use issues. Dehydration and suffocation of plants, nutrient deficiencies, heat suffocation, and sunburn are some of the most prevalent issues. To find changes in various parameters compared to programmed initial values, monitoring components can be utilized. A system made up of a central data gathering device, a network of sensors, and/or RFID tags can be used to monitor these factors in large part. Producers can: Get a real-time understanding of the many components of the agricultural system thanks to the information generated [16]. Generally, these computer systems are built with wireless connection technologies, offering low costs, simple installation, mobility, and agricultural systems. R, ZigBee, Bluetooth, Wibree, and WiFi Link are the four communication protocols that smart agriculture is most likely to use. Table 2.4 provides an overview of some of the variables that are typically examined in protected crop production systems. A central controller is necessary for a smart farm in addition to the monitoring system since it serves as a system for automation and control, processing data for all other components and delegating tasks to others. The decisions that the intelligent system makes in response to deviations that the sensors have observed are directly tied to these

Table 2.4 Environmental factors observed in smart crop systems [16].

Component	Environmental factor
Environment	Temperature and humidity
	Solar radiation
	Precipitation
	Wind speed and direction
Soil	Temperature and humidity
	Conductivity
	Salinity
Plants	Temperature and humidity
	Carbon dioxide
	Hydrogene

functions. Typical examples of functionality include the ability to switch irrigation systems on or off in response to environmental conditions, crop demand, and central controller decisions.

2.8 The APTs in SF and PA

Due to this conclusion, the discussion in this part will concentrate on APT as a case study for additional investigation. The remaining paragraphs of this section initially provide an overview of basic APT assaults before going into greater detail about APT attacks on SF and AP. We examined various behavioral factors in addition to assessing these attacks. The APTs typically represent risks where a perpetrator (or perpetrators) connects to the network regularly while remaining unnoticed in order to locate and steal strategic data [12]. Very strategic business, industrial, financial, government, or national security and defense organizations that store, transfer, and handle sensitive information frequently utilize APTs. Systematic methods for detecting and preventing APT attacks usually rely on intricate interactions between security service providers and end users with regard to information on military and political strategy, nuclear and space technologies, intellectual property, and more. Whitelisting (monitoring and simulating network access to permitted domains) and access control are common strategies for this and include traffic monitoring, access control, and these.

2.8.1 APT Attacks on SF and PA

The APT organizations attempt to attack high-value targets including greenhouses, zoos, and smart farms. The APT groups use cutting-edge strategies to accomplish its objectives, including phishing, social engineering, and zero-day assaults. According to the outcomes of the SF and PA's ATP attacks, theft of intellectual property (patents, etc.), stealing important data with food chains, controls, genetics, etc., changing database items or control parameters cause damage to important agricultural infrastructure.

2.8.1.1 The Anatomy of an APT Attack on SF or PA

Typically, there are three phases to an efficient ATP assault in SF or PA [30]. This one connects to the CKC step. Figure 2.8 depicts the anatomy of an APT attack.

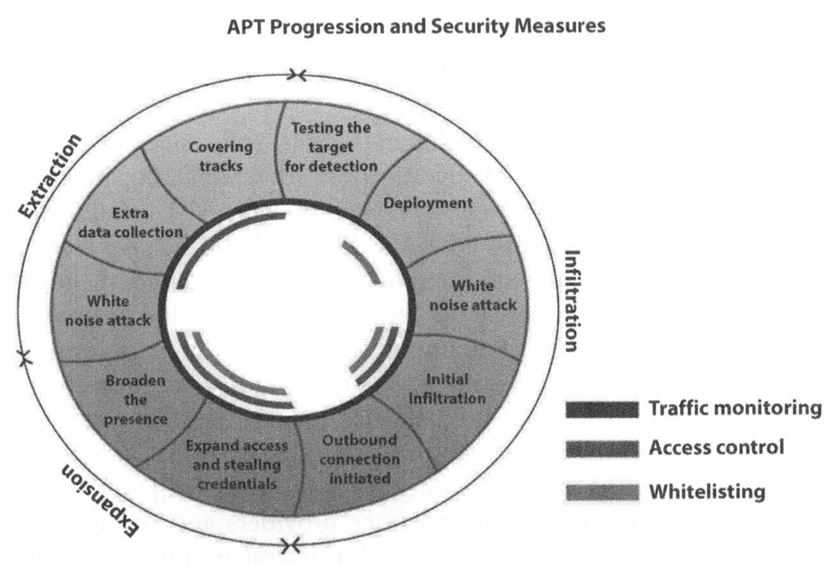

Figure 2.8 The anatomy of an APT attack.

2.8.1.1.1 Penetration (Infiltration)

To the first three CKC steps, this step can be mapped. This stage can be recognized through social engineering [29], which is a popular technique. Exploits and malware installation could be part of this process (SQL injection). Assailants may gain additional access to your network through backdoors created by snooping. The following are steps that are part of this step:

- Test target detection
- Composition
- White noise attack
- Initial penetration
- Initiate an outgoing connection.

2.8.1.1.2 Further Access (Expansion)

To stages 4–6 in the CKC, this step can be mapped. The attacker attempts to get long-term access to the network or other important resources during this phase. Mastering and using a crucial skill is the objective in order to set up the decisive action. You can conduct the following two tasks at this point:

- Use techniques similar to phishing to extend access and steal credentials.
- Expanding existence.

2.8.1.1.3 Information Stealth and Sabotage (Exploitation)

The last phase is comparable to the final CKC step. The APT groups can now harm systems, stop strategic operations, or steal important information. The APT organizations may now be interested in information on the value of raising animals and growing crops, as well as information on the cultivation process, which is crucial for business. The production process could be hampered by using this information to sell to other companies. On the victim's network, the stolen data might have been retained for potential sensitive transmissions in the future. To avoid being discovered by monitoring systems, APT attacks frequently employ white noise tactics [30]. The following actions are typically part of this step:

- White noise attack.
- Additional data collection.
- Cover tracks.

2.9 Challenges in the Implementation of Technologies in the Agricultural Sector

The demand for food generally rises as a result of economic and population growth. Food production is nevertheless impacted by climate change even with abundant human resources and fertile land. Uncontrolled temperature increases that have a direct impact on plant, animal, and human health are one of the variables that have had the biggest influence on the agricultural industry in recent years. Panama has many water resources, and the water there is of excellent quality, especially when it rains. The dry season has, however, recently been marked by seasonal water shortages. As an illustration, Panama went through one of the biggest water shortage problems in 2015–2016 as a result of the "El Nio" climatic phenomenon [11, 12]. Food production is mostly impacted by this. The decrease in watershed flow is what causes the loss of agricultural yield. Every year, water consumption rises in all areas of Panama's economy and population due to both factors. An estimate of 43% of Panama's total water use in 2010 (or 446 million cubic metres) was used for agricultural purposes. Inadequate regulation of urban development is another issue facing the nation. It impairs the soil's ability to absorb water, changes the water cycle, and increases runoff. Nevertheless, the groundwater of Panama has received very little study. This is a result of the absence of appropriate instruments and methods for keeping track of current aquifers, recharge zones, and potential output. Because of all this, there is a developing water issue that threatens systematic, effective, and sustainable water management. Land use modification is yet another problem in Panama. The roughly 1.8 million hectares of land in the nation offer perfect circumstances for agricultural growth, but due to the high food demand of the country's quickly expanding population, the actual land area utilized for agricultural operations has expanded to 2.9 million hectares. Land degradation is brought on by these regional disputes, which also cause widespread erosion, fertility loss, and pollution. In this situation, the challenge for Panama is to increase food production on the acreage used for agricultural operations in order to maintain the soil's fertility and the sustainability of the ecosystem. Lack of human resources to address agricultural issues is one of the main issues the Panamanian agricultural sector faces, along with issues like industrial waste.

2.10 Open Challenges and Research Areas

Concerning SF and PA security, this section contains open-ended research issues. To start the process of creating novel FS/PA mitigation measures, we analyze the existing research, construct a taxonomy, and look at APT case cases in this study. The information, resources, and guidelines of SF Maker can be used to produce future research through multidisciplinary study. When they create more specialized plans for SF and PA contexts, developers ought to test the mitigation methods discussed here. These issues could serve as a starting point for further inquiries into this subject. In order to enhance security and privacy in smart agriculture ecosystems, this section highlights public research challenges (see Figure 2.9).

- **Access Control from a Security Perspective:**

Owners, employees, greenhouses, and animal farms. Access control is traditionally covered. On their property, though, they ought to take security precautions. Unauthorized access, which is the basis for many harmful SF and PA attacks, must be stopped. Authentication, authorization, and reporting must be integrated. Despite the fact that the literature contains a large number of linked studies, more thorough investigation is still needed [30].

- **Data Protection:**

Bulk data collection, communication, and data processing are well-known features of SF and PA due to the powerful capabilities of smart devices and the IoT [1, 2].

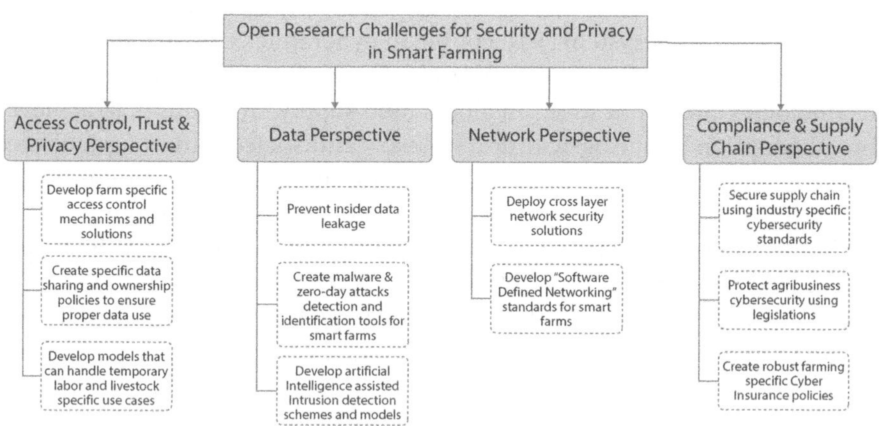

Figure 2.9 Open research challenges and future directions for cyber security in smart farming.

- **Network Infrastructure and Physical Layer Protection:**

Various attack surfaces [3]. This topic is covered by many research projects and related research is still needed.

- **Education as Risk Reduction:**

Researchers have not paid much attention to societal elements like education that lower Internet danger [4]. Farmers are not just concerned with lowering safety concerns on the work, but they also see education as a key tool in doing so [5]. Assessment of the current cybersecurity practices adopted by farmers is a prospective area of research. By doing so, we can identify the training's most valuable format, subject matter, and method of delivery.

- **Specific Secure Communication Protocols for SF and PA:**

Many network protocols, including Bluetooth, NFC, WiFi, SSL, TCP, UDP, 6lowPA, M2M (machine-to-machine), and IoT, have been incorporated into other technologies and are still being studied for personal security. Space arrangements are in place.

2.11 Conclusions

In order to reach their potential, SF and PA must be safeguarded from cyberattacks. These technologies can increase global food security and lessen the environmental effect of agriculture. The current state of security risks and cyberattacks is evident. An attack like that might have a significant impact on world markets, particularly those of poorer nations where agriculture is a major economic driver. Connected devices and data-focused applications in different industries compound security and privacy concerns are rendering these systems susceptible to cyberattacks. Cyber biosecurity threats and attacks are now more likely due to the quick development of technologies and applications in global food supply. In addition, to achieve broad adoption by the agricultural industry and society, smart agriculture implementation must take into account not just technological considerations but also environmental, economic, and social considerations. There are some crucial security factors to think about in SF and PA. A thorough threat taxonomy based on the CKC was developed as a consequence of research on cyberthreats in the SF and PA regions. Other from that, he writes about risk management and approaches to countering attacks by Palestinian Authority and Special Forces. Last but not least, we are creating a future research roadmap for this field.

References

1. Gupta, M.; Abdelsalam, M.; Khorsandroo, S.; Mittal, S. Security and privacy in smart farming: Challenges and opportunities. *IEEE Access*, 2020, *8*, 34564–34584.
2. Demestichas, K.; Peppes, N.; Alexakis, T. Survey on Security Threats in Agricultural IoT and Smart Farming. *Sensors*, 2020, *20*, 6458.
3. ia, T.N.; Qingqing, L.; Queralta, J.P.; Zou, Z.; Tenhunen, H.; Westerlund, T. Edge AI in smart farming IoT: CNNs at the edge and fog computing with LoRa. In *Proceedings of the IEEE AFRICON*, Accra, Ghana, 25 September 2019.
4. Jahn, M.M.; Oemichen, W.L.; Treverton, G.F. *Cyber Risk and Security Implications in Smart Agriculture and Food Systems*; Technical Report; College of Agricultureand Life Sciences, University of Wisconsin: Madison, WI, USA, 2019.
5. Kasten, J. Blockchain on the Farm: A Systematic Literature Review. *J. Strateg. Innov. Sustain.*, 2020, *15*, 129–153.
6. Vangala, A.; Sutrala, A.K.; Das, A.K.; Jo, M. Smart Contract-Based Blockchain-Envisioned Authentication Scheme for Smart Farming. *IEEE Internet Things J.*, 2021, *8*, 10792–10806.
7. Su, J.; Yi, D.; Su, B.; Mi, Z.; Liu, C.; Hu, X.; Xu, X.; Guo, L.; Chen, W.H. Aerial Visual Perception in Smart Farming: Field Study of Wheat Yellow Rust Monitoring. *IEEE Trans. Ind. Inf.*, 2021, *17*, 2242–2249.
8. Shadrin, D.; Menshchikov, A.; Somov, A.; Bornemann, G.; Hauslage, J.; Fedorov, M. Enabling Precision Agriculture Through Embedded Sensing With Artificial Intelligence. *IEEE Trans. Instrum. Meas.*, 2020, *69*, 4103–4113.
9. Sambo, D.W.; Forster, A.; Yenke, B.O.; Sarr, I.; Gueye, B.; Dayang, P. Wireless Underground Sensor Networks Path Loss Model for Precision Agriculture (WUSN-PLM). *IEEE Sens. J.*, 2020, *20*, 5298–5313.
10. Srinivas Kolli, M. Sreedevi. 2018. "Adaptive Clustering Approach to Handle Multi Similarity Index for Uncertain Categorical Data Streams". *J. Adv Res. Dyn. Control Syst.*, Vol. 10, 04-Special Issue, 2018.
11. Chukkapalli, S.S.L.; Mittal, S.; Gupta, M.; Abdelsalam, M.; Joshi, A.; Sandhu, R.; Joshi, K. Ontologies and Artificial Intelligence Systems for the Cooperative Smart Farming Ecosystem. *IEEE Access*, 2020, *8*, 164045–164064.
12. Sheikdavood K, Surendar P, Manikandan A. Certain Investigation on Latent Fingerprint Improvement through Multi-Scale Patch Based Sparse Representation. *Indian J. Eng.*, 2016; 13(31):59-64.
13. Amit Grover, A. Manikandan, Vivek Soi, Anu Sheetal, Mehtab Singh, Realisation of white LED using fiber based hybrid photonic structures, *Optoelectron. Adv. Mat.*, 15, 11-12, November-December 2021, pp.521-527 (2021).

14. Kolli, S., &Sreedevi, M "A novel index based procedure to explore similar attribute similarity in uncertain categorical data." *ARPN J. Eng. Appl. Sci.*, 14.12.
15. Yazdinejad, A.; Parizi, R.M.; Dehghantanha, A.; Karimipour, H. Federated learning for drone authentication. *Ad. Hoc. Netw.*, 2021, *120*, 102574.
16. Bahrami, P.N.; Dehghantanha, A.; Dargahi, T.; Parizi, R.M.; Choo, K.K.R.; Javadi, H.H. Cyber kill chain-based taxonomy of advanced persistent threat actors: Analogy of tactics, techniques, and procedures. *J. Inf. Process. Syst.*, 2019, *15*, 865–889.
17. Kolli, Srinivas and M. Sreedevi. "Prototype analysis of different data mining classification and clustering approaches." *ARPN J. Eng. Appl. Sci.*, 13.9 (2018): 3129-3135.
18. Balamurugan, D. & Seshadri, s.Aravinth & Reddy, P. & Rupani, Ajay & Manikandan, A.. (2022). Multiview Objects Recognition Using Deep Learning-Based Wrap-CNN with Voting Scheme. *Neural Process. Lett.*, 54. 1-27. 10.1007/s11063-021-10679-4.
19. Kamath, R.; Balachandra, M.; Prabhu, S. Raspberry Pi as Visual Sensor Nodes in Precision Agriculture: A Study. *IEEE Access*, 2019, *7*, 45110–45122.
20. Ahmed, B.K.I.; Prakash, A.; Husna, M.S.; Prakash, A. A Review on Smart Farming with Zinc-Fortified Sprouts. In *Proceedings of the Fourth World Conference on Smart Trends in Systems, Security and Sustainability (WorldS4)*, London, UK, 27 July 2020.
21. Sreedevi, T.R.; Kumar, M.S. Digital Twin in Smart Farming: A Categorical Literature Review and Exploring Possibilities in Hydroponics. In *Proceedings of the Advanced Computing and Communication Technologies for High Performance Applications (ACCTHPA)*, Kerala, India, 2 July 2020.
22. Venkataramanan, C. & Ramalingam, S. & Manikandan, A.. (2021). LWBA: Lévy-walk bat algorithm based data prediction for precision agriculture in wireless sensor networks. *J. Intell. Fuzzy Syst.*, 41. 2891-2904. 10.3233/JIFS-202953.
23. Kolli, Srinivas, AV Praveen Krishna, and XXXM. Sreedevi. "A Meta Heuristic Multi-View Data Analysis over Unconditional Labeled Material: An Intelligence OCMHAMCV." *Scalable Comput. Prac. Exper.*, 23.4 (2022): 275-289.
24. Manikandan, Annamalai, M,Ponni Bala. (2022). Intracardiac Mass Detection and Classification Using Double Convolutional Neural Network Classifier. *J. Eng. Res.*, 65. https://doi.org/10.36909/jer.12237
25. Idoje, G.; Dagiuklas, T.; Iqbal, M. Survey for smart farming technologies: Challenges and issues. *Comput. Electr. Eng.*, 2021, *92*, 107104.
26. Ferrag, M.A.; Shu, L.; Yang, X.; Derhab, A.; Maglaras, L. Security and privacy for green IoT-based agriculture: Review, blockchain solutions, and challenges. *IEEE Access*, 2020, *8*, 32031–32053.
27. Ametepe, A.F.X.; Ahouandjinou, S.A.R.; Ezin, E.C. Secure encryption by combining asymmetric and symmetric cryptographic method for data

collection WSN in smart agriculture. In *Proceedings of the IEEE International Smart Cities Conference (ISC2)*, Casablanca, Morocco, 19 October 2019.

28. Annamalai, Manikandan & Muthiah, Ponni. (2022). An Early Prediction of Tumor in Heart by Cardiac Masses Classification in Echocardiogram Images Using Robust Back Propagation Neural Network Classifier. *Braz. Arch. Biol. Technol.*, 65. 10.1590/1678-4324-2022210316
29. Kolli, S., Obaid, A.J., Saikumar, K., Reddy, V.S. (2023). An Accurate Swine Flu Prediction and Early Prediction Using Data Mining Technique. In: Rai, B.K., Kumar, G., Balyan, V. (eds) *AI and Blockchain in Healthcare. Advanced Technologies and Societal Change*. Springer, Singapore. https://doi.org/10.1007/978-981-99-0377-1_14
30. Sankar, S., Somula, R., Parvathala, B., Kolli, S., &Pulipati, S. (2022). SOA-EACR: Seagull optimization algorithm based energy aware cluster routing protocol for wireless sensor networks in the livestock industry. *Sustain. Comput. Inf. Syst.*, 33, 100645.

3

Precision Smart Farming and Cultivation with Virtual Reality/Augmented Reality Technology - Applications and Use Cases

Himani Sharma[1], Atin Kumar[1]* and Rohit Kumar[2]

[1]School of Agriculture, Uttaranchal University, Premnagar, Dehradun, Uttarakhand, India
[2]Faculty of Agricultural Sciences, GLA University, Mathura, Uttar Pradesh, India

Abstract

Prior to the introduction of technological advances in agriculture, a farmer's chances of producing quality crops were about equal to flicking a coin and hoping for heads. Smart farming uses new technologies that emerged at the start of the Fourth Industrial Revolution in the fields of agriculture and animal husbandry, precision farming aims to boost production quantity and quality while maximizing resource efficiency and reducing the negative environmental effects. A further benefit of using technology in agricultural and livestock production is that it will contribute to the worldwide availability of food. Traditional farming methods are regional in nature. There is a standard set of crops that are grown everywhere. Regarding the sowing, nourishing, watering, and harvesting periods, every farmer in that region uses the same practice. These behaviors lead to unpredictable outcomes and excessive resource utilization. There was no way for farmers to determine the reason for crop loss because they lacked information on the land they owned. The farmers were forced into debt and losses by this practice. Farming innovations increased the sector's glee, addressing the problem of uncertainty.

An example of artificial intelligence (AI) in use is virtual reality (VR). It exemplifies the use of computer technology. It aids in producing an artificial environment that is a replica of the actual environment and is comparable to it.

**Corresponding author*: atinchaudhary0019@gmail.com

Rajesh Kumar Dhanaraj, Balamurugan Balusamy, Prithi Samuel, Malathy Sathyamoorthy and Ali Kashif Bashir (eds.) Computer Vision in Smart Agriculture and Crop Management, (57–70) © 2025 Scrivener Publishing LLC

The application of augmented reality (AR) to farming will increase farm output. Agriculture will undergo a technological revolution as a result of AR. In the coming decades, the human population will grow steadily. As a result, AR will become more significant in agriculture.

Keywords: Precision farming, smart farming, augmented reality, virtual reality, artificial intelligence

3.1 Introduction

The third modern agricultural revolution is about to begin, and precision farming is a key component of it. The First Agricultural Revolution, which took place between 1900 and the 1930s, was mechanized agriculture, which enabled each farmer to create enough food for 26 people. The second movement, commonly referred to as the Green Revolution, started many years later, in the 1990s. Due to advances in science, newer varieties of pest-resistant, water-saving crops have been introduced, allowing each farmer to support 155 people [4].

By 2050, there will be 9.6 billion people on the planet, which means that food production will need to double in order to support everyone. The Third Revolution will be driven by cutting-edge analytical capabilities and continuously evolving IoT capabilities (Figure 3.1), such as drones for precision agriculture.

For farmers and the companies that represent them, it is essential to comprehend precision farming in its truest form.

The practice of collecting, analyzing and reacting to different intra- and inter-field variability inputs for contemporary agriculture is known as precision farming.

Figure 3.1 Internet of Things (IoT)-based smart farming.

According to common definitions, the word refers to "a technology-enabled strategy for agriculture administration that investigates, forms policies, and assesses the demands of particular farms and produce." Site-specific crop management (SSCM), satellite farming, and precision agriculture are other terms for the same practice [7].

By utilizing crop control software and scientific instruments, precision farming is practiced. Through sensor-equipped equipment positioned on the fields, thorough data gathering is carried out on the sampling of soil, plot evaluation, seasonal weather assessment, and crop research. A thorough and specific array of procedures can be implemented based on the conclusions that are developed using the data [10].

According to the National Academy of Science (NAS), precision farming should significantly contribute to improving farming's sustainability without reducing output or agricultural income. The World Economic Forum evaluated that if 15 to 25 percent of farms utilized precision agriculture, global food output could increase by 10 to 15 percent by 2030 while emissions of greenhouse gases and water use would both drop by 10 to 20 percent.

Precision agriculture's location-specific field supervision and variable-rate application (VRA) methods are two of the topics that are most frequently addressed. These steps are crucial to minimizing agriculture's negative environmental impact [4].

A farmer's chances of producing quality food prior to the introduction of modern technology for agricultural purposes were about the same as flipping a coin and getting result as per desire. There was little opportunity for farmers to determine the reason for losses to crops because they lacked knowledge about their farms. The producers were forced into debt and losses by this practice. The agriculture industry is optimistic as a result of developments in big data analytics, IoT, and readily available satellite imagery, which help to combat the problem of unpredictability [6].

Because it is customized to the requirements of each plot, precision agriculture and modern methods of agriculture put an emphasis on reducing the cost of production and waste. It focuses on gathering and analyzing data from farm IoTs (Figure 3.1), which includes sensors, drones, and robots to capture the data. Software as a service (SaaS) can be utilized to modify precision agricultural practices.

The worldwide population will reach 10 billion people by the year 2050! (World Bank) That is five times more than the 1.8 billion people we had in 1915. By 2050, the demand for sustenance is expected to nearly double. A novel possibility for addressing the problem of rising food demand in agriculture is augmented reality (AR) [8].

Indeed, farmers are severely distressed as a consequence of rising food demand. Farmers are holding out for better days. The inability to supply the rising demand for sustenance could have negative effects. Among them are only starvation, death, and political turmoil. A transitional shift in agriculture is therefore required. Farm output will increase as a result of the application of AR. Agriculture will undergo a revolution as a result of the AR technology surge. In the coming decades, the number of people will continue to increase. Consequently, AR becomes increasingly significant in agribusiness [1].

The AR continuum now encompasses agriculture, as well after revolutionizing numerous sectors! Consequently, learning about AR in agriculture would be fascinating.

Future food demand management strategies being used today are useless. Agriculture has undergone significant change over the years, but there is still room for growth [4]. Growing interest in smart farming is evidence that the agricultural sector is rapidly adopting cutting-edge technology. The new smart agricultural innovations are powered to a large degree by automated systems, artificial intelligence (AI), statistical analysis, and IoT. However, AR stands out among the numerous technological developments that are at present challenging agriculture. Agriculture can benefit from AR in a variety of ways, including field condition tracking, pest infestation detection, education for workers, etc. [6].

3.2 Advantages of Precision Smart Farming

In order to enhance food security, it has been suggested that agricultural production of food be increased given the world population's continued growth.

This is one of the issues farmers are currently dealing with that is challenging to fix, aside from the current worldwide problems related to agriculture like the effects of climate change, financial hardship, insufficient strategies towards the farming industry, and land scarcity.

For greater crop yield and control market demand, it is necessary to discover ways to create higher yields while taking into account elements that boost produce from agriculture. Precision farming can assist in achieving this. Utilizing precision agriculture technology has both financial and ecological advantages, and with this clever digital farming, food yields can be significantly improved. Precision agriculture helps to increase crop output, which eliminates global concerns about food security. Farmers can enhance their agricultural practices by considering the benefits of instantaneous evaluation, field visualization, and drone surveillance [9].

These innovations support better decision-making and increased output [9]:

3.2.1 Precision Agriculture Enhances Crop Cultivation

- Crop administration has gotten much simpler thanks to advances in agricultural technology like aerial vehicles and precision agriculture.
- Tracking mineral or water imbalances allows for the prompt application of pesticides and fertilizers when necessary.
- The pictures that are captured by these cameras will also have high clarity.
- Farmers will be able to see a bird's-eye picture of their crop thanks to this, which is not otherwise achievable.
- With this approach, any problem in the environment can be found right away.
- Since there are sophisticated sensors in position, the photographs will be extremely accurate, making it simple to identify any issues with the crops.
- It is possible to keep a running log of the data, which can be compared to other variable variables like water levels, potassium concentration, plant health, etc.
- Insects can be identified and exterminated as soon as they appear.
- With this method, fertilizers, as well as pesticides, can be used to their fullest potential.
- With the help of this clever digital farming technique, the general health of the product can be identified quickly and accurately [9].

3.2.2 Precision Farming Helps with Real-Time Data

In some circumstances where several choices must be made quickly, farming can become challenging.

- This frequently makes farm living "hard" because if such choices are made incorrectly, it could cost the farm's overall health and result in significantly lower yields in both quantity and quality.

- Precision farming and other cutting-edge agricultural technologies can be used to gather crucial information about the crops and analyze it to make timely, lucrative decisions.
- For instance, since weather and climate changes typically occur gradually and can be readily noticed when farmers take preventative action against them, their negative effects on the yield of crops can be lessened.
- With the help of the data provided regarding the development and general health of the fields and crops, choices can be made that are both accurate and timely.
- There are devices like robot sensors, which can be installed to monitor the crops and collect data about them.
- The farmer can get this data instantaneously.
- Because it is so useful for monitoring weather conditions, avoiding hazards, or rectifying them, technology like this can help with climate change's impacts on agriculture.
- It is possible to anticipate the seriousness of the environmental circumstances early, and farmers can develop mitigation measures to reduce the risks and hazards associated with weather-related conditions in agriculture.

3.2.3 Precision Farming Helps in Optimizing the Use of Resources

Farmers can lower their labor costs by using precision cultivation to make the best use of the resources at their disposal.

- Utilizing the available resources to their fullest will prevent the need for additional labor or resources.
- Additionally, an instrument for mapping is accessible, allowing for both field surveillance and the creation of the ideal planting strategy.
- As a result, it will be possible to choose the timing of when to sow and harvest crops.
- In addition to this, the choice of crop to grow can also be based on the time frame.
- Precision agriculture can be used to choose and apply fertilizers, as well as determine their amount and type.

3.2.4 Decrease in the Use of Pesticides, Fertilizers, and Water

In the past, farmers have applied pesticides, fertilizer, and water without first determining where they were required on the land. However, with smart farming, water and other substances are applied whenever, wherever, and in the right quantities. The expense of these chemicals decreases as their use is decreased since farming becomes less costly.

3.2.5 Reduce Strain on the Environment

The use of fertilizers, water, and other resources that are beneficial to farms has been decreased thanks to modern "smart farming" innovations, which have also increased output. The point of view is that hazardous substances can be utilized judiciously and only if they are required, rather than contaminating the natural environment.

3.2.6 Ability to Make Informed Decisions

Wisdom and expertise can only get farmers so far; data does not deceive. With the information they can gain from their range of agriculture-based technology tools, farmers can more easily plant trees, rotate their crops, conserve soil, and harvest. Farmers can now make better choices about what occurs on the farm because they will always have this knowledge at their fingertips.

3.3 Disadvantages of Precision-Smart Farming

1) Farmers might be deterred from using this agricultural technique by excessive investment expenses.
2) Before being used in practice, precision agriculture methods still need to be developed and require assistance from experts.
3) Before the system has collected enough data to be completely implemented, it may take a few years.
4) It is a very challenging job, especially the gathering and analysis of statistics [10].

3.4 How Could India Benefit from Precision Farming?

- In India, improving and expanding the use of precision agriculture technologies could contribute to lower costs of production, increased profitability, and greater efficient utilization of resources.
- It can fundamentally alter current farm management in India by boosting income generation, efficiency, long-term viability, crop quality, environmental preservation, on-farm overall standards of life, the nutritional value of food, and the financial prosperity of rural communities.
- Site-specific fertilizer, herbicide, and irrigation applications in coffee and tea estates in eastern India, oil palm groves in south India, and cotton fields; and fertilizer applications in cotton may all significantly decrease production costs and lower chemical burdens in the surrounding environment.
- When there is a shortage of water, irrigation can be more effective.
- Farmers can use forecasting and mitigation techniques to address issues like water depletion, nutrient shortages, and pests and diseases.
- It also raises the number of skilled job openings in the agriculture industry and gives new tools for assessing multifaceted factors, such as non-market functions.
- It plays a crucial part in keeping track of greenhouse situations in farming regions [2].

3.5 Challenges in Adopting Precision Farming in India

- Due to its distinct land ownership structure, inadequate infrastructure, farmers' lack of willingness to take on the hazards, economic and fiscal circumstances, and demographic factors, precision farming adoption in India is still in its infancy [8].
- The majority of Indian agriculture's small landholdings limit the economic benefits of precision farming technology that is presently available.

3.6 Cultivation with Virtual Reality/Augmented Reality Technology

An example of AI in use is VR. It exemplifies the use of computer technology. It aids in producing an artificial environment that is a replica of the actual environment and is comparable to it [3].

Software aids in the creation of virtual reality (VR). The person who uses it is exposed to it in a manner that makes them approve and think it is an actual environment. Two senses—vision and hearing—are used when experiencing VR. The application of AR to agriculture will increase agricultural output. Agriculture will undergo a technological shift as a result of AR. In the coming several decades, the human population will grow steadily. Consequently, the role of AR in farmland also becomes more significant. The AR spectrum has expanded to the world of agriculture after revolutionizing many other sectors! The application of AR technology to agriculture would therefore be fascinating to learn about.

Present approaches cannot meet the projected food demand. Despite recent substantial shifts, there is still room for growth in agriculture [8].

3.7 Benefits of Cultivation with Virtual/Augmented Reality Technology

The field-check and insect/pest detection procedures conducted on farmland are transformed by AR. Additionally, it establishes the characteristics of the soil and the crops that will be planted on a specific plot of ground. With just a quick glance at the application, AR aids in knowing pest and crop conditions [5].

3.7.1 Augmented Reality in Agriculture for Simulated Training

In reality, the implementation of AR is widespread in the manufacturing, medical care, and residential property industries, but its application in agriculture has not yet been studied. Because it requires the use of advanced machinery, cultivation is a complicated process. The most dynamic and secure type of coaching will be provided to farmers by AR. Using a tablet or smartphone, you can operate farm machinery! Farmers can now collect and communicate data on equipment, crops, and livestock in the best way they can.

3.7.2 Augmented Reality in Agriculture for Weather Updates

Additionally, weather predictions can be obtained using AR! Farmers can access accurate, real-time meteorological information with AR in agriculture. Farmers may find this knowledge useful when deciding which crops to plant. One of the industries with a significant impact on weather is farming.

In this scenario, AR provides cultivators with real-time weather information. Weather updates that are precise, timely, and relevant to a given place aid in reducing the threats involved with farming. The unpredictable character of the weather causes farmers to exhaust their efforts, precious time, and money. In that situation, AR will assist farmers in adopting some preventative measures [9].

3.7.2.1 Virtual Reality Saves Money on Plants Protection Tools

Knowing about pests, diseases, and the tools to use gives you the information you need to decide how to safeguard the crop and cut expenses.

A field study can be carried out in VR. It is simpler for farmers to assess the soil's effectiveness, choose the right crops, and efficiently utilize the soil's capabilities for crop output.

3.7.2.2 How Augmented Reality Works in Agriculture

Smart farming benefits greatly from AR. To disprove the idea, let us examine how AR functions. With the help of this technology, the present-day real-world environment is improved by the addition of text, audio recordings, illustrations, and audio-visual methods [2].

To provide this experience, AR technology utilizes a smartphone interface or accessories like AR glasses (Figure 3.2). Information from AR may

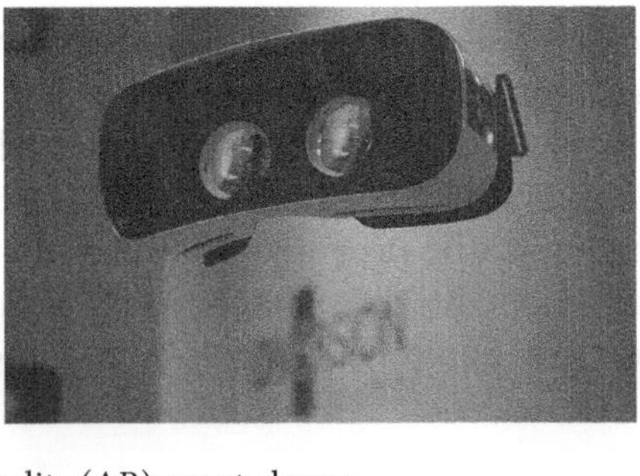

Figure 3.2 Augment reality (AR) smart glasses.

be displayed directly in the eyes of customers, allowing them to observe it, as well as to communicate with it. The material that is projected onto AR glasses or smartphone screens is made by AR software.

A farmer can more clearly analyze the health of crops and insect diseases when wearing AR glasses because they can identify pests, match them to those seen in the field, and show data displayed on the screen.

3.7.2.3 Is Augmented Reality the Future of Productive Farming?

With the help of digital visual elements, audio, or other forms of feedback provided by technological advances, AR is a better representation of what is really happening.

The global population is expanding rapidly, which has led to a sharp rise in consumer demand for commodities like food and farming products. Farmers will need to multiply their production by two. Farmers may benefit from emerging innovations in achieving their goals.

Apart from using AI, big data, blockchains, and the IoT, farmers may also use AR. By using AR in agriculture, growers can reduce crop waste products, increase output, and mentor other farmers (Figure 3.3).

The computerization of many farming operations, also known as "The Green Revolution," spurred by a population boom and greater need for nutritious food between the 1940s and 1980s benefited the entire agricultural sector.

This revolution had an impact on hybrid crops, agrichemicals, machinery, and other novel farm goods and machinery. The Green Revolution increases farming productivity while decreasing waste from operations and wastage, all of which contribute to an improvement in the availability of food [2].

Figure 3.3 Smart agriculture with AR technology futuristic concept, man holding tablet on the farm and use technology to predict the future agriculture growth.

There are currently no commercial products or solutions available for AR in agriculture, which is still in its nascent stages. As a consequence, even though COVID-19 is a momentary occurrence, the market may not immediately be impacted by a change in the economy. Therefore, COVID-19 had no effect on AR in the farming market.

According to the Business Intelligence and Strategy Research (BIS Research), the AR market for agribusiness is anticipated to grow at a Compound Annual Growth Rate (CAGR) of 37.4 percent from 2021 to 2026, reaching $4.43 million. It is anticipated that rising consumer hunger for precision agriculture and smart farming techniques will drive the market's rapid expansion and increase demand for other digital technologies, such as AR [9].

3.7.2.4 Why Does the Agriculture Sector Face Challenges with Augmented Reality?

There are some difficulties with AR in agriculture.

1. The bulky and pricey hardware components are among the main obstacles to the general adoption of AR. And most farmers do not have the financial means to implement such expensive technologies in their fields. In these circumstances, using AR in agriculture is a costly pipe fantasy. Farmers might be able to use AR in agriculture, though, as the price of AR hardware products is falling [7].
2. Farmers must first become thoroughly familiar with the concept of smart farming and the tools and equipment required before they can start using them. A lack of expertise could have catastrophic results.
3. There are many rural areas around the globe that lack reliable, strong internet access. Therefore, efforts to use intelligent agricultural methods are impeded in such regions.
4. The crop, farm zone, and crop/plant growth cycle all affect the production function. There is always a chance that the farmer will apply inputs incorrectly and let the crop suffer if they are not aware of this shifting production function [7].

3.8 Conclusion

Site-specific field supervision and VRA technology are two of the most usually argued aspects of precision agriculture.

These steps are necessary to lessen the impact of agriculture on the ecosystem. Area-centric agricultural methods are the norm. A region typically cultivates a certain collection of crops. Regarding the planting, nourishing, irrigation, and reaping periods, every farmer in that region adheres to the same practices. Unpredictability, excessive resource use, and unchecked waste output are all effects of these behaviors.

In order to handle all of the activities on the farm, smart farming concentrates on implementing the data that has been collected and incorporating it from multiple sources of information to show an overall picture. The next generation of agriculture will be shaped by smart farming, which represents a significant advance over conventional farming.

3.9 Summary

The broader reality marketplace is now led by VR and AR. The aforementioned methods are based on ideas from multidimensional processing, which is thought to be the subsequent phase in the development of computers.

The business development of AR and the corresponding VR ecosystem has been ongoing for ten years. These technologies have seen development over the last five years due to the continuing trends of mechanization and Industry 4.0, and this trend is predicted to continue over the following five years. But only a few apps are seeing progress at this time.

Due to the enormous demand, technological advancements such as AR are expected to grow at a faster pace during the projected time frame of 2021–2026.

References

1. Dhanaraju M, Chenniappan P, Ramalingam K, Pazhanivelan S, Kaliaperumal R (2022) Smart Farming: Internet of Things (IoT)-Based Sustainable Agriculture, *Agriculture.* 12(10):1745. https://doi.org/10.3390/agriculture12101745

2. https://www.agrifarming.in/smart-farming-in-india-challenges-techniques-benefits/
3. https://www.analyticssteps.com/blogs/how-virtual-reality-revolutionizing-agriculture/
4. https://www.iberdola.com/innovation/smart-farming-precision-agriculture/
5. https://www.researchdive.com/blog/6-significant-benefits-of-precision-farming/
6. https:www.queppelin.com/augmented-reality-in-agriculture/
7. Javaid M, Haleem A, Singh RP, Suman R (2022) Enhancing smart farming through the applications of Agriculture 4.0 technologies. *Int. J. Intell. Netw.*, Volume 3, Pages 150-164, https://doi.org/10.1016/j.ijin.2022.09.004
8. Monteiro A, Santos S, & Gonçalves P (2021) Precision Agriculture for Crop and Livestock Farming-Brief Review. *Animals*: An open access journal from MDPI, *11*(8), 2345. https://doi.org/10.3390/ani11082345
9. Shrivastava A, Nayak CK, Dilip R, Samal SR, Rout S, Ashfaque SM (2023) Automatic robotic system design and development for vertical hydroponic farming using IoT and big data analysis, *Mater. Today Proc.*, Volume 80, Part 3, Pages 3546-3553, https://doi.org/10.1016/j.matpr.2021.07.294.
10. Terence S, Purushothaman G (2020) Systematic review of Internet of Things in smart farming. *Trans. Emerg. Telecommun. Technol.*, 31, e3958.

4

Stereo Vision Subsystem and Scene Segmentation Self-Steering Tractors in Smart Agriculture

Dileep Pulugu[1]*, Revathy Pulugu[2], K. Muthumanickam[3], S. Gopinath[4] and A. Manikandan[5]

[1]*CSE, Malla Reddy College of Engineering and Technology, Hyderabad, India*
[2]*CSE, Narsimha Reddy Engineering College, Kompally, Hyderabad, Telangana State, India*
[3]*Information Technology, Kongunadu College of Engineering and Technology, Thottiam, Tamil Nadu, India*
[4]*ECE, Karpagam Institute of Technology, Coimbatore, India*
[5]*ECE, SRM Institute of Science and Technology (Deemed to be University), Kattankulathur, Chennai, India*

Abstract
A key component of smart agriculture is the automatic navigation of agricultural equipment. Machine vision algorithms are being used more and more in smart agricultural machinery applications to pinpoint the precise position of crop rows and plan vehicle navigation routes in real time. This problem describes a tractor system based on vision. This research focuses more precisely on the system design, user safety, typical navigational mistakes, tractor navigation control systems, and presentation. Current research has demonstrated that stereo vision systems offer stronger control stability and are more effective at real-time on-site navigation than monocular systems. It is very accurate for many different crops, such as cotton, sunflower, and corn. By combining backdrops of foliage and sky, it is able to successfully derive orchard navigation pathways. This article offers a quick overview of the state of autonomous agricultural vehicles and lays out all the fundamental principles for integrating computer vision into field autonomous driving. Most essential, agricultural vehicles can adapt to unforeseen field changes.

**Corresponding author*: p.dileep@mrcet.ac.in

Rajesh Kumar Dhanaraj, Balamurugan Balusamy, Prithi Samuel, Malathy Sathyamoorthy and Ali Kashif Bashir (eds.) Computer Vision in Smart Agriculture and Crop Management, (71–92) © 2025 Scrivener Publishing LLC

Automation of farm vehicles is anticipated to boost output and safety while lowering operating costs by requiring less labor and upkeep. Tractors, like other autonomous vehicles, employ a number of digital cameras to monitor attitude, calculate speed, and change direction while the driver is not there. We can provide efficient solutions for scene segmentation and real-time barriers thanks to our experience with computer vision and deep learning algorithms. A stereo vision subsystem for calculating distance is part of the vision control system. Two GPS devices are in charge of driving the car. Absolute positioning, relative positioning, local signal recognition, line cutting, obstructions are all possible with the help of GPS, magnetic compasses, and other tools. Automatic picture segmentation, which identifies lines and cuts for exact navigation, is the fundamental technology. The tractor can avoid collisions with other cars, pedestrians, and animals with the use of an object detection and classification subsystem that is part of the computer vision system.

Keywords: Machine vision algorithms, scene segmentation, stereo vision

4.1 Introduction

By crop monitoring, good management enables you to reach lucrative judgements. Agricultural data is now a crucial consideration when making crucial choices on behalf of farmers thanks to recent advancements in data analysis and management in agriculture-related technology intensification [1, 2]. Farming 5.0, the next generation of farming, is being facilitated by the integration of robots and artificial intelligence (AI) technology with data-driven farm management. Precision farming and data processing are key components of Agriculture 4.0, commonly referred to as digital or smart farming, which aims to assist farmers in making business choices [3, 5]. For vast farms with inexperienced personnel or unforeseen issues that are hard to manage, smart farms offer methodical and practical techniques. Figure 4.1 depicts the Agriculture 5.0 idea, which includes the crop management cycle. Harvesting is where crop management systems begins. Mobile site observation platforms are part of space return surveys. Via remote sensing, the platform gathers information about plants, soil, and the environment in order to offer spatial data to decision-making algorithms. It employs artificial intelligence algorithms to make wise judgements in real time, and sensory feedback controls the system-based decision-making process. Continue doing this all the way through the plant's life cycle [4–6].

An agricultural truck has a movable monitoring platform connected to it. Smart sensors offer existing agricultural vehicles, such as tractors, a higher level of self-awareness, and they may be expanded to integrated

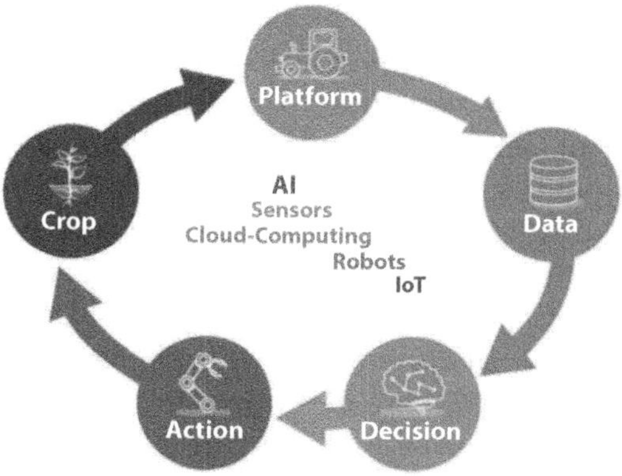

Figure 4.1 The agricultural 5.0 concept and the crop management cycle.

smart and self-driving vehicles that run independently in the field [8–10]. Thus, autonomous tractors or autonomous robots are what are used as agricultural vehicles in Agriculture 5.0 [11]. Tractors and other agricultural equipment must cover enormous regions for extended periods of time while doing monotonous chores. A high level of automation in agriculture is made possible by the autonomous navigation of agricultural tractors. It is used for task automation, more accurate crop structure routing, elevated task safety, and decreased personnel and running expenses [12]. Sensing the surroundings is how autonomy is attained. Local or global location is frequently used for autonomous navigation in agricultural areas [13]. Global information is the GPS-provided precise location of the tractor in the field. This will concentrate on utilizing newly acquired local geographic data from self-guiding tractor vision sensors. An overview of vision-based guiding systems for autonomous tractors is what this paper aims to do. extensive research community for several scenarios, including the navigation of autonomous off-road vehicles.

4.2 Global Positioning System

The only commercially available navigation technology for agricultural vehicles is GPS-based systems. Real-time kinematic (RTK) GPS-based automated steering systems are now available from several tractor manufacturers. Several applications, including seed routing, mapping, traffic control, and crop management, can make use of GPS RTK position data. The GPS guidance systems offer absolute guidance, in contrast to the complementing direction offered by machine vision, which necessitates the use of a GPS-guided food system to grow crops or create crop rows using specific tracking technologies [14]. A GPS base is also necessary for a GPS guiding system. The location and the RTK GPS-guided tractor or agricultural robot should be separated by around 10 kilometers. Nevertheless, GPS systems are unaffected by weed density, shade, loss of vegetation, or any other factors since they do not rely on the visual look of crops. It makes machine vision guiding systems function worse. Another benefit of GPS navigation systems is that curves can be simply encoded into them [15]. There seem to be three restrictions while using GPS while driving. First off, locations that are shielded from microwaves do not have GPS. Also, under a variety of topographical circumstances, GPS does not guarantee a continuous centimeter position accuracy (buildings, trees, steep and winding terrain, satellite leaks or differential correction signals). The inherent delay (data delay) needed to analyze the signal and decide whether space velocities exceeding the second limit could point to a control system issue. The high price of agricultural application is the third (although the cost is low due to widespread use). The use of GPS RTK as a single location sensor in the autopilot system of a self-propelled cannabis harvester was studied by Stoll and Kutzbach [16]. They discovered that the steering deviation was always less than 100 mm. Depending on the displacement speed, the lateral displacement variation (error) along the linear trajectory ranges from 25 to 69 mm. Investigations were made with the use of an RTK GPS control system to steer a tractor from a curving track to an autonomous vehicle [19, 20]. The root mean square error measured on a sinusoidal route with a speed of 6.5 km/h, an amplitude of 2.5 m, and a wavelength of 30 m is 6 cm, while the greatest error measured is 13 cm [17]. The agricultural machinery is guided by an integrated navigation system; the tracking error is less than 5 cm, and the travel speed may exceed 3 m/s. On both straight and curving roadways, all field activities, including batching, cultivation, and spraying, may be completed. The greatest variation for a perfect trajectory is less than 12 cm.

4.3 Self-Steering Tractors with Vision Have Evolved

The 1980s saw a tremendous advancement in computers, electronic sensors, and computation that inspired interest in self-driving cars. Several different bootstrapping methods, such as optical, mechanical, and ultrasonic, have been put forth [19, 23]. From the early 1990s, numerous agricultural applications have made extensive use of the GPS system as a relatively new and precise orientation sensor enabling completely autonomous navigation [24]. The adoption of dependable GPS sensors is nonetheless constrained in agricultural navigation applications due to their expensive cost. When the crop row structure can be seen, machine vision technology based on nearby optical sensors may also be utilized to identify agricultural vehicles. In order to steer the vehicle between the machines for field operations, the camera system can then calculate the relative positions of the machines based on the cutting line. Local characteristics aid in modifying the trajectory of the vehicle across the terrain. The latter is the major justification for why ordered fields with predictable crop behavior have been the main focus of most tractor visual guiding research. To establish directed routes for photos containing crop lines, a number of image processing techniques have been presented. Just a few vision-based guiding systems, nevertheless, have been created to use real-world coverage [25]. The bearing angle and offset for the tractor should be determined using principal component analysis, according to Pinto and Reid's [26] suggestion. The pose recognition issue, where location is established by combining azimuth and offset, addresses this [27]. The author created a machine vision method for crop edge identification with Benson *et al.* The tractor has algorithms that automatically identify domain borders and locate guidance points. The same authors implemented a fuzzy logic-based vision-based guiding system to automate maize harvesting [28]. Assuming accuracy, like GPS, harvesting at night. Osório and Pinto created a tiny agricultural grain harvester with an automated vision system [29]. The suggested approach calculates the guiding signal after separating the uncropped crop line from the backdrop using a black-and-white camera. The most popular form of agricultural navigation technology is automatic tractor steering. Since roughly 20 years ago [30], autonomous tractors have been available for purchase. Sensor fusion is a part of commercial tractor navigation technology, which does not only rely on machine vision. A tractor navigation system based on vision has been created, however due to stability problems that have since been discovered, commercial use is still in its early phase. The promise of vision-based automated control of agricultural machinery has, however, been shown in related studies. Hence, it is anticipated that in the coming 10 years, vision-based self-driving tractors would significantly alter the agricultural industry. Also gaining popularity are agricultural robots and agricultural robots that may launch a revolution that promises to replace the tractor, which is currently the craze in agriculture. Based on the same concepts and sensors, agricultural robots may autonomously explore fields while operating accurately and technically at the crop size [5]. Agricultural robots can function in adverse weather circumstances and can flexibly apply multiple tools (topping tools, lawn mowers, atomizers, etc.). As a result, tractors are essential equipment on every farm, large or small, and while we do not currently have any plans to replace them, we do have plans to grow.

4.4 Safety Issues

The usage of farm tractors is a factor in the majority of accidents connected to agricultural work [7, 8, 21]. The latter might result from the following factors. In addition, (1) many small farms are lacking in operators and sophisticated machinery. (2) Labor-intensive agricultural tasks needing mechanical aid; (4) seasonal employment in a region that varies with the seasons, making it difficult for users to become used to the environment. Self-driving tractors utilizing outdated sensors in an automated steering system must be safer and more accurate than a human driver for a tractor to be completely autonomous. Studying tractor safety concerns can therefore aid in the development of the safety measures needed for completely autonomous driving. Yet, autonomous tractors are intended to supplement human operators rather than to take their place. Long trips over expansive fields are draining and confusing. To conduct and troubleshoot other agricultural chores (such as trimming and spraying), tractors must travel autonomously around the field while human drivers are present [11]. Several academics have looked at tractor safety difficulties, but they have mostly concentrated on problems with technical aspects like vibration [13], rollover protection structures (ROPS) [14], operator ergonomics [16], and technology [18]. A gadget to track the condition of tractor parts is suggested as a potential remedy for mechanical risk monitoring [21]. We also suggest cutting-edge options for intelligent tractor guidance, such virtual reality [25]. Because faults cannot always be avoided, the system frequently needs to monitor and promptly notify system issues. Emergency notification systems [27], as well as warning and alarm systems [28], are designed to assist drivers in waking up and realizing their surroundings. Programs and significant interventions focused at assisting and enhancing farmers' situational awareness can help avert fatal incidents [18]. The most frequent root causes of tractor safety issues are depicted in Figure 4.2 [17].

The dependability of the steering system architecture's hardware and software is a key factor in self-steering tractor safety. The latter is based on how autonomous the system is. Even steering units with the same degree of autonomy might vary greatly, though. The three types of sensor-based

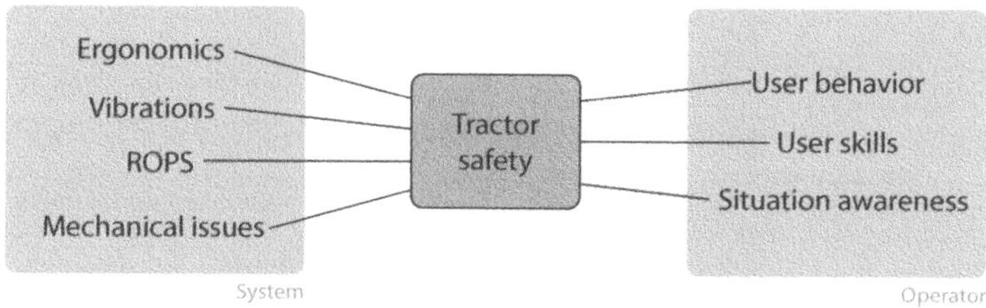

Figure 4.2 The primary causes of tractor safety hazards.

autonomous driving systems are actuator-based, algorithm-based, and sensor-based recognition systems. Motion problems include high-altitude driving, driving speed, high-speed rotation and slippage of tractors, electrohydraulic control system instability, and more. Misjudgments on the slope account for the majority of decision-making errors. Measurements made by the sensory systems lead to perception mistakes. All recognition mistakes in an autonomous tractor system based on machine vision happen throughout the tools and techniques used for picture processing [19]. Real-time picture acquisition and processing must be quick and precise when making choices. It might be challenging to balance detection algorithms' speed and accuracy in dynamic situations like fields. Crop line routing is the name for automatic field routing. The state of the crop will have an impact on how well the system performs; if the crop cannot be readily spotted, the system will not function as intended. Hence, under various climatic circumstances, crop lines ought to be distinguished. The most frequent issues with detecting agricultural lineage are missing vegetation, seedlings at various development stages, uneven densities of foliage, and weeds. Weed characteristics are very similar to those of certain immature crops. Beets and greens, for instance, are identical in size and form. As a result, vision algorithms need to be resistant to weed growth at all phases [26]. More perplexity brought on by lights. The algorithm may be impacted by variations in the field strength. Moreover, the tractor is shadowed by direct sunshine, which yields subpar detection results [22]. Several cameras that are positioned on the tractor's head may be seen on a camera mounted on the nacelle. Planting height may potentially restrict the system's ability to recognize rows. The obtained sensor data is crop-scale and has excellent accuracy and precision because the tractor is a land vehicle. Because to the high data quality, environmental factors like illumination and shadows can dramatically decrease system accuracy [10]. The primary issues with the optical identification system for autonomous steering tractors are outlined in Figure 4.3.

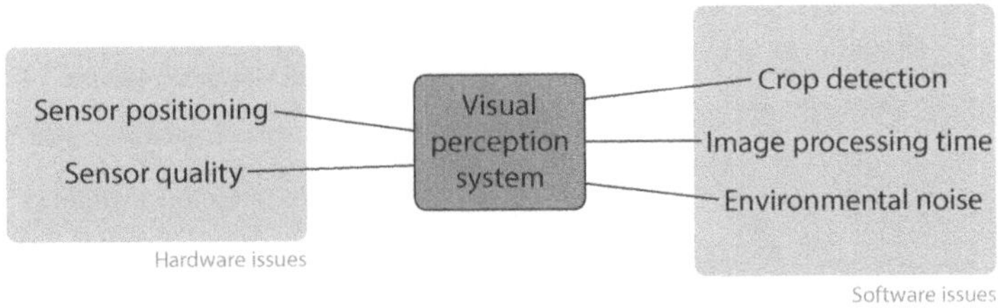

Figure 4.3 Significant issues with the self-steering tractor's optical sensing system.

In conclusion, vision-based steering is adaptable yet influenced by environmental circumstances. Steering precision can be significantly increased by multi-sensor systems that incorporate data from several sensors. The absence of purely visual navigation methods in contemporary literature is mostly due to this. Multi-sensory fusion technology is the key foundation for future automated control systems.

4.5 The System Architecture of Self-Guiding Tractors

Below is a rudimentary illustration of the self-driving tractor. It provides a vision-based system architecture and considers the necessary elements needed to make autonomous driving a reality.

4.6 Basic Modeling

A consistent flow of information is necessary for the creation of autonomous machines. The Sense-Perceive-Plan-Act cycle (SPPA) cycle is used to describe this [11]. By closed-loop connection, the SPPA loop links detection, awareness, planning, and action. Sensors gather (detect) physical data from the environment, receive and evaluate the data (perception), choose (plan) a suitable exploration trajectory, and operate (control) the tractor to follow the chosen trajectory. The fundamental modeling of an autonomous tractor is shown in Figure 4.4.

Basic mechanics and cognitive driving intelligence are two fundamental components needed for autonomous tractor navigation (CDI). The hardware and software used for platform navigation and control must incorporate CDI. Although control comprises all steering characteristics including steering speed, angle, and speed, navigation includes location, mapping, and route planning. CDI can make use of ride control software,

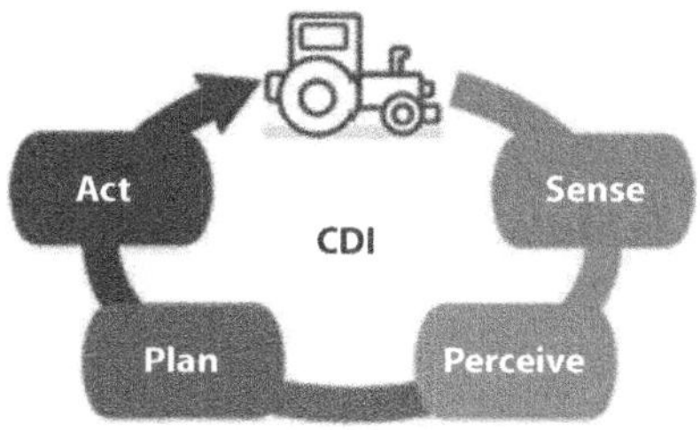

Figure 4.4 Simple self-steering tractor models.

route planning algorithms, and sensor data from navigation and location devices. The tractor utilising CDI is the host.

4.7 Building with a Vision

Figure 4.5 depicts the suggested vision-based navigation system's flowchart.

In this design, an optical sensor serves as the main sensor. Optical sensors that record images that a personal computer (PC) analyzes and uses to extract the directional signal receive RTK GPS data. In order to assess the system, we put the prototype shown in Figure 4.5 on a real tractor and ran a number of autonomous driving experiments.

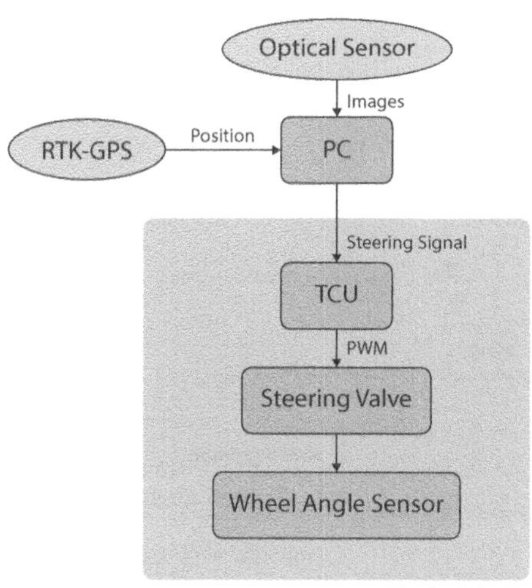

Figure 4.5 Basic tractor navigation system functioning using eyesight.

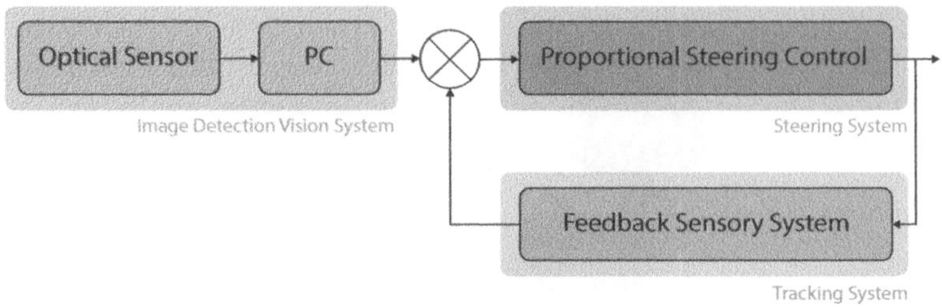

Figure 4.6 A route tracking control system's fundamental structure.

4.8 Path Tracking Control System

The image sensor, tracking system, and direction control are the three primary systems that make up the tracking control system's fundamental design, as seen in Figure 4.6.

4.9 Development of a Tractor-Based Agricultural Row Detection System Using Stereovision

4.9.1 System Architecture

The main goal is to develop the core technology for an agricultural machine vision control system that uses a crop row number recognition system based on stereo vision. Figure 4.7 depicts a signal flow diagram for an agricultural

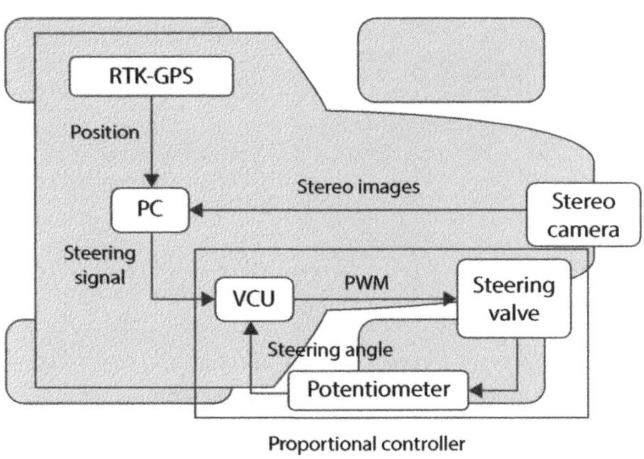

Figure 4.7 The RTK signal GPS's flow diagram for a tractor navigation system based on stereo vision.

machine navigation system based on stereo vision and machine independence. The present tractor steering mechanism is changed, an electro-hydraulic steering system is added in parallel, and the research platform is updated with a stereo vision navigation system to accomplish autonomous steering.

The STH-MD1 stereo vision camera is an essential part of the stereo vision navigation system (VidereDesign, CA). The navigation computer utilizes a personal computer to process stereo images, determine tractor navigational parameters, and provide turn signals. The automated steering system consists of an electrohydraulic directional control valve with proportional solenoid actuator and a potentiometer-based wheel angle sensor. In automated driving, the navigation system uses the RS232 serial interface to communicate steering control signals to the vehicle control unit.

4.9.2 Stereo Vision Crop Row Detecting System

A stereo vision camera serves as the foundation for the crop line detecting technology. Based on two simultaneous planar pictures taken by a binocular stereo camera, this stereo system simulates a three-dimensional scene as shown in Figure 4.8.

The image's horizontal distance R is specified as uR in uL. Calculating the depth of interest at the stereo camera lens's center Zc is one of the key goals of stereo view image processing. The depth of an item may be

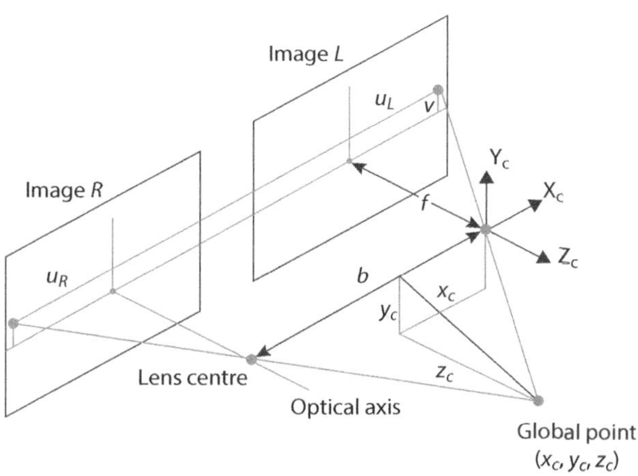

Figure 4.8 Stereo vision camera.

determined using the formula below utilizing two picture planes that are horizontally aligned and put in the same plane.

$$uL - uR = d = bf / zc \tag{4.1}$$

The stereo image difference may be thought of as the value of the variable d in Equation 4.1. The following steps can be used to get a point's 3D location in left camera coordinates when the tilt is known:

$$xc = - uL/f\ zc = -buL/d \tag{4.2}$$

$$yc = - v/f\ zc = -bv/d \tag{4.3}$$

$$zc = bf / d \tag{4.4}$$

The STH-MD1 stereo camera (VidereDesign, CA) is used by this stereo navigation system to record stereo pictures and estimate the distance between the tractor and the crop line. Two steerable 1.3MP progressive CMOS pictures make up the stereo camera independently with a 12.0mm lens through an IEEE 1394 interface. The stereo camera may record seven frames (1280 x 1032 pixels), 25 frames (640 x 512 pixels), or 80 frames depending on the picture resolution acquired (320 x 256 pixels). A near-infrared (NIR) filter with an 880 nm center wavelength was put in front of each scene picture. A 381-vertical field of view cameras and 501 horizontal field of view cameras were positioned 1.7 meters in front of the search platform at a 311-degree slope (Figure 4.9). This camera can capture a live

Figure 4.9 Placement of cameras on the research tractor platform.

scene up to 1 to 8 meters in front of the tractor using the configuration and settings given. For stereo computations, stereo cameras transform the raw pictures they record into modified images. The geometric link between the parameters is established by this rectification procedure associated with the left and right pictures, as well as the camera settings. The first one comprises the focal length, lens distortion, and optical center of the picture. The camera maker's commercial software is used for calibration and stereo computations.

4.10 Creation of a Crop Row Detecting Method Using Stereo Vision

4.10.1 Overview of the System and Picture Preparation

In this work, we create a crop row recognition system based on a stereo vision that gives agricultural machinery navigational information to identify along crop rows during various trimming operations. The algorithm's information flow is depicted in Figure 4.10. Stereo image processing, elevation mapping, and track point selection are the three steps used to find an accurate and trustworthy line tracing solution in stereo images. On the system, stereo cameras are always installed at an awkward angle b. Sector angle and roll angle a are frequently included in mounted cameras. The transformation to XYZ axes defines the camera coordinates from the center of gravity (CG) at the bottom of the tractor to the origin of the vehicle, as well as the XcYcZc axis from the left camera's focal point to the origin (Figure 4.11).

The stereo vision system can turn the 3.0m x 3.0m landscape image in front of the machine into a topographic map based on the aforementioned camera arrangement and configuration. The 3D points obtained by the stereo vision system must first be converted from camera coordinates to vehicle coordinates in order to recreate the 3D terrain map from those data. The camera's rotation angle is one of six factors used in this transformation to set up the camera. 1. Tendency Angle of the propeller, B. Using the following formula, the position offset (xd, yd, zd) represents the direction and location of the camera in vehicle coordinates:

$$[x\ y\ z] = Ry\ (g)\ Rx\ (b)\ Rz\ (b)\ [zc\ xc\ yc\ ] + [xd\ yd\ zd] \qquad (4.5)$$

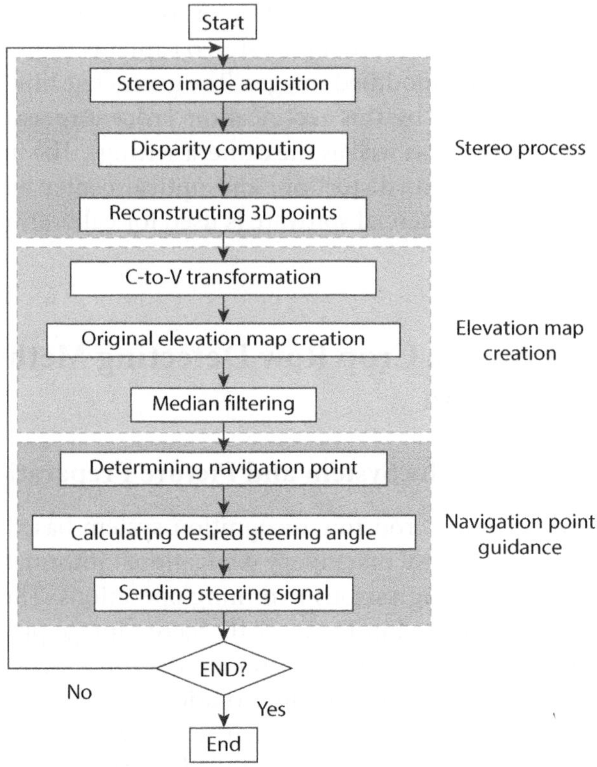

Figure 4.10 Stereo vision-based crop line detection method.

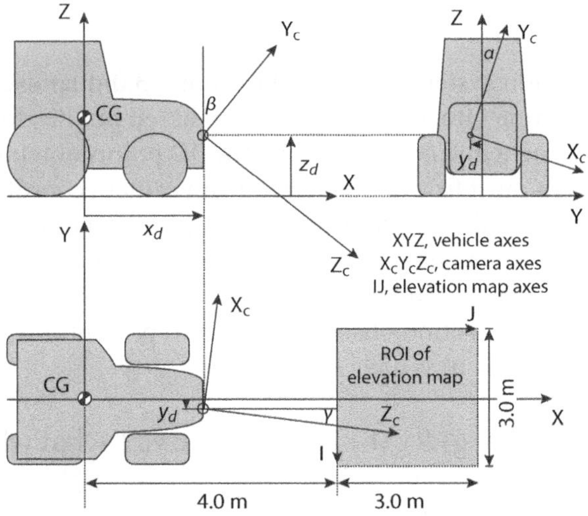

Figure 4.11 Camera's center of gravity.

$$Rx(y) = (\cos_0 y_0 - \sin_1 y_0 \sin y_0 \cos y) \tag{4.6}$$

$$Ry(y) = (\cos y \sin y_0 - \sin_0 y_0 \cos_1 y_0) \tag{4.7}$$

$$Rz(y) = (\cos^1 y^0 \sin^0 y^0 - \sin y \cos y) \tag{4.8}$$

The roll angle, tilt angle, and translation angle of these six parameters for the tractor platform employed in this study are 1.1, 31.0, and -1.3, respectively, and the position offsets xd, yd, and zd are 2.66, 0.04, and 1.70, respectively.

4.10.2 Visual Odometry Based Relative Localization

One of the most difficult problems for robots in the field is self-localization. The relative positioning system gradually updates the position during movement and is used in situations where high precision and short-term accuracy are required. Major relative positioning systems use wheel odometers and inertial measurement units (IMUs) [6, 7]. Wheel odometry relies on kinematic model integrity and accuracy, which is hard to assume in uneven out-door terrains. Also, its accuracy degrades rapidly when wheels tend to slip. On the other hand, inertial measurements cannot be used reliably when the robot dynamics in the field are slow. Therefore, these vehicles move slowly in the field and cannot be used for self-localization of autonomous vehicles in agriculture. Furthermore, the dynamic influence of uneven terrain can easily distort the forward dynamics. Alternatively, a visual odometer can be used, which is not dependent on the vehicle's kinetic model [24]. This means that one or more cameras are used to take pictures of the surrounding environment to estimate relative changes in the position of the robotic car's body. The two most common VO estimators use optical flow-based feature tracking [26] and planar transformation between images using frequency-domain phase correlation [24]. In this experiment, we had to determine the optimal visual odometry measurement method available for self-localization of a two-wheel tractor. We therefore performed a basic exploratory experiment involving loop closure and error propagation to analyze errors in terms of frame size and frame rate.

4.10.3 Visual Odometry Using Feature Tracking

Other common visual odometry estimations based on feature tracking use optical flow fields. In this approach, we identify relatively distinct features in the environment and track them across video frames to estimate robot motion. The computational cost for the resource tracking technique is mostly determined by the quantity of monitored resource points. The best method for real-time location updates is optical flow field estimation with feature tracking. As each attribute is monitored separately, statistical data filtering procedures may be used with ease. Over time, a variety of optical flow estimating methods have been created, each with unique benefits and drawbacks. Among these, it has been demonstrated that Lucas Candy's [28] approach is more effective and precise than other ways. Moreover, the OpenCv image processing package has an enhanced version of this approach. Similar to the frequency-domain method, a video sequence is captured while the robot is travelling with the camera fastened to the ground. The statistics are then filtered using the Lucas–Kande optical flow estimation method to gradually update the robot's location. The optical flow produced by Lucas Candy's technique is seen in Figure 4.12.

4.10.4 Visible Odometry Using a Single Camera

Using a commercial toy vehicle moving at up to 30 cm/s over a series of studies outside, we evaluated the effectiveness of the two visual odometry methods. On the rover's back, 50 cm above the surface, is a digital camera. The speed of picture registration is also significantly accelerated by image processing software. We discovered that utilizing a C++ compiler and the Intel OpenCv Image Processing Library results in acceptable performance.

Figure 4.12 The figure depicts the optical flow field that was produced on a video frame that was taken from the ground.

Figure 4.13 Rover platform with downward faced camera.

The rover's position is updated during each frame using a straightforward range measurement model. Understanding the robot's location and orientation in relation to the fixed world structure's centre of mass, P(x0, y0, 0), is critical to understanding how the robot will behave as shown in Figure 4.13.

For rovers, the position is from a known position, it is estimated by adding the incremental distances traveled in the X direction (Δx) and Y direction (Δy). If the calculated change of the heading ($\Delta\theta$) is then the updated, position P′ can be obtained as follows:

$$P' = P + [\Delta s \cos(\theta + \Delta\theta) \Delta s \sin(\theta + \Delta\theta) \Delta\theta]$$

4.11 Stereo Vision for Absolute Localization

In the previous chapter, visual odometry measurements were briefly introduced as an alternative to wheel odometry measurements for outdoor localization of autonomous vehicles. Navigating and controlling a robot requires precise positioning in its environment, and the required accuracy is determined by the application. In many mobile robotics environments, pose can be simplified by estimating only Cartesian position and orientation. Therefore, traditional mobile robots are equipped with wheel encoders to measure the distance traveled by the robot, gyroscopes, and accelerometers to measure incremental changes in the robot's orientation, and cameras to track the path the robot travels. These are called proprioceptive measurements because the onboard sensors estimate the robot's relative motion, rather than making measurements at absolute reference points. The process of integrating these proprioceptive measurements over time to maintain an estimate of the robot pose is called inferential navigation. Dead reckoning localization is easy to implement and can provide high accuracy for short paths. But dead reckoning fails in most real-world applications. This is because it inherently provides no means of correcting errors. This is clearly illustrated by the visual odometry results discussed in the previous chapter. To improve the results of visual odometry, we need a way to absolutely detect the pose of a robot relative to one or more fixed points. Absolute sensor measurements are known as exteroceptive, and to determine absolute positions, the robot detects features in the environment and poses in direct world coordinates relative to these (usually fixed) reference positions. In this model, the localization of points does not require knowledge of the robot's previous position, and the localization error is determined only by sensor noise and bandwidth. The GPS is a common means of absolute positioning. Many outdoor autonomous robots, including agricultural autonomous vehicles, rely primarily on GPS. There are many low-cost GPS devices available, but even with many satellites in view, they are accurate to within 5 meters. Therefore, it is suitable for commercial applications such as map navigation, but it cannot be used for fully autonomous navigation. Previous experiments combining differential GPS and an inertial navigation system for precise positioning yielded errors of a few centimeters and fractions of a degree [6]. Beacon-based positioning, including RF or optical beacons, is used as an alternative to GPS in relatively small areas of operation. But these systems need to fix beacon emitters or reflectors in several locations which require excessive rearrangement when the operating environment change [29]. Distance measurements using time-of-flight can reduce the number of beacons required, but require expensive hardware and are not recommended for low-cost applications. Due to this limitation, we found a way to absolutely detect the pose of the robot using stereo vision from a reference fixed point. Stereoscopic vision has long been used to measure the depth of a pair of images taken by two cameras placed at a fixed distance from each other. Since distance measurements for stereo camera systems are designed to limit incremental errors due to visual error measurements, an accuracy of ±1 m is considered sufficient for stereo rang measurement units. Another advantage is that the same hardware used in the visual odometer system can also be used in the stereo odometer system at no additional cost, thus reducing the overall cost of the system. Stereo measuring systems can measure the distance to objects smaller than 15 meters with an acceptable accuracy. In this experiment, we used two inexpensive webcams as a stereo pair and used Intel's OpenCv image processing library to track and calculate object distances.

4.12 Multi-Vision Methods

Several vision sensors provide input pictures for multipart systems. Three pictures and three cameras' worth of GPS data are combined to make GPS. The complexity of the system is reduced since there is no need to spin the camera to collect various views when using numerous cameras to combine photos to create the final image. The suggested multi-camera system's perspective is four times quicker than that of existing stereo vision systems. Later, we will publish a comparison table of every vision-based autonomous tractor navigation algorithm that was taken into consideration. The same table includes information on the performance of the algorithm and the visual system that was employed. The approach makes use of a camera as the primary navigation sensor. Additional sensors can be used as complements or to compare various sensors and methodologies.

4.13 Conclusions

Future sustainable agriculture is being paved by data-driven farming together with the proper sensor systems, including artificial intelligence techniques. To do this, the vision-based technology offers accurate navigational data and autonomously steers the tractor along the crop line. In this approach, the operator is free to participate in activities like steering while the autonomous driving system handles the tiresome work of crop follow-up, enhancing management effectiveness, work efficiency, and the general safety of agricultural operations. This study demonstrates that problems like vision-based tractor systems engineering, safety and navigational mistakes, and vision-based navigation control systems enhance the significance of machine vision systems in autonomous tractors. With systematic evidence of trends and difficulties, this work seeks to increase our understanding of vision-based search techniques in agriculture. Future studies can build on this work to better understand the architectures, sensors, algorithms, and security concerns of vision-based machines. The information offered is transferable to other circumstances where the employment of autonomous guided vehicles in agricultural navigation operations is a priority.

References

1. Saiz-Rubio, V.; Rovira-Más, F. From Smart Farming towards *Agriculture 5.0*: A Review on Crop Data Management. *Agronomy*, 2020, *10*, 207.
2. Martos, V.; Ahmad, A.; Cartujo, P.; Ordoñez, J. Ensuring Agricultural Sustainability through Remote Sensing in the Era of Agriculture 5.0. *Appl. Sci.*, 2021, *11*, 5911.
3. Shanmugapriya, P.; Rathika, S.; Ramesh, T.; Janaki, P. Applications of Remote Sensing in Agriculture-A Review. *Int. J. Curr. Microbiol. Appl. Sci.*, 2019, *8*, 2270–2283.
4. Fan, J.; Zhang, Y.; Wen, W.; Gu, S.; Lu, X.; Guo, X. The future of Internet of Things in agriculture: Plant high-throughput phenotypic platform. *J. Clean. Prod.*, 2021, *280*, 123651.
5. Amit Grover, A. Manikandan, Vivek Soi, Anu Sheetal, Mehtab Singh, Realisation of white LED using fiber based hybrid photonic structures, *Optoelectron. Adv. Mater. Rapid Commun.*, 15, 11-12, November-December 2021, pp.521-527 (2021).
6. Dr. S. Vijayalakshmi. Early detection of breast cancer using robust back propagation neural network classifier. *Rom. Biotechnol. Lett.*, 2022; 27(2): 3407-3415 DOI: 10.25083/rbl/27.2/3407.3415.
7. V. A.R, S. David, E. Govinda, K. Ganapriya, R. Dhanapal and A. Manikandan, "An Automatic Brain Tumors Detection and Classification Using Deep Convolutional Neural Network with VGG-19," *2023 2nd International Conference on Advancements in Electrical, Electronics, Communication, Computing and Automation (ICAECA)*, Coimbatore, India, 2023, pp. 1-5, doi: 10.1109/ICAECA56562.2023.10200949.
8. Ali, R., Manikandan, A. & Xu, J. A Novel framework of Adaptive fuzzy-GLCM Segmentation and Fuzzy with Capsules Network (F-CapsNet) Classification. *Neural Comput. Appl.*, (2023). https://doi.org/10.1007/s00521-023-08666-y.
9. Franceschetti, B.; Rondelli, V.; Ciuffoli, A. Comparing the influence of Roll-Over Protective Structure type on tractor lateral stability. *Saf. Sci.*, 2019, *115*, 42–50.
10. Balamurugan, D. & Seshadri, s.Aravinth & Reddy, P. & Rupani, Ajay & Manikandan, A. (2022). Multiview Objects Recognition Using Deep Learning-Based Wrap-CNN with Voting Scheme. *Neural Process. Lett.*, 54. 1-27. 10.1007/s11063-021-10679-4.
11. ha, H.; Lodhi, V.; Chakravarty, D. Object Detection and Identification Using Vision and Radar Data Fusion System for Ground-Based Navigation. In *Proceedings of the 2019 6th International Conference on Signal Processing and Integrated Networks (SPIN)*, Noida, India, 7–8 March 2019; pp. 590–593.
12. Karur, K.; Sharma, N.; Dharmatti, C.; Siegel, J.E. A Survey of Path Planning Algorithms for Mobile Robots. *Vehicles*, 2021, *3*, 448–468.

13. Ge, J.; Pei, H.; Yao, D.; Zhang, Y. A robust path tracking algorithm for connected and automated vehicles under i-VICS. *Transp. Res. Interdiscip. Perspect.*, 2021, *9*, 100314.
14. Schouten, G.; Steckel, J. A Biomimetic Radar System for Autonomous Navigation. *IEEE Trans. Robot.*, 2019, *35*, 539–548.
15. da Silva, S.P.P.; Almeida, J.S.; Ohata, E.F.; Rodrigues, J.J.P.C.; de Albuquerque, V.H.C.; Reboucas Filho, P.P. Monocular Vision Aided Depth Map from RGB Images to Estimate of Localization and Support to Navigation of Mobile Robots. *IEEE Sens. J.*, 2020, *20*, 12040–12048.
16. Venkataramanan, C. & Ramalingam, S. & Manikandan, A. (2021). LWBA: Lévy-walk bat algorithm based data prediction for precision agriculture in wireless sensor networks. *J. Intell. Fuzzy Syst.*, 41. 2891-2904. 10.3233/JIFS-202953.
17. Li, Y.; Wang, X.; Liu, D. 3D Autonomous Navigation Line Extraction for Field Roads Based on Binocular Vision. *J. Sens.*, 2019, *2019*, 1–16.
18. Manikandan, Annamalai, M, Ponni Bala. (2023). Intracardiac Mass Detection and Classification Using Double Convolutional Neural Network Classifier. *J. Eng. Res.*, 11(2A). 272-280. 10. 36909/jer.12237.
19. Wang, Q.; Meng, Z.; Liu, H. Review on Application of Binocular Vision Technology in Field Obstacle Detection. *IOP Conf. Ser. Mater. Sci. Eng.*, 2020, *806*, 012025.
20. Annamalai, Manikandan & Muthiah, Ponni. (2022). An Early Prediction of Tumor in Heart by Cardiac Masses Classification in Echocardiogram Images Using Robust Back Propagation Neural Network Classifier. *Braz. Arch. Biol. Technol.*, 65. 10.1590/1678-4324-2022210316.
21. Zhang, X.; Dai, H.; Sun, H.; Zheng, N. Algorithm and VLSI Architecture Co-Design on Efficient Semi-Global Stereo Matching. *IEEE Trans. Circuits Syst. Video Technol.*, 2020, *30*, 4390–4403.
22. Rabab, S.; Badenhorst, P.; Chen, Y.-P.P.; Daetwyler, H.D. A template-free machine vision-based crop row detection algorithm. *Precis. Agric.*, 2021, *22*, 124–153.
23. Manikandan, A., & Nirmala, V. (2015). A Low Cost Thermoelectric Refrigerator. *Int. J. Appl. Eng. Res.*, *10*(55), 3097–3101.
24. Corno, M.; Furioli, S.; Cesana, P.; Savaresi, S.M. Adaptive Ultrasound-Based Tractor Localization for Semi-Autonomous Vineyard Operations. *Agronomy*, 2021, *11*, 287.
25. X. Di and R. Shi, "A survey on autonomous vehicle control in the era of mixed-autonomy: from physics-based to AI-guided driving policy learning," *Transp. Res. Part C Emerging Technol.*, vol. 125, Article ID 103008, 2021.
26. Bommaraju, K., Manikandan, A., & Ramalingam, S. (2017). Aided System for Visually Impaired People in Bus Transport using Intel Galileo Gen-2: Technical Note. *Int. J. Veh. Struct. Syst.*, *9*(2), 110–112. https://doi.org/10.4273/ijvss.9.2.09

27. H. Khayyam, B. Javadi, M. Jalili and R. N. Jazar, "Artificial intelligence and internet of things for autonomous vehicles," *Nonlinear Approaches in Engineering Applications*, vol. 1, pp. 39–68, 2020.
28. M. Cunneen, M. Mullins and F. Murphy, "Autonomous vehicles and embedded artificial intelligence: the challenges of framing machine driving decisions," *Appl. Artif. Intell.*, vol. 33, no. 8, pp. 706–731, 2019.
29. A. Osório and A. Pinto, "Information, uncertainty and the manipulability of artificial intelligence autonomous vehicles systems," *Int. J. Hum. Comput. Stud.*, vol. 130, pp. 40–46, 2019.
30. Balamurugan, D. & Seshadri, s.Aravinth & Reddy, P. & Rupani, Ajay & Manikandan, A. (2022). Multiview Objects Recognition Using Deep Learning-Based Wrap-CNN with Voting Scheme. *Neural Process. Lett.*, 54. 1-27. 10.1007/s11063-021-10679-4.

5

Vision-Based Image Classification and Image Segmentation Algorithms for Plant Disease Diagnostics

N. Ashokkumar[1]*, A. Manikandan[2], S. Hariprasath[3] and P. Vijayalakshmi[4]

[1]*Department of ECE, Mohan Babu (MBU) University, Tirupathi, India*
[2]*ECE, SRM Institute of Science and Technology (Deemed to be University), Kattankulathur, Chennai, India*
[3]*ECE, Saranathan College of Engineering, Trichy, India*
[4]*CSE, Knowledge Institute of Technology, Salem, India*

Abstract
One of the main sources of income is agriculture. Many peasants in developing nations like India find work in agriculture. Many crops are grown in India's agriculture, which provides a living for about 70% of the country's people, according to a study. Indian farmers primarily employ manual planting due to outdated technologies. What crops will grow on a farmer's property is unknown. Exotic plant diseases that harm plants through their leaves have an impact on agricultural productivity and decrease in profitability. Moreover, it lowers the level and quality of agricultural output. For accelerated plant development and increased agricultural yields, leaves are crucial. Farmers and academics have had trouble detecting illnesses on plant leaves. Nowadays, farmers apply pesticides on plants that may have a negative impact on human health and economy. These plant diseases need to be found quickly, using a variety of ways. A variety of plant diseases are covered in this article, as well as cutting-edge methods for finding them.

Keywords: Image segmentation, plant disease, plant village dataset, image classification

**Corresponding author*: ashoknoc@gmail.com

Rajesh Kumar Dhanaraj, Balamurugan Balusamy, Prithi Samuel, Malathy Sathyamoorthy and Ali Kashif Bashir (eds.) Computer Vision in Smart Agriculture and Crop Management, (93–122) © 2025 Scrivener Publishing LLC

5.1 Introduction

Agriculture's production is heavily influenced by the economy. The factory's productivity, product quality, and quantity would all suffer if the proper steps are not taken in this area. As an illustration, the shear disease that can be fatal to American pine trees exists. In India, the agricultural sector provides a living for close to 70% of the people [1]. For the purpose of avoiding crop loss, identifying plant diseases is crucial. Plant disease detection manually is challenging. It takes a lot of time, effort, and understanding of plant diseases to complete this. In light of this, plant disease detection can be performed using image processing and machine learning models [2]. Using images of leaves, learn how to identify plant illnesses. The information and image quality present in photographs can be produced through "image processing," a subset of signal processing [3]. The majority of plants have bacterial and fungal disease subspecies. Plant diseases can also be caused by climatic factors due to the exponential population expansion. Biological (living) agents including fungus, bacteria, viruses, nematodes, parasitic plants, air pollution, and other illnesses are studied by phytopathologists to understand plant diseases and nutritional imbalances [4]. We also research abiotic (non-living) plant harm brought on by things like poor growth circumstances. It takes more than just comparing plant samples with ill humans whose symptoms correspond to plant symptoms on the internet or in books to learn how to identify plant diseases. Similar symptoms might be caused by a variety of plant ailments. Plant issue diagnosis is a technological advancement that has developed over time [5]. It is true that learning about plants and plant diseases involves a lot of trial-and-error and iteration. Gardner Masters will have a better chance of diagnosing the issue accurately and resolving it at the extended client's location if they comprehend the procedure [6]. The main objective as a Gardner Master is education. With the diagnostic you get, you have the chance to explain to extension consumers the right conditions for plant development and offer long-term solutions to avert plant illnesses. Plant diseases undoubtedly lower the production and quality of agricultural goods. Research on plant diseases focuses on visually discernible patterns in plants. We suggest and employ the support vector machine (SVM) classification method in this study [7]. The health and illness of plant leaves have a significant impact on the yield of crops on farms. Plant disease analysis used to only be done manually by subject-matter specialists. It takes a long time to process all this work. The accompanying publications give image processing techniques [8]. The illness often manifests as symptoms on the fruit, stems,

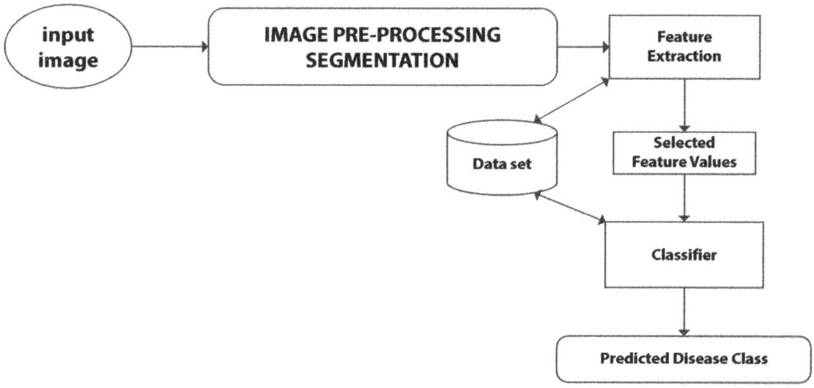

Figure 5.1 Basic diagram for plant diseases diagnosis.

and foliage. One of today's fastest-growing technologies, image processing primarily involves the conversion of pictures into signals while utilizing signal processing techniques. It has been used in many facets of business [9]. Crop cultivation is crucial to the agricultural industry. The major cause of the current food losses is a fall in the production of diseased crops. It has not been investigated how to detect plant illnesses early. The primary issue is to decrease the usage of pesticides while increasing the amount and quality of agricultural products [10]. To estimate the leaf infection zones, we enhanced the k-means clustering technique. To divide up contaminated areas into pertinent groups and segment them, a color-based segmentation model is developed [11]. On example photographs, an experimental examination of the infected area and time complexity is conducted. Image processing methods can be used to identify plant diseases [12]. Disease detection involves a number of steps, including picture acquisition, image preprocessing, image segmentation, feature extraction, and classification as shown in Figure 5.1.

5.2 Signs and Symptoms of Plant Disease

Fungi or organisms that resemble fungus are responsible for the majority (about 85%) of plant illnesses. Some severe ailments affecting food and feed crops, however, are brought on by bacteria and viruses [13]. Plant diseases are caused by certain nematodes. Abiotic, or non-infectious, plant illnesses include harm from air pollution, nutritional deficits, or toxicity from growing in unsuitable conditions [14]. The form of the plant can be an excellent predictor of the kind of pathogen if plant disease is detected.

Plant disease symptoms are the pathogen's outward physical manifestations. For instance, fungus that produces fruiting bodies is ill. Tripomycosis, the bacterial exudate of the canker, is caused by nucleated bacterial cankers. Although the ulcer itself is made of plant tissue, the illness is indicated by a viscous fluid discharge that is mostly made of bacteria [15].

Plant disease manifestations on the outside are known as symptoms. There might be observable alterations in the color, shape, or functionality when plants react to infection. Fungus plant disease and dahlias are both produced by the verticillium wilt, which is typically identified by leaves with a diamond shape [16]. The typical symptoms of bacterial lozenges are brown necrotic lesions with vivid yellow haloes on the inner and exterior of leaves of soybean plants. Not the actual cause of the ailment, but something to pay attention to, are symptoms brought on by a causative agent. Typical indications and symptoms of bacterial, viral, and fungal plant diseases are illustrated here.

A. Fungal Illness

Fungal infections may be transported by wind and water and impact infected seeds, soil, yields, and weeds. It happens in low or competent clear water with wet, grey blotches in initial tissues [17]. The environment then turned gloomy, as a white fungus began to grow over the floor. On the upper surface of the highly treated transparent wool, the heaped wool creates streaks that range from yellow to white. Outside of the leaves, it spreads and turns yellow.

Indications of fungal illness:

- Rust on leaves (common leaf rust in corn)
- Stemmed rust (wheat stem rust)
- Sclerotinosis (white mold)
- Particulate mildew

Fungal disease symptoms:

- Berries with a birds-eye spot (anthracnose)
- Seedlings that have dried out (phytophthora)
- Leaf blight (septoria brown spot)
- Chlorosis (yellowing of leaves)

B. Bacterial Diseases

The term "bacterial leaf spot" is frequently used to describe bacterial infections. Little, greenish-yellow lesions that deform and distort on young leaves or older leaves that have dark, watery, oily lesions are often where it starts.

Indications of bacterial illness (difficult to observe, but can include):

- Infected ooze
- Lesions bathed in water
- Bacteria from a severed stem are flowing into the water.

Symptoms of bacterial diseases:

- Sheperd's crook stem ends on woody plants
- Fruit spot
- Canker
- Crown gall
- Leaf spot with a yellow halo

All viral illnesses cause some yield loss, and virus-infected plants typically live less time [18]. Although certain viruses can also show up on leaves, fruit, and roots, the leaves of virus-infected plants often exhibit the symptoms more frequently. Analysis of viral illnesses can be challenging. The virus stunts development, causing the leaves to droop and dry up are shown as Figures 5.2 and 5.3.

Indications of viral diseases:

- None because the viruses themselves are invisible.

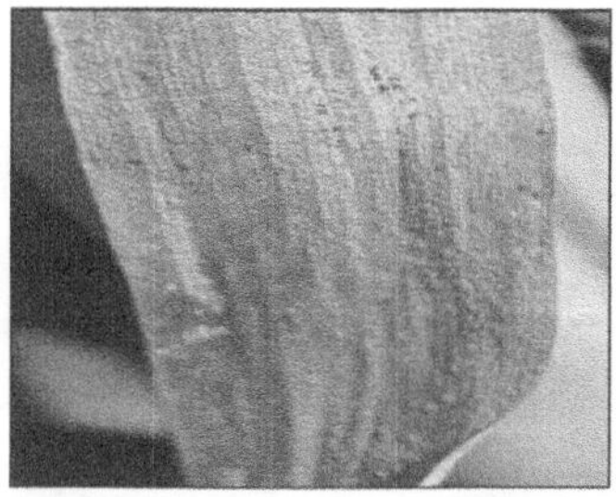

Figure 5.2 Sign of stripe rust is pustules on a winter wheat leaf.

Figure 5.3 Kidney bean leaf viral illnesses.

Viral illness signs:

- Crumpled leaves
- Strange leaves
- an inlaid leaf design
- plant stupor, crinkled leaves

Several viral, bacterial, and fungal infections share the same symptoms [19]. When unexplained plant issues occur, abiotic disease, herbicide damage, and worm issues should also be taken into account. This is only a sample and neither exhaustive nor exhaustive in any way.

5.2.1 The Importance of Plant Diseases in the US

- Night lozenges, which were first introduced in 1904 and essentially abolished night usage in North America, are some examples of plant disease infections that result in catastrophic plant losses in the United States [20].
- The introduction of white pine green cans in 1912 caused enormous financial damage to the timber sector as shown in Figure 5.4.

Figure 5.4 Viral disease.

- Citrus canker (and a closely related bacteria called citrus leaf spot, identified in 1984) was introduced in 1910 and wiped off millions of citrus plants. Since their introduction in 1930, Dutch elms have killed out numerous elms from the East Coast to the Pacific Northwest [21].

5.2.2 Different Types of Plant Disease

We created this section so that researchers can look at different diseases to understand the types of imaging tasks and the types of features to consider [22]. Plant diseases occur when viruses, bacteria infect plants and prevent them from growing normally. Effects on plant foliage range from discoloration to death [23]. Fungi, microorganisms, viruses, nematodes and other diseases occur. Some common diseases of corn, betel nut, coconut palm, papaya, cotton, pepper, tomato and eggplant are discussed here as shown in Figure 5.5. I will detail several subspecies of the disease.

- **Rust:** In mature plants, it typically develops on the underside of the leaves. Initially, the underside of the leaves had a few dots. These blotches develop into clusters of reddish-yellow spores over time. The petioles eventually become yellow-green and then black. Leaf curling, yellowing, and shedding are symptoms of severe infections [24].
- **Kole Roga:** It is the primary betel nut ailment. The fungus *Phytophthora palmivora* is the cause [25].
- **Yellow leaf disease:** A disease brought on by the betel nut pathogen phytoplasma that causes green leaves to become yellow and eventually decreases production.
- **Leaf rot:** Ailments that affect coconut plants. Bacteria and fungus are the culprits. The size, shape, and color of leaf markings vary [26].
- **Leafy leaves:** The illness can be identified by its leafy leaves. It may be brought on by viruses, fungi, or tafrina [27].

Since they are submerged in water after first developing on the leaves, leaf disease kills most cotton plants in this container. Ultimately, it develops holes in the leaves and becomes black [28].

- **Streak disease:** *Xanthomonas pvvesicatoria*, a dangerous bacterial disease, is detected in peppers [29] with symptoms like leaf patches or tiny yellow-green lesions.

Figure 5.5 Diseases of different plants.

Plague: The transmission of plague is rapid. Mildew may develop in damp, cool climates. evidence of the emergence of grey patches with erratic shapes on the leaves. The area is surrounded by a ring of white mold [30].

- **Bacterial depletion:** This phenomenon lowers the production of growing eggplant. The plant collapses and the leaves wither [31].

5.3 Techniques and Algorithms for Detecting Plant Disease

5.3.1 Detection Techniques Used Today for Agricultural Diseases

Crop diseases can be discovered and identified using direct or indirect approaches. When a large number of samples need to be evaluated, high-throughput analysis techniques such as molecular and serological approaches can be utilized [32]. For precise disease/pathogen identification, these techniques directly identify disease-causing organisms, such as bacteria, fungi, and viruses. Conversely, indirect approaches use a variety of factors, including morphological alterations, temperature changes, variations in growth rates, and volatile organic chemicals emitted by sick plants, to diagnose plant illnesses [33].

5.3.2 Clustering Algorithms

K-means grouping, based on a collection of characteristics, objects are categorized into K-classes using k-means clustering. By reducing the sum of squared distances between an object and its cluster, objects are classified [34]. Next, consider how the k-means clustering method works.

(1) Heuristically or at random choose the Kth cluster center.
(2) Minimize the distance between each pixel and the cluster's center as you assign each pixel to a cluster.
(3) Determine the cluster's centroid by averaging each pixel. The next step is to repeat steps 2 and 3 until convergence is achieved [35].

The user's expertise determines the K-value and region of interest (ROI) selection in the generally used procedure, and occasionally the user does not choose the ROI appropriately. As a result, we must manually choose the ROI value for each image in the database. Identifying the appropriate disease area takes effort and is prone to mistakes. Thus, plant leaf lesions may be automatically diagnosed via automated clustering. In this study, illness symptoms were automatically separated from "healthy" leaf areas using k-means clustering. As a result, the user must select another technique that automatically calculates the ROI for k-means clustering. This is achieved by setting a threshold between the color of healthy leaf regions and the color of sick areas [36]. In order to do this, we mask pixels whose red values are lower than the leaf image's blue and green values. Only one pot of leaves is left for the remainder of the picture as a result. The procedures for separating pictures from their surroundings and getting them ready for grape leaf DA detection are shown in Figure 5.6.

5.3.3 Image Preprocessing

Real-world locales were used for the image, however there was noise in the form of spores, watermarks, and dust. Data preparation's objectives are image denoising and pixel value correction [37]. In order to recognize and categorize plant leaf diseases, computer vision techniques have been thoroughly investigated and put to use in agricultural applications. In this article, we provide fresh research demonstrating the use of image processing methods for locating and categorizing plant leaf diseases in various outdoor settings [38].

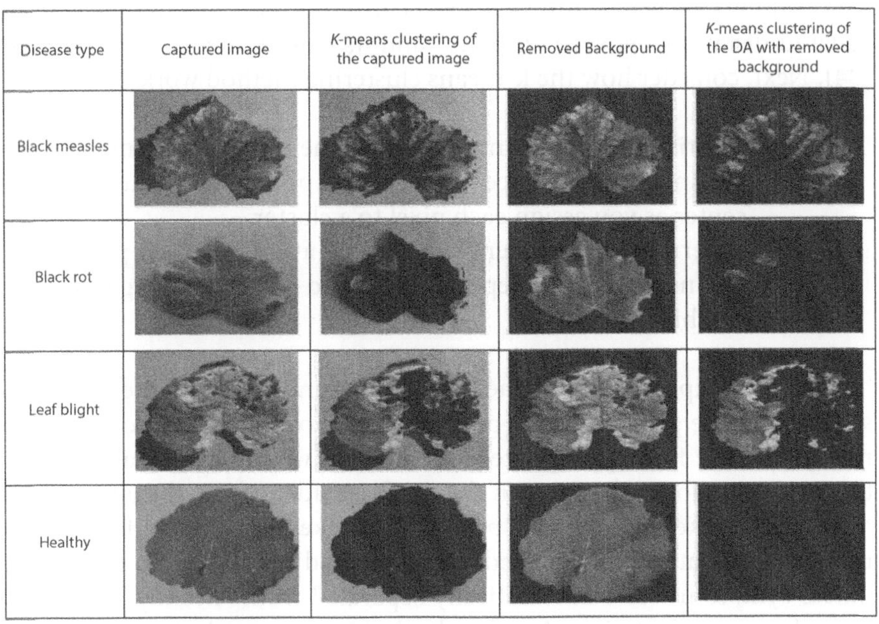

Figure 5.6 Create backdrop photos to isolate and identify DA in grape leaves.

5.4 Dataset for Diagnosis Plant Disease

The main force underlying these technical advancements is datasets. We provide DiaMOS Plant in this work, a field dataset gathered for the analysis and tracking of plant ailments. In order to highlight the importance of the data gathered and to fulfil its main purpose of maximizing information content, it is particularly created for the categorization and detection of leaf diseases [39]. The research community can utilize this study's recommendations for selecting and organizing datasets. The completeness and representativeness of the datasets used to develop the algorithms have a significant impact on the precision and dependability of integrated AI systems. In order to train an intelligent neural network on known examples and improve generalization, a large amount of data must be collected [40].

Implementing the dataset is not a simple and uncomplicated process and requires subject matter specialists owing to the time and money spent obtaining, annotating, and categorizing the photographs. Scientific advancement is slowed down by the availability of digital agriculture (DA) datasets, an issue that has been documented in the literature as shown in Figure 5.7 [41]. We have been focusing on data collecting in recent years.

Figure 5.7 Sample dataset images.

The most well-known application in this sector is the Plant Village [42], which paints a consistent backdrop on the underside of leaves using 54,000 pictures. The environmental circumstances under which the model can function are not accurately reproduced by datasets produced under controlled settings.

5.4.1 DiaMOS Plant Dataset

DiaMOS Plant is the name of the extended dataset tested [43] for identifying and tracking plant issues, a key component of plant culture. Using this dataset, classification and detection procedures using machine learning and deep learning algorithms can be tested. A number of 3505 images were gathered in total, comprising 499 images of fruit and 3006 images of leaves. Fruit, nut, growth, and ripening comprise the fourth stage of a fruit's development. Leaf desiccation, slug damage, leaf spot, and healthy leaves were the four main categories of biotic and abiotic stresses. The DiaMOS Plant refers to the group of 3505 images of fruits and leaves.

The three trees in this picture are from the same Italian land. Many cameras were used to take the pictures, including a smartphone (Honor 6x) and a DSRL (Canon EOS 60D). As a result, photos are offered in two resolutions: 3456 x 5184 and 2976 x 3968. The usage of two devices simultaneously was necessary due to the vast number of persons participating in the data collecting. Also, different resolutions add value and make the dataset more complicated. In this topic of literature, employing several devices is a common strategy since it may give the model inputs that are diverse and representative. Smartphones for agricultural and non-agricultural operators have different technological features in real-world circumstances, including resolution.

5.4.2 RoCoLe Dataset

A number of 1,560 photos of leaves are included in the Robusta Coffee Leaf Image Dataset (RoCoLe). Make an assessment by combining them. Using a smartphone camera at working distances of 200 mm and 300 mm, the axial (top) and abaxial (bottom) surfaces of the blades were captured. This data also contains in-depth item ratings gathered using the Labelbox online tool [44]. Specifically for binary and multi-classification challenges, the RoCoLe is utilized for segmentation operations, as well as for the training and validation of machine learning algorithms. Make a list of plant diseases. The presence of disease (green disease 1 and red spider mite 2) and varied circumstances (healthy vs. unhealthy) on the front and back sides of coffee leaves are represented in the RoCoLe dataset, which also reflects the severity of the infection.

5.4.3 Plant Pathology Dataset

The Phytopathology dataset [45] includes 3651 RGB images of several apple leaf diseases collected during the 2019 growing season and offered for sale in unsprayed apple orchards at Cornell AgriTech (Geneva, NY, USA). A total of 1200 apple scabs, 1399 cedar apple greens, 187 complex disease presentations (several diseases on the same leaf), and 865 healthy leaves were found in 3651 RGB images. Mobile devices frequently operate in busy, open-air settings. Healthy leaves, apple scab, various illnesses, and cedar apple green disease were three of the categories that were manually added to the dataset. These statements have the backing of a board-certified plant pathologist.

5.4.4 Plant Village Dataset

An image-based compilation of 54309 images dubbed "Village of Plants" displaying ventricular leaf disease on leaves with a constant backdrop (black or grey). A standard point-and-shoot Sony DSC-Rx100/13 camera was used by the authors to take 4–7 photographs of each leaf. Only a handful of the 14 crops shown include tomatoes, apples, blueberries, cherries, corn, grapes, oranges, peaches, potatoes, raspberries, beans, squash, and tomatoes. There were 17 diseases associated with fungi, four diseases associated with bacteria, two ovarian diseases associated with fungi, two viral diseases, and one disease transmitted by ticks. Pictures of healthy, disease-free leaves from 12 different crops are also shown in Figure 5.8.

Figure 5.8 Robusta coffee leaf image dataset.

5.5 Segmentation

Although processing a picture using that requires segmenting it into many segments with comparable qualities or features, segmenting the leaf image is still crucial. A number of techniques, including Otsu's approach and k-means clustering, can be used to split data [54]. Processes like picture segmentation, image enhancement, and color space conversion fall under the category of segmentation. Initially, filters are used to improve digital photos. The backdrop image is filtered to remove the leaf images. The filtered image's RGB colors are then translated into color space parameters [55]. The best color identification technique is Hue Saturation Value (HSV). Based on the discontinuity and similarity qualities, image segmentation algorithms may be split into two types [46]. These characteristics give rise to three primary methods for segmenting images: theory-based segmentation, edge-based (contour-based), and hybrid segmentation [56].

The goal of region-based segmentation is to divide a picture into distinct pixel sections that have the same properties. For area-based segmentation, you may utilize split and merge, k-means clustering [57], region expanding, thresholding, etc.

- Edge-based segmentation, which is distinct from approaches based on regions of an image since it entails establishing boundaries between areas of an image. There are three different kinds of borders, according to [47]: points, lines, and contours. Contours are abrupt tone shifts in a picture that have a certain amplitude, slope, or variation in ambient intensity. It can alternatively be explained as the line separating two homogeneous areas [48].
- Methods that combine edge-based and region-based techniques utilizing particular theories. We made use of ideas such artificial neural networks, fuzzy logic, evolutionary algorithms, and clustering [49].

5.5.1 Segmentation using the RCNN

Using the recurrent convolutional neural network (RCNN) approach, we segment leaf disease using enhanced photos. In order to separate the foreground and background of these photos of leaves, CRUN initially employs a region-based convolutional neural network. Diseased leaf portions were separated from healthy leaves using U-Net. In this stage, the RCNN is used to separate foreground elements in the picture, such as leaves, from background elements, such as black and white. As seen in Figure 5.9, the layers of the RCNN structure resemble those of the AlexNet.

The input, convolutional, and maxpool layers of RCNN are analogous to those of AlexNet. But, instead of SoftMax from the RCNN to the final layer, we have bounding boxes and SVM output, and we are completely linked as indicated as Figure 5.10. The ability to update all network layers during training is the main benefit of RCNN. There is no need for disc space for the cache. It has great detecting quality, as well.

To split the leaf picture and find leaf and sick zones throughout the image, use this bounding box. Regions of a picture such as the foreground,

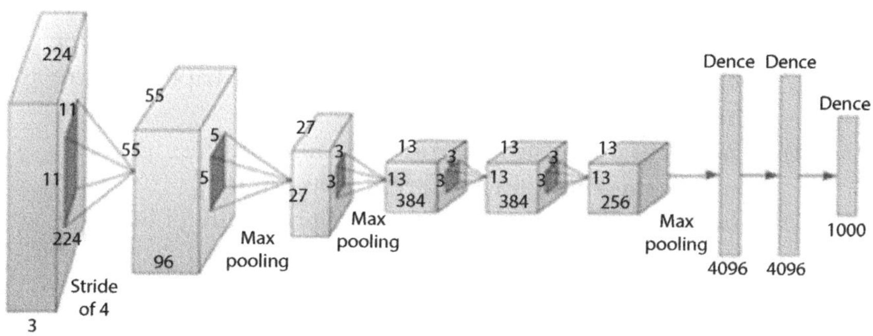

Figure 5.9 AlexNet architecture.

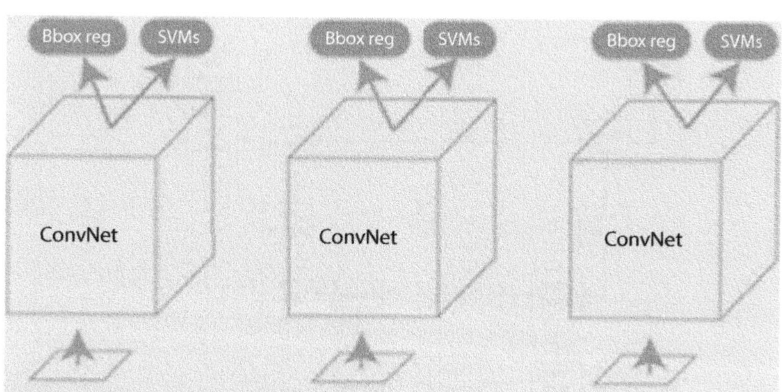

Figure 5.10 Final layer of RCNN.

foliage, background colors, and color transitions are represented by bounding boxes. The disease area is then finalized by applying this bounding box result to U-Net [50].

The two SoftMax and completely connected layers are not utilized architecturally by this architecture, which is why RCNN and U-Net are used to divide the lesion area. The procedure states that U-Net separates the boundary region output from the lesion region. The U-shaped design of U-Net, which has more convolutional and max-pooling layers, gave rise to the name as shown in Figure 5.11 (Bounding box RCNN outputs for detecting disease areas using U-Net-based medical image segmentation characteristics). These layers are swapped out with a significant number of convolutional and max pooling layers.

(1) Upsampling the picture by a factor of two is referred to as phrasing up.
(2) down denotes a ratio of 2 downsampling of the pixel value.
(3) The term "conv3" refers to the [128 128] convolution filter being used to conduct convolution on all triplet picture channels.
(4) The act of duplicating a prior outcome and fusing it with the present condition is referred to as joining.
(5) The term "max pool" describes the procedure of bringing the outcomes of the preceding step into line.

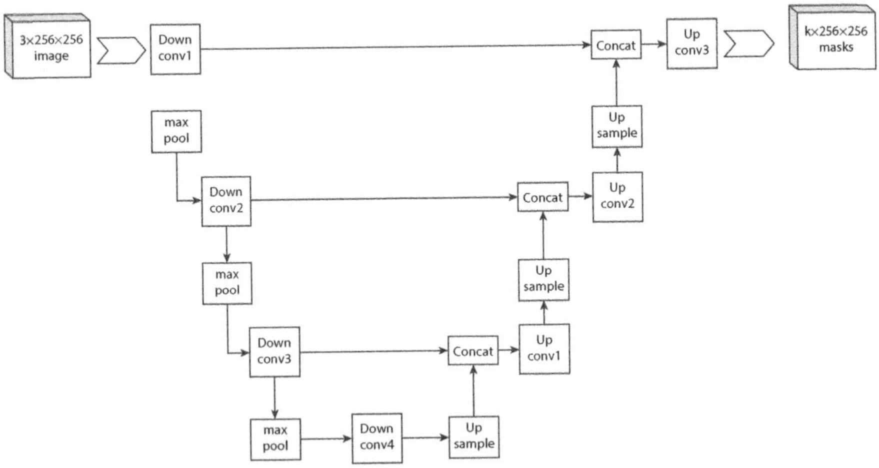

Figure 5.11 Architecture diagram of U-net.

Figure 5.12 Sample segmentation image.

Images are downscaled in the manner mentioned above before being convolved with convolutional layers. The picture is then split into normal and abnormal parts using convolutional filtering once the convolutional output has been upsampled. For the amended original dataset, the previously described U-Net architecture method was applied to separate sick areas from leaves.

5.5.2 Segmentation using TGVFCMS

A hybrid partition named total generalized variation fuzzy C-Means (TGVFCMS) was used to divide the affected leaf sections. The data is separated, and then it is sent to CNN for classification. The TGVFCMS technology excels in edge preservation and noise reduction. In order to reduce undesirable noise and artefacts in FCM-based approaches, our TGVFCMS was created to incorporate TGV normalization into the smoothing cycle. The TGV framework is shown below.

This is due to the fact that the parent and initial gradients produced in (4) are both sparsity constrained. Voici the second TGV.

The segmented sample image shown in Figure 5.12.

5.6 Classification

Formerly, researchers concentrated on raising the reliability of automated plant leaf disease detection systems. We examine methods for categorizing plant diseases using various classifiers, such as convolutional neural networks, support vector machines, K closest neighbors, and artificial neural networks.

5.6.1 Convolutional Neural Network (CNN)

A deep feed-forward neural network that can handle multidimensional data is a convolutional neural network. The goal of a CNN is to condense a picture into a format that is digestible while maintaining the elements required for accurate prediction. The CNN supports the AlexNet, Google Net, and VGGNet architectures. Researchers in several computer science domains have expressed a great deal of interest in it [51]. It is employed in agriculture to categorize plant diseases. The input layer, the convolutional layer, the pooling layer, the fully connected layer, and the output layer make up the CNN model, as illustrated in Figure 5.13. Provide input in the form of photos to correctly categories plant diseases. To extract characteristics from pictures, convolutional layers are employed. From the retrieved features, the pooling hierarchy calculates feature values. More convolutional and pooling layers may be utilized to extract more detailed information, depending on the complexity of the picture. The output of a completely linked layer is transformed into a single vector that may be utilized as the input of the following layer. Plant diseases will eventually be categorized in the output layer.

For identifying plant leaves with a CNN approach, the following activities were carried out. AlexNet and Google Net are the two CNN designs

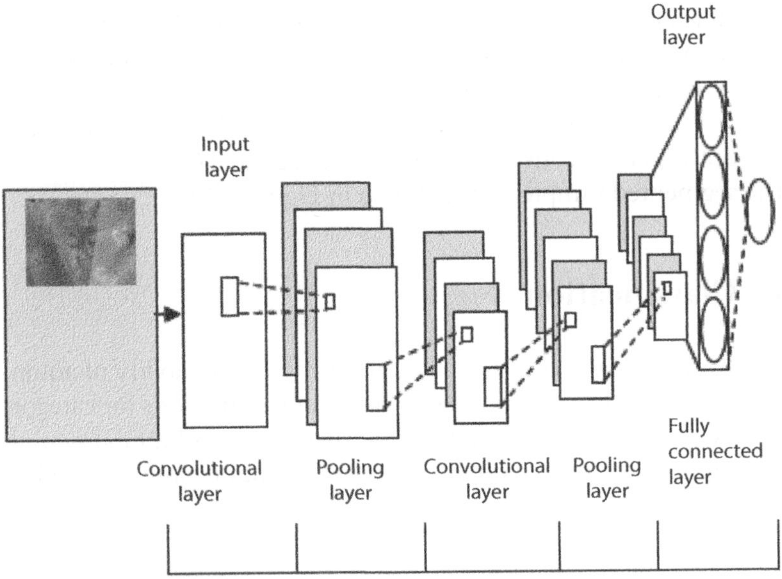

Figure 5.13 Architecture of deep CNN.

that receive the majority of attention in Mohanty *et al.* 2016 [10], while transitional learning and training from scratch are employed as the training mechanisms. The Plant Village collection, which includes around 54306 photos of various plants with 38 disease classifications, was utilized to create the images. The model is optimized, and predictions are made using the downsized, 256x256-pixel versions of every image in the dataset. The CNN architectures employed by Ferentinos, K. P. [52] in 2018 included AlexNetOWTBn, Over feat, AlexNet, VGG, and Google. The net was trained using several metrics. Torch7, a machine learning computer framework, is used to implement the training and testing of these models. Around 87848 photos from the Plant Village collection, which include 25 plant species in 58 distinct disease categories, were used in this study. The architectures with the greatest success rates are VGG and AlexNetOWTBn.

5.6.1.1 *Convolutional Layer*

The first layer in a CNN arrangement is always a convolutional layer. The MN1 is a valid input layer for CNN. The 2D size of a 2D picture with a single layer is MxN. This filter is convolved with the picture and has the same depth as the input image. The outcome is the final picture, which is produced by convolving the input image with this curve or form. Since we approximate the input image's curve during convolution, the form revealed by the filter ends up having greater values. The convolution process can be represented as follows using equation 5.8.

$$s\,(t) = (x^{*}w)(t) \tag{5.8}$$

5.6.1.2 *Pooling Layer*

The data size is reduced by using pooling layers. This entails segmenting the matrix data into individual values and replacing each fragment. The illustration demonstrates how the extremum or mean of all values in the current segment replaces the partitioned matrix.

5.6.1.3 *Fully Connected Layer*

In order to conform to the network layer design, dimensions are modified in the fully linked layers. With a completely connected layer, every input and output dimension is linked to every other dimension. Like a typical neural network, all activations from the preceding layer are passed on to the following layer.

5.6.2 Classification Using Artificial Neural Networks

A neural network (ANN) is a type of information processing model that takes its cues from how biological systems, particularly the brain, process information. It is made up of synthetic neurons, also known as processing elements (PEs), which are linked together by coefficients to form a neural network. Find links and patterns in data to learn from experience rather than code. Artificial neural networks may be used to extract patterns since they can make sense of complicated data. Feed-forward ANNs and feed-back ANNs are the two forms of ANNs, where a signal is created and the behavior of any layer does not impact that layer. Bidirectional network loop propagation is requested. An artificial neuron, as seen in Figure 5.14, has several inputs that can each have a value between 0 and 1, but only one output. Neurons contain weights and a general bias for each input [53].

5.6.3 Classification Using K-Nearest Neighbor (K-NN)

This nonparametric statistical classifier assigns neighbors the proper weights. The calculated Euclidean distance metric is used in this instance to categorize the data. It also goes by the moniker "lazy learner" since it stores all input values for learning tuples without computing them during the learning phase. As a result, it cannot be employed in domains that need dynamic categorization of big databases. Blue stars in Figure 5.15 denote test data points. Two classes are provided by the red circle and green square

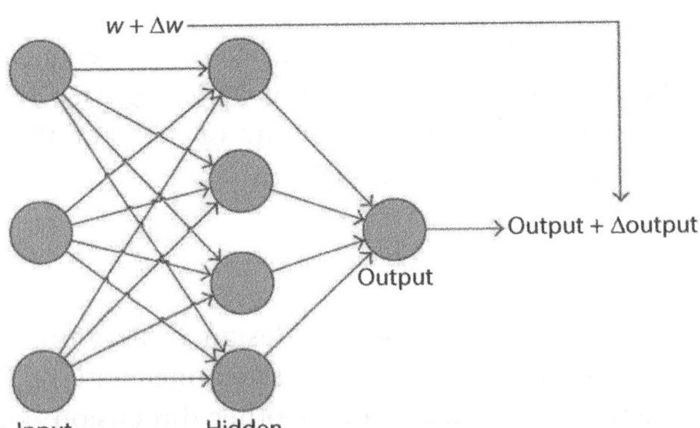

Figure 5.14 Classification using ANN [29].

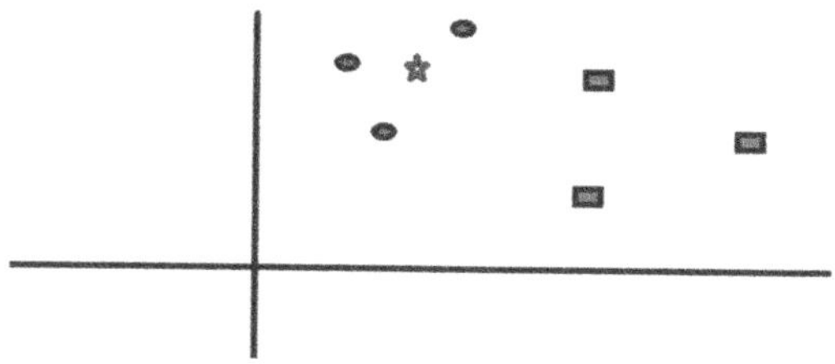

Figure 5.15 K-NN classifier.

that surround the test point. In Figure 5.15, there are six points, hence six distances must be determined.

Three crucial factors for K-NN categorization are:

- It will appear as shown.
- Easy to understand result output; quick computation; high prediction rate; and short calculation time.

Text mining, pattern recognition, predicting stock market trends, categorizing numerous plant diseases in agriculture are all common uses for this technique. The next step is to use the K-NN approach to find plant illnesses. The trained Inception v3 model's convolutional layers were used by Ram Charan *et al.* in 2017 [25] together with a transition learning mechanism based on SVM and KNN. Diseases are categorized in three distinct ways. Models for the SVM, KNN, and Inception SoftMax Layer V3. Using a method, Sureza *et al.* 2017 [26] suggested detecting illnesses like brown spot and rice blast transmission. In the Shivamogga area of Karnataka, the camera recorded roughly 300 photographs that were used in this article. With a global threshold, we partition data using Otsu's approach. With a global threshold, we partition data using Otsu's approach. For classification, the KNN approach is used, and for feature extraction, linked components.

A subgroup of deep neural networks (DNNs), known as DBNs (Hinton *et al.*, 2006), includes multilayer restricted Boltzmann machines (RBMs) and multilayer perceptrons (MLPs). Connected basic units and visible units make up an RBM. Used for connections with weights. An MLP is an input-, hidden-, and output-layered feed-forward network. Multiple-layer networks are capable of handling challenging problems and producing more effective picture categorization for identifying plant diseases. The DBN architecture is seen in Figure 5.16. The attribute data gleaned from

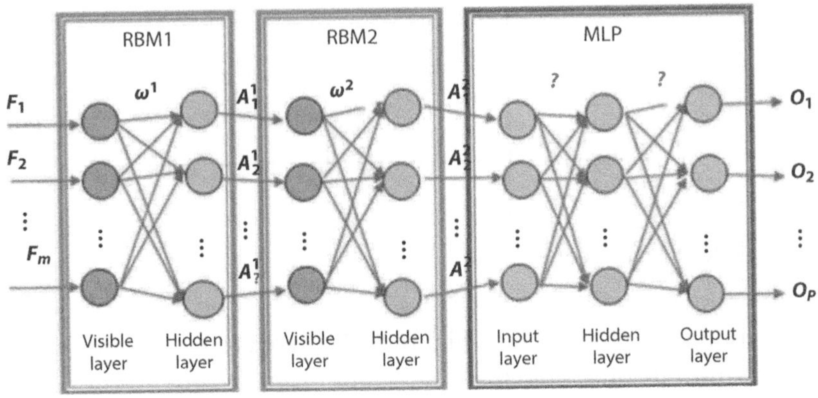

Figure 5.16 Structural design of a DBN classifier.

analyzing the features serves as the input of the visible layer, while the first RBM's hidden layer is stated as follows.

$$F1 = \{F11, F21, \ldots, Fx1, \ldots, F31\}\ ; 1 \leq x \leq 3$$

where F1x represents the first RBM's xth visible neuron. The first RBM's secret levels are described as follows:

$$A1 = \{A11, A12, \ldots, A1o, \ldots, A1e\}\ ; 1 \leq o \leq e$$

A1o here stands for a fully concealed neuron as well as another hidden neuron. Polarizations are present in both the surface and deep levels of every neuron. The deflections of the exposed and covered layers are represented by y and z, respectively.

$$y1 = \{y11, y12, \ldots, y1x, \ldots, y13\}$$

where the xth observable neuron's deflection is indicated by the symbol y1x.

$$z1 = \{z11, z12, \ldots, z1o, \ldots, z1e\}$$

5.6.4 Classification Using an Adaptive Neuro-Fuzzy Classification Model (ANFIS)

A "fuzzy reasoning system" was proposed [20]. Certain input/output operations make it simple to create adaptive neuro-fuzzy inference system (ANFIS), a fuzzy inference system. A mixed learning technique was used to build and construct the ANFIS architecture in the study. For the categorization of plant diseases in this work, a linear Takagi Sugano-type adaptive neural purification architecture was employed. The major objective of adopting ANFIS is to enhance the decontamination system's capacity for self-learning in order to achieve the categorization of plant diseases. A multilayer feed-forward network is what ANFIS is. Thirteen color feature vectors were retrieved for this investigation. 13 input variables and one output variable, a categorical category, are present. According to Figure 5.17, the ANFIS architecture for classifying plant diseases using the color moment model is composed of five layers.

The first layer's nodes are all adaptive nodes with node functions.

$$L_{1,i} = \mu_{Ai}(x_k)$$

Ai is the label given to the node if L1 is layer xk, where k=1, 2,...13. Use Gaussian membership functions for membership functions (MF).

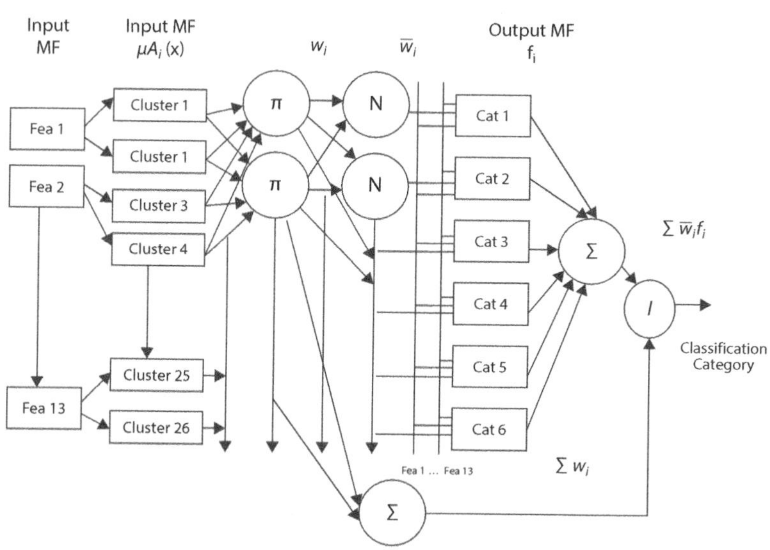

Figure 5.17 ANFIS architecture-based tomato plant disease classification.

$$\mu_{Ai}(xk) = e^{\frac{1}{2\left(\frac{x_k - c}{\sigma}\right)}^2}$$

where "c" stands for the center and "c" is a collection of parameters. shows the membership function's width. The term "prerequisite parameters" refers to these variables.

Layer 2: Each node in this layer that is indicated by the letter P is a fixed node that is the same as the sum of all the inputs.

$$L_{2,i} = w_i = \mu_{Aj}(x_1)\mu_{Bk}(x_2)\ldots\ldots\mu_{Mn}(x_{13})$$

Where i= 1,2,3…..26; j,k=1,2; n=n-1

Layer 3: The tier is represented by the letter N and contains only fixed nodes. The i-th node determines the proportion of the i-th rule and determines the total strength of a certain rule as follows:

$$L3,I=wi=wi/\Sigma i\ w\ i :=1,2,\ldots\ldots 26$$

The output of this layer is normalized.

Layer 4: The fourth layer's nodes are all adaptive Node

$$L_{4,i} = \bar{w}_i f_i = w_i(p_i x_1 + q_i x_2 + \cdots\ldots\ldots + s_i x_{13})$$

Where $\bar{w}_i$ is the normalized output of layer 3, i = 1,2,3… ..26 and $\{p_i, q_i, s_i\}$ is the "parameter set" of present the node. The parameter of the layer refers as "consequent parameters".

Layer 5: The output from the sum of all inputs is calculated by this layer as follows:

$$L_{5,i} = \sum \bar{w}_i f_i = \frac{\sum_i w_i f_i}{\sum_i w_i}$$

In this work, a variety of learning principles are used to teach ANFIS. According to hybrid learning, there are two unique steps: the first is the forward pass of the input, where the pertinent feature signal is sent to four layers and the "result parameters" are determined using the "least squares estimate technique." The reverse pass is the second. Gradient descent is used to update the "assumption parameters," which represent the total of all nodes' error rates from output to input. The hybrid approach, which is 20 times quicker than the original backpropagation, decreases the dimensionality of the search space.

5.7 Conclusion

The agricultural industry has incorporated data mining technologies. To identify impacted crops and aid agriculture, the initiative will put creative ideas into practice. The k-means clustering technique was used to separate and assess regions of leaf infection. Publish photos for apps that recognize diseases. It provides a fantastic alternative, particularly for isolated rural farming areas. The system created here is used to diagnose plant illnesses, and accurate characterization and well-established classification techniques are crucial for operating the system in real time. As a consequence, the suggested strategy, which uses an ANN classifier for classification and a Gabor filter for feature extraction, produces better results, with a 91% recognition rate. For different plant diseases, an ANN-based classifier that combines color and texture data can be employed. The findings are instructive and encouraging for the creation of cutting-edge machine vision systems for the detection and categorization of plant diseases. The suggested approach is highly useful for distinguishing between regular and damaged items. We take into account a literature dataset created for the classification and detection of leaf diseases, offering a performance benchmark when the collection is available. The amount of information that a piece of data may convey in terms of a feature that defines how it will be shown or displayed. The suggested dataset was built taking these aspects into account. It is a good strategy for decreasing the duration and extent of diseased areas that cluster. Techniques for feature extraction aid in the extraction of diseased leaves and classification of plant diseases. We were aided in navigating the procedure by an integrated voice navigation system. Creating an open multimedia (audio/video) for illnesses that are automatically treated if recognized is a potential upgrade for this project.

References

1. Ali, R., Manikandan, A. & Xu, J. A Novel framework of Adaptive fuzzy-GLCM Segmentation and Fuzzy with Capsules Network (F-CapsNet) Classification. *Neural Comput. Applic.*, (2023). https://doi.org/10.1007/s00521-023-08666-y
2. Annamalai, Manikandan & Muthiah, Ponni. (2022). An Early Prediction of Tumor in Heart by Cardiac Masses Classification in Echocardiogram Images Using Robust Back Propagation Neural Network Classifier. *Braz. Arch. Biol. Technol.*, 65. 10.1590/1678-4324-2022210316.
3. Manikandan, Annamalai, M,Ponni Bala. (2023). Intracardiac Mass Detection and Classification Using Double Convolutional Neural Network Classifier. *J. Eng. Res.*, 11(2A). 272-280. 10. 36909/jer.12237
4. Sheikdavood K, Surendar P, Manikandan A. Certain Investigation on Latent Fingerprint Improvement through Multi-Scale Patch Based Sparse Representation. *Indian J. Eng.*, 2016; 13(31):59-64.
5. S. Dhanasekaran, Dr. P. Mathiyalagan, Rajeshwaran, A. Manikandan, "Automatic Segmentation of Lung Tumors Using Adaptive Neuron-Fuzzy Inference System ", *Ann. RSCB*, 2021; pp. 17468–17483.
6. Manikandan, A., & Sakthivel, J. (2017a). Recognizable Proof of Biometric System With Even Distorted And Rectification States. *J. Adv. Res. Dyn. Control Syst.*, *9*(2), 1393–1398.
7. Manikandan, A., & Jamuna, V. (2017). Single Image Super Resolution via FRI Reconstruction Method. *J. Adv. Res. Dyn. Control Syst.*, *9*(2), 23–28
8. Manikandan, A., Suganya, K., Saranya, N., Sudha, V., & Sweetha, S. (2017). Assessment of Intracardiac Masses Classification. *J. Chem. Pharm. Sci.*, 5, 101–103.
9. Ashokkumar, N. & Meera, S. & Anandan, P. & Murthy, Mantripragada & K S, Kalaivani & Alahmadi, Tahani & Alharbi, Sulaiman & Raghavan, S. & Jayadhas, s. (2022). Deep Learning Mechanism for Predicting the Axillary Lymph Node Metastasis in Patients with Primary Breast Cancer. *Biomed Res. Int.*, 2022. 10.1155/2022/8616535.
10. K. N. R. C. K. D. D. T. "Survey on 2D-DCT Based Image Watermarking With High Implanting Limit and Robustness". *Int. J. Recent Innov. Trends Comput. Commun.*, vol. 4, no. 10, Oct. 2016, pp. 161-4, doi:10.17762/ijritcc.v4i10.2576.
11. Karpagalakshmi, RC, Tensing, D & Kalpana, A.M. (2016). Image Localization using Deformable Model and its Application in Health Informatics. *J. Med. Imaging Health Inf.*, vol. 6(8), pp. 1972-1976. https://doi.org/10.1166/jmihi.2016.1959.
12. Namrata, K, Karpagalakshmi, R, C, Manikandasaran, S, S. (2017). Implementation of Novel Technique for Image Watermarking Using 2D-DCT. *Int. J. Pure Appl. Math.*, volume 117(16), pp. 221-226.
13. T. Ramalingam, R. Umamaheswari, R. C. Karpagalakshmi, K. Chandramohan, M.S. Sabari. (2021). Location of plant Leaf maladies utilizing picture

division. *J. Image Process. Artif. Intell.*, vol. 7(3). http://dx.doi.org/10.46610/JOIPAI.2021.v07i03.002.

14. Gopalan, S. H., Vignesh, V., Mahendran, N., & Dinesh, M. T. P. P. (2021). Dynamic Clinical Trials Management in Anunreliable Environment using Blockchain. *Des. Eng.*, 817-822.
15. S. S, N. H. A. Rufus, D. Anand, R. S. Rama, A. Kumar and A. S. Vigneshwar, "Evolutionary Optimization with Deep Transfer Learning for Content based Image Retrieval in Cloud Environment," *2022 International Conference on Augmented Intelligence and Sustainable Systems (ICAISS)*, Trichy, India, 2022, pp. 826-831, doi: 10.1109/ICAISS55157.2022.10011122.
16. Ashokkumar, N Ashokkumar and P, Krishnagandhi and Kannan, B and Raju, Y David Solomon, Smart Farming Field Observation Using Embedded Systems (2020). *J. Electr. Eng. Technol.*, 11(4), 2020, pp. 241-245.
17. Kavitha, A., N. Ashok Kumar and M. Revathy. (2018). Automatic Identification of Maritime Boundary Alert System using GPS. *Int. J. Eng. Technol.*, 7.3.1 pp.20-22.
18. Ashokkumar, N., B. Kannan and Y. Raju. (2020). Smart Farming Field Observation Using Embedded Systems. *Int. J. Electr. Eng. Technol.*, 11(4).
19. Nagarajan, Ashokkumar, A. Kavıtha and S. Devı. "Wireless Sensor Data Fusion Techniques in Estimating Temporal Resource Attributes in Scenarios of Intermittent Connectivity." *El-Cezeri*, 9.2: 413-423.
20. Umamaheswari, R. Karpagalakshmi, R.C. Kavinkumar. (2020). An Android Application For Smart Attendance Management System By Using Face Recognition. *J. Eng. Sci.*, vol 11(11).
21. Chandramohan, K. Karpagalakshmi, R. C. Sumaiya, Subhashini and Sowmiya. (2020). Design and Implementation Of Weeding Robot For Organic Farming. *J. Eng. Sci.*, vol 11(11).
22. Karpagalakshmi, R.C. Gobinathan, Umamaheswari, R. Siva, A. (2020). Reviewing And Modelling Of Seed Sowing Multipurpose Robot Towards Smart Farming. *J. Eng. Sci.*, vol 11(11).
23. Gopalan, S. H., Suba, S. A., Ashmithashree, C., Gayathri, A., & Andrews, V. J. (2019). Digital forensics using blockchain. *Int. J. Recent Technol. Eng.*, *8*(2), 182-184.
24. Venkataraman, Vignesh & Suryanarayanan, Harihara Gopalan & Mohan, M & Ramya, R S & Ananthakumar, R. (2021). A Quantum-Based Blockchain Approach to Voting Protocol Using Hyperledger Sawtooth. *J. Phys. Conf. Ser.*, 1916. 012088. 10.1088/1742-6596/1916/1/012088.
25. Ali, R., Manikandan, A., Lei, R. *et al.* (2024). A novel SpaSA based hyper-parameter optimized FCEDN with adaptive CNN classification for skin cancer detection. *Sci. Rep.*, 14, 9336. https://doi.org/10.1038/s41598-024-57393-4.
26. Vignesh, & Suryanarayanan, Harihara Gopalan & Mary, Celsia & Indumathy, N & Harida, A. (2021). Certain Investigation on IOT Based Smart Pill Carafe in Health Care. *J. Phys. Conf. Ser.*, 1916. 012152. 10.1088/1742-6596/1916/1/012152.

27. Manikandan, A., & Sakthivel, J. (2017). Solid Waste Bin Management System. *J. Adv. Res. Dyn. Control Syst.*, *9*(2), 29–33.
28. Bommaraju, K., Manikandan, A., & Ramalingam, S. (2017). Aided System for Visually Impaired People in Bus Transport using Intel Galileo Gen-2: Technical Note. *Int. J. Veh. Struct. Syst.*, *9*(2), 110–112. https://doi.org/10.4273/ijvss.9.2.09
29. Nirmal Kumar, P., & Manikandan, A. (2017). Zigbee Based Online Air Pollution Monitor. *J. Chem. Pharm. Sci.*, 230–232.
30. Manikandan, A., & Pradeep, S. (2017). IoT Based Electricity Theft Detection. *Int. J. Control. Theory Appl.*, *10*(12), 211–215.
31. Manikandan, A., & Tamilselvan, S. (2017). RFID Based Voice Bank Alert System for Blind People. *J. Chem. Pharm. Sci.*, 308–310.
32. Manikandan, A., & Sakthivel, J. (2016). IoT Based Real Time Traffic Analyzer. *J. Chem. Pharm. Sci.*, 139–142.
33. Manikandan, A., Ramalingam, S., & Aathi bhagavan, V. (2016). Potholes Alert System for Riders. *Int. J. Adv. Nat. Appl. Sci.*, *10*(9), 440–444.
34. Manikandan, A., & Nirmala, V. (2015). A Low Cost Thermoelectric Refrigerator. *Int. J. Appl. Eng. Res.*, *10*(55), 3097–3101.
35. T. Thamaraimanalan, M. Mohankumar, S. Dhanasekaran, "Experimental analysis of intelligent vehicle monitoring system using Internet of Things (IoT)," *EAI Endorsed Trans. Energy Web*, p. 169336, Jul. 2018
36. V. V. Teresa, K. Rajeshwaran, S. S. Kumar, S. Vishnupriyan and S. Dhanasekaran, "IoT-based Underground Cable Fault Detection," *2022 International Conference on Augmented Intelligence and Sustainable Systems (ICAISS)*, Trichy, India, 2022, pp. 1094-1100, doi: 10.1109/ICAISS55157.2022.10010910.
37. Ashokkumar, N., Nagarajan, P., Venkatramana, P. (2020). 3D(Dimensional)—Wired and Wireless Network-on-Chip (NoC). In: Ranganathan, G., Chen, J., Rocha, Á. (eds) Inventive Communication and Computational Technologies. *Lecture Notes in Networks and Systems*, vol 89. Springer, Singapore. https://doi.org/10.1007/978-981-15-0146-3_12.
38. Ashokkumar, N. and A. Kavitha. "A novel 3D NoC scheme for high throughput unicast and multicast routing protocols." *Tech. Gazette*, 23.1 (2016): 215-219.
39. Ashokkumar, N. and A. Kavitha. "Network on Chip: A Framework for Routing in System on Chip." *J. Comput. Theor. Nanosci.*, 12.12 (2015): 6077-6083.
40. Kumar, N. A., Kavitha, A., Venkatramana, P., & Nandan, D. (2022). Architecture Design: Network-on-Chip. In *VLSI Architecture for Signal, Speech, and Image Processing* (pp. 147-165). Apple Academic Press.
41. Gopalan, S.H. ZHRP-DCSEI, a Novel Hybrid Routing Protocol for Mobile Ad-hoc Networks to Optimize Energy Using Dynamic Cuckoo Search Algorithm. *Wireless Pers. Commun.*, 118, 3289–3301 (2021). https://doi.org/10.1007/s11277-021-08180-1

42. Gopalan, S., & Radhakrishnan, R. (2016). Improved Cuckoo Search Optimisation Based Energy-Delay Aware Routing Algorithm in Manet for Rescue and Emergency Applications. *Int. J. Comput. Technol. Appl.*, 9, 20.
43. Gopalan, S. H., & Krishnan, R. R. (2016). Trust based fuzzy aided ACO for optimal routing with security in MANET. *Asian J. Res. Social Sci. Humanit.*, *6*(cs1), 529-544
44. Gopalan, S. H., & Radhakrishnan, R. (2014). Probability Based Optimized Energy Efficient Routing Algorithm for Mobile AD-HOC Network. *Middle East J. Sci. Res.*, *22*(4), 591-595
45. Kavitha, T. & Pandeeswari, N. & Shobana, R. & Vinothini, V.R. & Karuppanan, Sakthisudhan & Jeyam, A. & Malar, A. (2022). Data congestion control framework in Wireless Sensor Network in IoT enabled intelligent transportation system. *Meas.: Sens.*, 24. 100563. 10.1016/j.measen.2022.100563.
46. Natarajan, Vignesh & Thandapani, Kavitha. (2022). Reliable efficient cluster routing protocol based HTDE scheme for UWSN. *Indones. J. Electr. Eng. Comput. Sci.*, 28. 498. 10.11591/ijeecs.v28.i1.pp498-507.
47. Natarajan, V.P., Thandapani, K. (2021). Adaptive time difference of time of arrival in wireless sensor network routing for enhancing quality of service. *Instrum. Mes. Metrol.*, Vol. 20, No. https://doi.org/10.18280/i2m.200602
48. Natarajan, Vignesh & Thandapani, Kavitha. (2022). An improvement of communication stability on underwater sensor network using balanced energy efficient joining distance matrix. *Int. J. Syst. Assur. Eng. Manag.*, 13. 10.1007/s13198-021-01593-y.
49. S. Ahankari, M. Rajmohan, A. PruthaRani, D. Yeshasree and T. Kavitha, "Wireless Underwater Communication: A Networking Approach for Estimating First Order Lag in Routing Data," *2022 International Conference on Electronics and Renewable Systems (ICEARS)*, Tuticorin, India, 2022, pp. 743-749, doi: 10.1109/ICEARS53579.2022.9751824.
50. Karpagalakshmi, R.C., Vijayalakshmi, P., Gowsic, K., & Rathi, R. (2021). An Effective Traffic Management System Using Connected Dominating Set Forwarding (CDSF) Framework for Reducing Traffic Congestion in High Density VANETs. *Wirel. Pers. Commun.*, *119*, 2725 - 2754.
51. R. C. Karpagalakshmi and D. Tensing, "Vehicle object observation using position based local gradient model," 2012 International Conference on Radar, *Commun. Comput. (ICRCC)*, Tiruvannamalai, India, 2012, pp. 293-298, doi: 10.1109/ICRCC.2012.6450598.
52. V. A.R, S. David, E. Govinda, K. Ganapriya, R. Dhanapal and A. Manikandan, "An Automatic Brain Tumors Detection and Classification Using Deep Convolutional Neural Network with VGG-19," *2023 2nd International Conference on Advancements in Electrical, Electronics, Communication, Computing and Automation (ICAECA)*, Coimbatore, India, 2023, pp. 1-5, doi: 10.1109/ICAECA56562.2023.10200949.
53. S. Suvitha, R. C. Karpagalakshmi, R. Umamaheswari, K. Chandramohan, M. S. Sabari, (2021). An Estimation and Evaluation of Network Availability

in Link State Routing Networks. *J. Netw. Secur. Comput. Netw.*, https://doi.org/10.46610/JONSCN.2021.v07i03.003.

54. K. Suganyadevi, V. Nandhalal, S. Palanisamy and S. Dhanasekaran, "Data Security and Safety Services using Modified Timed Efficient Stream Loss-Tolerant Authentication in Diverse Models of VANET," *2022 International Conference on Edge Computing and Applications (ICECAA)*, 2022, pp. 417-422, doi: 10.1109/ICECAA55415.2022.9936128.
55. Manikandan, A, Pradeep,S. (2017). Quantitative Analysis of Network Arrangement in Randomized Appropriation in WSN" *J. Chem. Pharm. Sci.*, pp. 181-184, 2017.
56. Manikandan, A., & Rajarajachozhan, C. (2017). Artificial Bee Colony for Socially Aware Networking. *J. Chem. Pharm. Sci.*, 2, 299–301.
57. Venkataramanan, C. & Ramalingam, S. & Manikandan, A.. (2021). LWBA: Lévy-walk bat algorithm based data prediction for precision agriculture in wireless sensor networks. *J. Intell. Fuzzy Syst.*, 41. 2891-2904. 10.3233/JIFS-202953.

6

Smart Dust Technology for Monitoring and Control Systems in Smart Agriculture and Crop Surveillance Systems

M. Yogeshwari[1]* and A. Prasanth[2]

[1]*Department Information Technology, School of Computing Sciences, Vels University, VISTAS, Chennai, Tamilnadu, India*
[2]*Department of Electronics and Communication Engineering, Sri Venkateswara College of Engineering, Sriperumbudur, India*

Abstract

Developments in information and communication technology (ICT) have brought a huge impact on the social and economic development of the world. In India, most of the rural community depends on farming for their lives. Modern farming practice is necessary to satisfy the food grain requirements of the growing population. The ICT sector plays a vital role in imparting sustainable knowledge to farmers by offering a wide range of solutions to some of the challenges with an objective of improving crop productivity. Wireless sensor networks (WSNs) is one of the enabling technologies and is successfully used for efficient and profitable precision agriculture.

A sensor network is a highly distributed, self-organized system consisting of a set of battery-operated small sensor nodes. It is used to monitor the various location-specific parameters related to soil, crop, and climate. The sensor nodes are mostly deployed in open and harsh environments and should operate in an unmanned manner for an expected duration. These nodes rely on battery power for their functioning, hence energy efficiency is of paramount interest in these resource constraint networks. It is identified that the grid topology-based sensor network is well suited for agriculture applications. The various field-specific parameters collected by the sensor networks are reported to a base station (sink) for further processing.

**Corresponding author*: myogeshwari.scs@velsuniv.ac.in

Rajesh Kumar Dhanaraj, Balamurugan Balusamy, Prithi Samuel, Malathy Sathyamoorthy and Ali Kashif Bashir (eds.) Computer Vision in Smart Agriculture and Crop Management, (123–138) © 2025 Scrivener Publishing LLC

The data stored in the sink are used for the mission of soil moisture forecasting and sugarcane yield classification problems. Sugarcane is one of the profitable and indigenous crops cultivated in India. Being a long-term irrigated crop, it does not need a specific soil type but needs a high water requirement. Therefore, adequate soil moisture should be available throughout the growing period, as the cane growth is directly proportional to the water transpired. This makes the work discussed in this chapter to give focus to development of smart agriculture.

Keywords: Smart agriculture, computer vision, internet of things, wireless sensor networks, crop management, machine learning, information and communication technology, smart dust

6.1 Introduction

The economics of a nation is significantly influenced by agricultural production [1]. Lack of appropriate detection might seriously harm the amount and quality of the agricultural harvest [2]. One of the most basic needs for a person to live and survive is food. The provision of food around the world is greatly aided by agriculture [3]. Food scarcity in developing countries can be reduced by increasing the yield in agriculture. Three major reasons for low yield are because the farmers are unaware of the fertility of the land, excess use of fertilizer as it is randomly applied in the field without soil testing and the use of old methodology in farming that has no monitoring system in the agricultural field.

Human life depends on agriculture, which are also important sources of food, medicine, and other things. However, biodiversity is fast vanishing globally [4]. First and foremost, the decision on crops to be planted in the agricultural land should be based on soil fertility. The erroneous decision of the crop for a particular land causes loss in agricultural production. Secondly, during the growth of the plants, it absorbs nutrients from the soil that leads to a lack of nutrients for further growth.

Thirdly, the absence of checks regarding soil nutrients through traditional methodology adopted by farmers might cause improper usage of fertilizer. Fourthly, nutrients are lost due to environmental hazards like soil erosion. As the productivity of agriculture depends on the soil nutrients of the land, the research ponders on monitoring the soil nutrients in an automated, accurate, and economical method. The prevalence of several leaf diseases in the agricultural sector lowers the quality of the crop [5].

The manual measuring of plant features used in traditional plant phenotyping is a laborious, expensive process [6]. Mint leaves are rich in vitamin A, manganese, and fiber. It protects the body from oxidative stress, improves brain functions, acts on digestive disorders and prevents cold effects. Curry leaves are rich in vitamin A, B, C, calcium, protein, iron reduces weight, for vision improvement, cure anemia, improves skin and hair health. Neem leaf is bitter in taste but it is of high medicinal value for headache, hair, skin, removing toxic effects in blood and treats ulcer too.

Leaf rust has the potential to be less harmful to sensitive kinds than stripe and stem rust, although it may still spread over most seasons [7]. Nerium leaf is used in wound healing, diabetes, cancer medicines, cardiac illness, asthma, etc. It is toxic when used in excess. The crown flower leaf has high medicinal value to treat diabetes, treat ear pain, natural anti-inflammatory disease healer, reduces the red blemish in the skin, etc. Jack fruit leaves relieves tooth aches and controls blood circulation. Ashoka leaf, barks, flowers are used for preparing capsules. It is used for gynecological relate disease and cures many disorders. Since every part of Ashoka has medicinal value, it is known as universal plant.

The most important commercial crop, cotton, has a tremendous impact on how our agricultural production structure is developing. Therefore, we need to increase cotton output since it has become a heated topic in our nation [8]. Peepal leaf treats cold, fever, treats asthma, provides relief from eye pain, eye pain, nosebleed, Jaundice, heart diseases, dysentery, constipation etc. It also controls cholesterol level, powerful antioxidant, natural solution to damaged liver etc., Mango leaves are antimicrobial antioxidants. They help cure diabetes, treats vascular problems and eye-related problems.

Lemon leaves, which are rich in vitamin C are good energy boosters; they strengthen blood vessels. Their oil extract is used for skin care. They can treat insomnia and stress relate disuses. Jack fruit leaf stabilizes body resistance. It treats up on wounds and speeds up the healing process. Also, it is used in skin care in removing dead skin cells. Plant diseases come in a variety of forms. The illnesses may result in substantial yield losses, a reduction in the quality of agricultural goods, and even pose a danger to food security. Taking disease control measures is based on early detection and diagnosis of plant diseases [9].

Plant management involves close observation, particularly for the control of diseases that can drastically reduce productivity [10]. Finding plant diseases early on is essential for reducing losses in agricultural output productivity and quantity. In order to better understand physiological aspects of plant development, photosynthetic activity, and transpiration, it is vital to measure the leaf area [11]. The ability to monitor huge fields of crops and automatically identify disease signs as soon as they emerge on plant leaves makes automated plant disease detection a crucial study area [12].

Since most of our terrestrial ecosystems depend on plant biodiversity, all life on earth is dependent on plant species in one way or another since they offer a variety of natural sources of energy [13]. Image identification of plant diseases is helpful for the growth of plant protection information technology and might lessen the reliance of agricultural output on experts and technicians in the field of plant protection [14]. The two major factors that affect both the quantity and quality of cotton output are illnesses and insects that affect cotton. The key to efficient cotton protection is the early discovery of pests and practical management strategies [15].

Soil nutrients are classified as macronutrients and micronutrients. Macronutrients are substances present in high proportion in the soil namely Nitrogen (N), Magnesium (Mg), Calcium (Ca), Phosphorous (P), Potassium (K). and Sulfur (S) micronutrients that present in the soil are less in proportion and less significant for plant growth. Chlorine (Cl), Manganese (Mn), Boron (B), Zinc (Zn), etc., are microelements also known as trace elements. Major consumption of nutrients required for plant growth are N, P and K. Hence, the determination of soil nutrients level is significant for the farmer in all processes starting from crop plantation till harvesting using smart dust technology. Monitoring crops for disease detection is crucial to successful production [16].

6.2 Smart Dust Technology in Smart Agriculture

6.2.1 WSN-Based Precision Agriculture

In wireless sensor networks (WSN), numerous dust sensor nodes are interconnected, limiting the processing capability. A considerable number of sensor observations gets transmitted to the sink node, increasing network complexity. The significant drawback in everyday internet usage, as the internet is a virtual network of devices, plays a significant role in any application considered.

For the effective growth of the Internet of Things (IoT), communication and connectivity are seen as the two biggest challenges [17]. The IoT enabled with several intelligent embedded sensor nodes are installed in the designated monitoring zone, WSN offers user-friendly communication [18]. For the purpose of detecting, measuring, and reporting on environmental changes, the sensor devices are interconnected with the internet and interact with one another wirelessly [19]. The IoT makes a network of internet-connected objects capable of collecting or changing information that uses embedded dustsensors or devices. It primarily connects a machine with an IP address somewhere into the world. It can be linked in conformity with any other device, anywhere within the world known as the internet. The IoT requires a unique paradigm to provide optimal network performance in sharing accurate data in the wireless dustsensor network.

Since the WSN is data-centric, data processing is a critical problem in the WSN network, which determines the performance of the WSN. Usually, the dustsensor node is battery powered with limited energy. In precision agriculture, an enormous amount of data is acquired and processed. This idea of precision agriculture using WSN will help the farmers increase the crop yield by applying modern technologies as a replacement for conventional farming practices. Agricultural yield is incredible. The goal of offshore ocean monitoring and exploration applications for underwater acoustic sensor networks is variety. In the majority of these applications, the network is made up of several sensor nodes that are placed at various water depths [20].

The IoT sensors are connected to the electric vehicles to gather data on the EVs' location, charging status, and trip distance [21]. Smart irrigation using IoT with machine learning techniques plays a significant role in optimum water-resource utilization. It causes a severe effect on plants if the plant is not adequately cared for frequently. As a result of this, respective product quantity, productivity, or quality will be affected. A large number of sensor nodes with limited battery power make up a WSN[22]. Because sensor nodes are limited by non-replaceable batteries, energy consumption will generally be a major challenge for WSN. The suggested method is beneficial in monitoring large crop areas. This device aids in forecasting irrigation planning based on irrigation needs and sensing factors. Widely dispersed sensor nodes in a terrestrial environment are used in WSNs to perceive and assess the environment's physical conditions. In comparison to other nodes, the sensor node close to the sink will lose energy more quickly [23]. Diseases are the main factor causing an agricultural product's decreased productivity and efficiency [24].

6.2.2 IoT in Agriculture

Smart farming and precision agriculture is a futuristic and effective technique for practicing agricultural and sustainably generated food. It has converged the linked devices and new technology into agriculture. Deployment of IoT-based smart farming removes or reduces the physical labor involved and thus enhances production in all possible methods. An overview of the IoT-based precision agriculture environment is outlined in Figure 6.1.

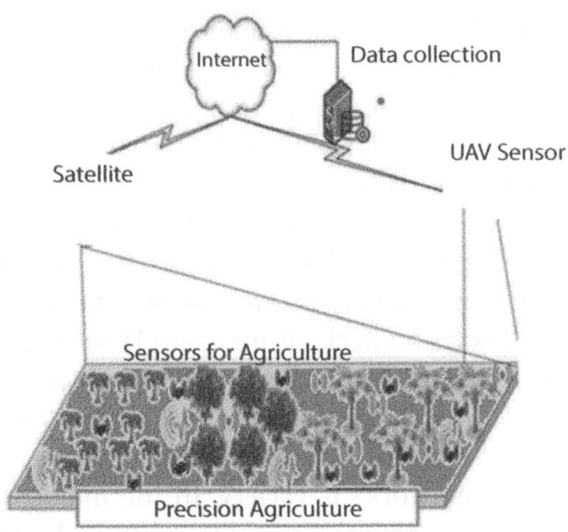

Figure 6.1 The IoT-based precision agriculture environment.

The overall agriculture system is enhanced through IoT-based smart farming involving real-time area surveillance. The integration of dust sensors and device interconnectivity allowed smart agriculture to not only save farmers' time but also to reduce the wasteful use of resources such as water and electricity. The records maintain many critical data, such as soil moisture, humidity, and temperature. Through the data observed, numerous insights can be drawn for the future.

6.2.3 Machine Learning in WSN and Its Agriculture-Based Applications

The ability to learn from experience is machine learning. The algorithms learn from the input datasets based on the refined equations, progressively train the samples, and enhance performance. The ML tools' advantage is that it is more powerful in solving the dust sensor data-based nonlinear problems or other sources. This is useful for smart agriculture technology.

Based on the real-world scenarios, it facilitates the decision-making capability with minimum human assistance. Today, machine learning algorithms are used in all domains. However, based on the data quality, the accuracy is affected. The dataset representations and target variables play a significant role in machine learning techniques. The techniques

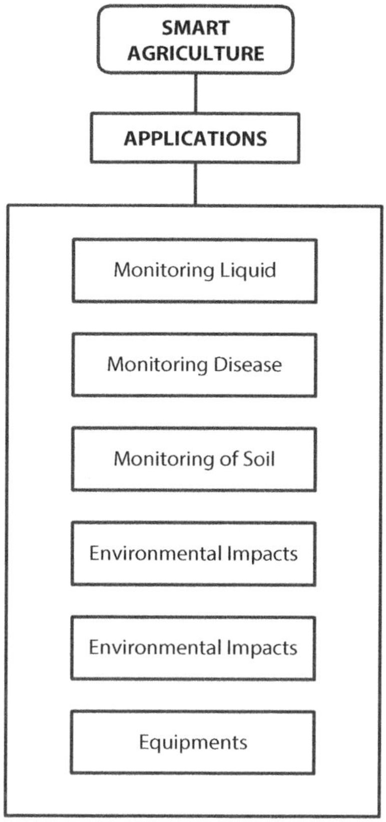

Figure 6.2 Applications in smart agriculture.

are divided into two categories: i) supervised learning; and ii) unsupervised learning. The set of labels are known for the data in supervised learning.

In contrast, unsupervised learning relies on hidden patterns. The selection of an algorithm depends on data insights. However, both learning techniques used in today's IoT intelligent data analysis. The WSN- and IoT-enabled smart farming is the emerging ML technique used to quantify and understand big data applications. The various forms related to machine learning in intelligent agriculture depicted in Figures 6.2 and 6.3.

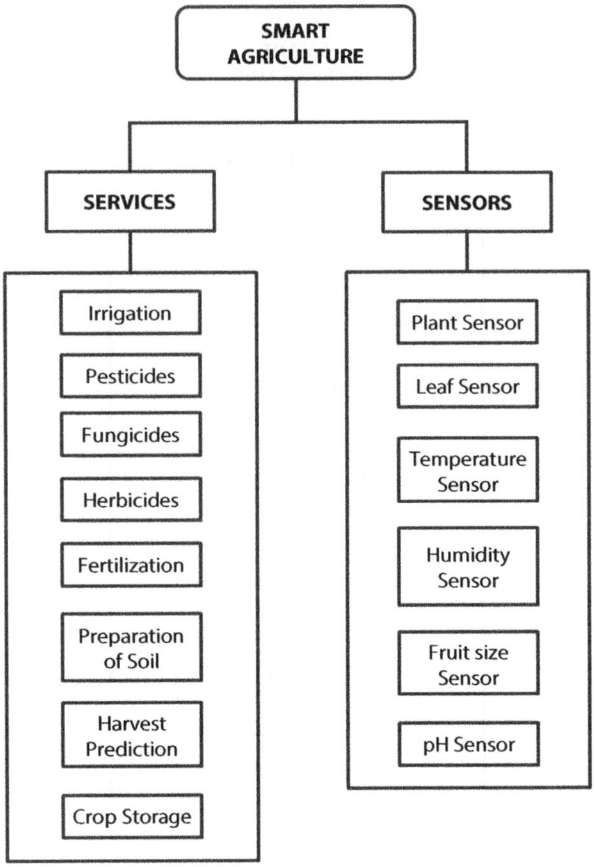

Figure 6.3 Services and sensors in smart agriculture.

6.3 Precision Agriculture and Its Functional Elements

A precise time schedule for irrigation actions, based on anticipated meteorological factors, such as soil and plant traits and water level, is necessary for precision agriculture. The sort of water allocation was highlighted by scheduling irrigation based on the system's interstitial time usage. When planning watering, soil wetness is a crucial consideration. The flooding in the area is initiated when the predetermined cutoff number is reached. In drought circumstances, conventional irrigation cultivation consumes 90% of the water. Precision agriculture measures the accurate estimation of water usage in farming for efficient water management. Because of the vast variety of products grown and the plurality of farms, describing irrigation methods over a large region is challenging.

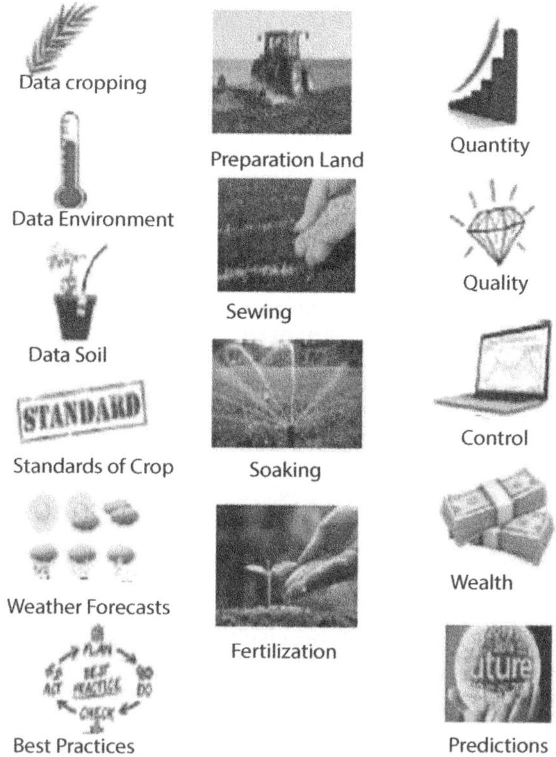

Figure 6.4 Functional elements of precision agriculture.

Because of their efficiency and ease of field upkeep, advanced farming systems aid in the IoT. Creating this type of automatic system also aids in the watering of vegetation without the need for human intervention. It will also be able to place seedlings and measure soil moisture. Figure 6.4 depicts the different functional components of precision agriculture. Monitoring pests that cause agricultural diseases mainly involves aspects such as detecting, assessing, and treatment. Image processing has recently emerged as a crucial technology in which raw images taken from the agricultural area are evaluated to identify insect types for pest management. By assisting in the restoration of the natural environment, these IoT-based pest control devices have reduced total expenses.

6.4 Yield Monitoring and Forecast

Yield monitoring analyzes the range of factors corresponding to the agricultural yield, as shown in Figure 6.5, such as moisture content and the

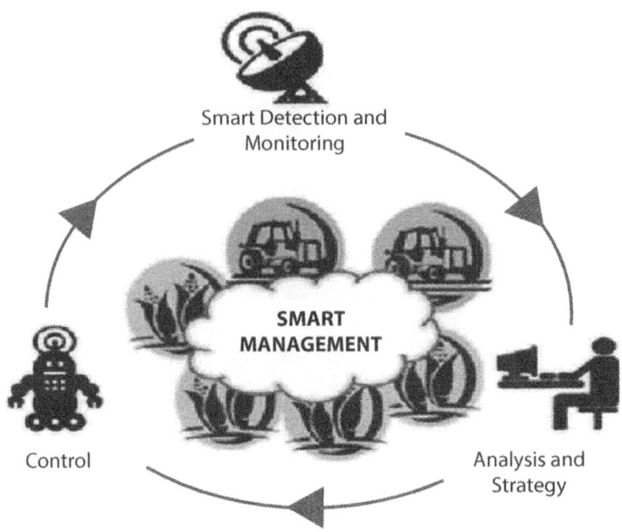

Figure 6.5 Yield monitoring in smart farming.

quantity of harvested crops. It precisely validates and checks the crop records to provide the wet-level estimates. It also determines how well the plant has performed and what it needs to do next. Crop forecasting is a work possible based on the yield harvest, which supports the farmers in planning their future requirements and making appropriate decisions.

Measuring yield maturity and quality is another challenge that exists in harvest prediction. The prediction requires an understanding of fruit size and color variations. The correct prediction of the harvest period maximizes the production and adjust the policy mechanisms. Hence, obtaining a real benefit ultimately depends on the harvest period and procedure and the understanding of the crop conditions, whether they are ready for the harvest.

Anticipations such as the crop sugar beet have a harvest of 140 tons throughout the year. This task depends on modeling or mapping the discipline variability in sugar beet production. Simultaneously, the uniformity of the crop in the yield depends on the environmental factors, reducing the effect of water stress over the crop yield.

6.4.1 Information Sources Required for Yield Monitoring

The information required for a farmer to effectively use precision agriculture technology is beyond the knowledge level of a conventional farmer. The end-users will only have a certain level of knowledge about precision agriculture in most cases. Nevertheless, their experience is not sufficient to transform a uniform crop production technology into precision agricultural technology. Therefore, the potential benefit associated with a given precision agriculture information source depends on how it can bridge the potential information gap between the information threshold required to use precision agricultural technologies effectively. This benefit is indirect, and it is highly dependent on how the adoption of precision agriculture through an information source will maximize farm profits.

Besides, there is a potential cost that could be associated with the choice of an information source. People choose information because there is a potential benefit to be derived from using them. In the case of precision agriculture, it helps farmers use the spatial and temporal variability within a field, and the array of technologies that come with precision agriculture, to optimize their input use. Thus, the information required for precision agriculture can be intense and, in most cases, beyond average farmers' mental ability.

Therefore, it is not surprising that the quality of human capital endowment of the economic agent and the type of enterprise operated greatly influence the choice and use of information. Endogeneity is a pervasive problem in precision agriculture technology. Besides, the motive of enhancing activities like access to capital, off-farm income, and farm income by potential end-users of innovation can positively affect their adoption of a novelty.

6.4.2 Demand for Irrigation Planning

Due to freshwater demand and the increasing world population, water scarcity affects the central part of the world. The world population keeps increasing from 7.2 billion current communities and will reach around 9 billion by 2050. The domestic utilization sector and agricultural sector consumes the central part of the freshwater. Due to advancements in technology, an efficient, smart irrigation system is the hourly need in most developed countries.

Developing countries focus on a cost-effective smart irrigation system and need for countries like India, where 4% of freshwater is serving 17% of the world population. Henceforth, the smart irrigation system with an efficient strategy for water resource utilization is modeled based on the latest technologies. Most of the irrigation system depends on a few essential parameters such as precipitation, evaporation, and soil moisture for designing the smart irrigation system.

By imposing the brand-new sensing of IoT in agriculture, the standard farming elements substantially changed. Currently, the integration of wi-fi sensors unimaginably promotes agriculture with smart farming. Its limitations on dense farming issues, yield optimization, drought response, land suitability, irrigation, and pest control are the challenges.

6.5 Advanced Agricultural Practices

Food quantity and quality enhanced by adopting the novel methods. The methods practiced for centuries. To increase crop production, focusing on fertilizers, seeds, etc., is required. This methodology is not possible with conventional strategies, and there exists an adequate demand in meeting this gap. The scientists think alternate approaches on food supply such as bioengineered, genetically engineered foods challenge DNA and genetic issues.

Several studies highlight that these foods cause severe human effects such as infertility, human health disruptions, ageing, and fault insulin regulations. Similar technologies acceptance in society leads to bio-organic food products. Research conducted for decades helps in conventional agriculture processes improving the yield with minimum effect. Sophisticated and controlled techniques are required to tackle these issues and move towards cultural and urban farming, for which the usage of advanced techniques based on dustsensor technologies is the mandatory requirement today.

The old method of agricultural farming is greenhouse farming. The idea of growing plants gained its popularity in the 19th century. The same was promoted in countries having extreme weather conditions. Environment effects are less in indoor-grown crops. More importantly, they receive less light during daytime conditions. Traditionally, the crops are grown only in suitable parts, and the actual time required is for the communication devices to set up and enhance their applications. The success of crop production varies in controlled environments such as accuracy, material, ventilation, and a decision-support system. Environment parameters play a critical task in measuring climate change conditions.

An IoT-based prototype monitors environmental parameters such as soil temperature, soil humidity, and water level. In precision agriculture, on-time information communication and reporting are considered a primary function that requires a dependable and safe relation between the participating objects. Reliable communication, the operators in agricultural sectors play a crucial role. The agricultural sector requires a stable and grand architecture concerned with energy consumption, reliability, cost, and energy networks connected in remote areas where the farmers located.

6.6 Conclusion

Today, agriculturalists are involved in maximizing production with minimum costs. This method requires a novel approach to address the issue and increase yield and profitability, reducing the environmental impacts of food production. To incorporate precision agriculture in the cultivation field requires various ecological parameters that indirectly assess the above issue. However, the intra-field variability in sugar beet yield and quality is generally unknown; therefore, it needs to quantified concerning soil properties and micro-climate conditions.

The observed variability in the crop environment can then be managed by adjusting the resources to the specific places where they are required. Therefore, the precision agriculture model has been utilized and advanced here following the crop growth accounting the field variations such as ambient temperature, soil moisture, and soil type.

Precision agriculture mitigates the conventional management strategies to increase agricultural profits and accurate resource utilization. Precision farming, according to the soil types, influence the production yield-based crop development. The potential techniques of precision agriculture improve the Indian agricultural sectors. The study proposed focuses on the identification of limiting factors that affect the crop production.

- Providing accurate predictions on the water-based soil conditions for stem.
- Suggestions of various irrigation schemes concerning socio-economic parameters.
- Providing accurate predictions on the water-based soil conditions for leaf.
- Providing accurate predictions on the water-based soil conditions for plant.
- The primary objective of this research, "ACRIS: Agriculture Cultivation Recommender and Smart Irrigation System," is to:
- Make beneficial suggestions to farmers who use IoT in precision agriculture for increasing their cultivation yield by predicting the various soil parameters.
- Provide accurate information on climate, moisture, and water factors.
- Predicts the soil moisture and plan the irrigation accordingly.

- Mitigate water scarcity by optimizing irrigation system in precision agriculture.
- Provide timely, accurate, reliable information, and farmers can benefit from this knowledge to improve their productivity.

References

1. M Yogeshwari, G Thailambal, Automatic feature extraction and detection of plant leaf disease using GLCM features and convolutional neural networks, *Mater. Today Proc.*, Elsevier, 2021/5/11https://doi.org/10.1016/j.matpr.2021.03.700.
2. Dr. M. Yogeshwari, Dr.R.Varalakshmi, A review on plant leaf disease identification and classification image, *JARDCS*, 2019,volume 11, issue 8, pp 1463-1475.
3. Basvaraj S. Anami, J.D. Pujari, Rajesh Yakkundinath, Identification and classification of normal and affected agriculture / horticulture produce basedon combined color and texture feature extraction, *Int. J. Comput. Appl. Eng.*, vol 1, Sept 2011, pp.356-360.
4. Pang, Po Ken, Lim, King Hann, *Review on Automatic Plant Identification Using Computer Vision Approaches*, 10.1088/1757-899X/495/1/012032, June, 2019.
5. M Yogeshwari, G Thailambal, Automatic segmentation of plant leaf disease using improved fast fuzzy C means clustering And adaptive Otsu thresholding (IFFCM-AO) algorithm, 2020, *Eur. Mol. Clin. Med.*, volume 7, issue 3, pp 5447-5462.
6. ShubhraAich, I. Stavness, Leaf Counting with Deep Convolutional and Deconvolutional Networks, *IEEE International Conference on Computer Vision Workshops (ICCVW)* 2017.
7. Grant Hollaway, Senior Plant Pathologist Cereals, Leaf rust of wheat, *Victorian Government Department of Environment and Primary Industries Melbourne*, May 2014, pp.1-2.
8. DiaoZhihuo, Wang Huan and Song Yinmao, Image segmentation method for cotton mite disease based on colour feature and area threshold, *JATIT*, vol.48, Feb 2013, pp.527-533.
9. Haiguang Wang, Image Recognition of Plant Diseases Based on Principal Component Analysis and Neural Networks, *International Conference on Natural Computation (ICNC)*, IEEE, May 2012, pp. 246-251.
10. Hrushikesh Marathe, PrernaKothe, Leaf disease detection using image processing technique, *IJERT*, vol 2, Issue 3, March 2013, pp.1-4.

11. Anand Kulkarn & Ashavin Patil R. K, Applying image processing technique to detect plant diseases, *Int. J. Modern Eng. Res. (IJMER)*, Vol. 2, Issue 5, Sep-Oct 2012, pp.3661-6264.
12. Al-Hiary. H, Bani-Ahmad, M-Reyalat and M. Braik, Fast and accurate detection and classification of plant diseases, *IJCA*, vol.17, -no. 1, March 2011, pp. 31-38.
13. Ali Jan Ghasab, Feature decision-making ant colony optimization system for an automated recognition of plant species, DOI:10.1016.2014.11.011,2015.
14. Haiguang Wang, Application of Neural Networks to Image Recognition of Plant Diseases, *International Conference on Systems and Informatics (ICSAI 2012)*, IEEE, June 2012, pp. 2159-2164.
15. Hui Li, RonghuaJi, Jianhua Zhang, Xue Yuan, Kaiqun Hu, Web-Based Intelligent Diagnosis System for Cotton Diseases Control, IFIP Springer-Advances in Information and Communication Technology, *AICT*, 2011, pp. 483-490.
16. Dheeb Al Bashish, Malik Braik and Sulieman Bani Ahmad, Detection and classification of leaf disease using K-mean based segmentation and neural networkbased classification, *Int. Technol. J.*, vol.10, Issue 2, 2011, pp. 267- 275.
17. Prasanth, S.Jayachitra, A Novel Multi-Objective Optimization Strategy for Enhancing Quality of Service in IoT enabled WSN Applications, *Peer-to-Peer Netw. Appl.*, Vol.13, 2020.
18. P. Paruthi Ilam Vazhuthi, A.Prasanth, SP. Manikandan, KK Devi Sowndarya, A hybrid ANFIS reptile optimization algorithm for energy-efficient inter cluster routing in Internet of Things-enabled Wireless Sensor Networks, *Peer-to-Peer Netw. Appl.*, 2023, 1-29.
19. P. Vazhuthi, S.P. Manikandan, A. Prasanth, Concurrency and Computation: Practice and Experience, 34 (2022) 1-28.
20. A. Prasanth, Certain Investigations on Energy-Efficient Fault Detection and Recovery Management in Underwater Wireless Sensor Networks, *J. Circuits, Syst. Comput.*, Vol. 30, 2021, 2150137:1-18.
21. K.B. Bhaskar, A. Prasanth, P.Saranya, An energy-efficient blockchain approach for secure communication in IoT-enabled electric vehicles, *Int. J. Commun. Syst.*, 35 (2022) e5189:1-27.
22. A. Prasanth, S. Pavalarajan, Implementation of Efficient Intra- and Inter-Zone Routing for Extending Network Consistency in Wireless Sensor Networks, *J. Circuits, Syst. Comput.*, Vol.29, 2020.
23. A. Prasanth, S. Pavalarajan, Zone-Based Sink Mobility in Wireless Sensor Networks, *Sensor Rev.*, Vol.39, pp.874–880, 2019.
24. AsmaAkhtar, Automated Plant Disease Analysis (APDA): Performance Comparison of Machine Learning Techniques, *11th International Conference on Frontiers of Information Technology*, December 2013, pp.60-65.

7

An Advanced Application of UAV – Drone Technologies in Precision Agriculture for Seed Dropping, Fertilizers and Pesticides Spraying and Field Monitoring

Daniel Lawrence I.[1]*, A. Rehash Rushmi Pavitra[2], Ragupathy Karu[3] and M.P. Saravanan[4]

[1]*Agni College of Technology, Chennai, Tamil Nadu, India*
[2]*SRM Institute of Science & Technology, Chennai, Tamil Nadu, India*
[3]*Bannari Amman Institute of Technology, Sathyamangalam, Tamil Nadu, India*
[4]*Saveetha Engineering College, Chennai, Tamil Nadu, India*

Abstract
In precision agriculture, too much advancement has been made to boost crop output. Almost 70% of rural residents in nations like India depend mostly on agriculture for their income. Sensors and the Internet of Things are crucial for creating a more environment-friendly and fruitful global agriculture. Large volumes of data may be intelligently monitored and analyzed to provide deeper understanding and information that can then be used to improve forecasting, judgment, and sensor management. Unmanned aerial vehicles (UAVs), that could potentially quickly be deployed for monitoring reasons, have considerably benefitted precision agriculture. The World Health Organization (WHO) has reported adverse effects from manual fertilizer and pesticide spraying, causing a million cases. To address this issue, drones are being used to spray fertilizers and pesticides at different densities. These UAVs can assist farmers in increasing crop production by monitoring crop health and soil health, reducing environmental harm, and using resources efficiently. Finally, the UAVs are used to spray fertilizer and pesticides as per the observed data results. Drones are used in the majority of applications as a movable aerial platform for high-resolution image capture to determine the impact of pest, disease and weed patches in a cultivated field. The temperature, humidity, light intensity, pressure, and soil moisture of an agriculture field

**Corresponding author*: danilawernce@gmail.com

Rajesh Kumar Dhanaraj, Balamurugan Balusamy, Prithi Samuel, Malathy Sathyamoorthy and Ali Kashif Bashir (eds.) Computer Vision in Smart Agriculture and Crop Management, (139–158) © 2025 Scrivener Publishing LLC

environment is collected through sensors. Importantly, the former can identify the crop health and weed patches through aerial captured images and soil moisture and environment conditions by sensors. Especially, the drone use for precise crop health monitoring is economical, less time-consuming process. The study emphasizes the importance of novel technologies in achieving sustainable pest management in modern agriculture and how this will require collaboration between experts from multiple disciplines, including agronomy, ecology, software development, and engineering.

***Keywords*:** UAV, agriculture, database, cloud, prediction, sustainable

7.1 Introduction

India depends heavily on agriculture, but it is still far from adopting the most recent technologies to create high-quality farms. Developed nations have already begun using unmanned aerial vehicles (UAVs) for remote sensing, photogrammetry, and precision agriculture. It moves quickly and can lighten a farmer's workload. The UAVs typically come with cameras, sensors, and sprayers to monitor crops and apply pesticides. Both military and commercial applications have previously been made of a range of UAV types [1]. Precision agriculture involves the analysis of data collected by UAVs to evaluate their usefulness in performing agricultural tasks such as crop monitoring, estimating crop height, applying pesticides, and conducting soil and field analyses. Researchers investigate the technology, methods, systems, and limitations associated with UAVs. To create a micro autonomous mini unmanned rotorcraft vehicle, hardware components, software integration, autonomous flight, modeling, design, and implementation are used for drone. Precision in agriculture aims to develop technology that enables agricultural operations to succeed economically. The UAVs are a cutting-edge technology that has led to the emergence of various UAV-based applications, which have driven industry development and increased service demand. Aerial photographs with high spatial resolution can be obtained using UAVs at a lower cost than other approaches [2]. Governments and the corporate sector both heavily rely on remote sensing to map soil characteristics, categorize different types of crops, identify water stress in crops, keep track of agricultural diseases, and map crop production [3]. Farmer observation of the harvest from the air is made easier by drone surveillance equipment. The water system, types of soil, pests, and fungus infections are all included in this material. The drones' cropped photographs provide spectral data in the infrared and visible ranges. Farmers may make necessary adjustments for improved crop management since crop information is often available [4]. The system offers resources for the assessment of irrigation possibilities under various soil, climatic, and management variables, as well as capabilities for exploring geographically referenced datasets useful for irrigation. The irrigation system for water management utilizes Arc view geographic information systems (GIS) and the Avenue computer program for customizing GIS applications, designing new tools, and finding locations with water deficits [5]. These photos can reveal information about plant health, which is not visible to the naked eye, and the ability to use this technology for routine assessment of crop yield, even up to once per week or hour, is important. The irrigation systems used for water management in agriculture can vary greatly in size, from individual farms to larger collaborations and administrative bodies [6, 7]. Unfortunately, there is a high probability of agricultural inputs being wasted during procedures, increasing the cost of production. In recent years, automation has become more prevalent in agriculture for tasks such as harvesting, spraying, and drip irrigation, leading to increased efficiency and reduced waste of water. Ideas and techniques to deliver fertilizers to specific locations at appropriate times are also encouraged. They are especially valuable for mapping and analyzing specific areas. In precision agriculture, it is crucial to quickly identify the best path and route to efficiently spray pesticides and fertilizers in targeted and stressed locations [8]. A variety of farming situations, meteorological circumstances (from dry to rainy years), the irrigation technologies used to water supplied for irrigation, and the hydraulic properties of the water supply systems may all be taken into consideration when estimating irrigation needs. The variety of approaches and procedures used in agriculture to boost agricultural production and output has increased during the last several decades [9]. Although fertilizers are used to increase the soil's fertility by providing important nutrients to boost crop output, pesticides have been utilized to control insects and other pests [10]. Now, the population is projected to increase from seven billion to nine billion. To fulfill the population's ever-growing requirements, agricultural systems will be under a lot of strain, which leads to increase in an environmental impact by humans. By shielding crops from pests, insects, disease, and other stresses when used in agricultural techniques, there is little question that agricultural output will grow. The UAVs will transport fertilizer spray or pesticides spray in accordance with user requests or specifications with the lowest possible flight duration, covering all sites in a single visit [11].

7.2 Irrigation Management

Drones have thermal, multi-spectral camera that can detect areas which need modifications in irrigation management level. These sensors can determine the vegetation index, a health indicator for crops, once they have begun to grow by analyzing the heat signature of the crop. Before the planting of crops over a field, drones may collect and generate an accurate map of the soil for soil analysis. This information is used to guide the pattern of seed planting [12]. Water shortage in irrigation is a problem that must be dealt with, and supply management systems are crucial in this regard, as both the research and practical experience demonstrate. Because supply and demand management depend heavily on one another, it is mentioned here. On the one hand, better agricultural irrigation is necessary for the correct use of supply measures like those pertaining to water of subpar quality [13]. Water shortage in irrigation is a problem that must be dealt with, and supply management systems are crucial in this regard, as both the research and practical experience demonstrate. Because supply and demand management depend heavily on one another. On the one hand, better agricultural irrigation is necessary for the correct use of supply measures like those pertaining to water of subpar quality [14]. Creative and environmentally conscious research, as well as the proper transfer of technology, is needed to utilize water for agricultural output in water-scarce locations. The utilization of recovered wastewater and saline waters for on-farm irrigation management is a major topic of discussion in this essay [15]. First, the research offers a number of hypotheses on the concepts of water shortages, including aridity, droughts, degradation, and shortage of water, as well as approaches for coping with these liquid regimes. The efficacy of irrigation, water use, and water efficiency conceptual frameworks are investigated from many perspectives. This topic discusses ways to address water scarcity in the supply chain, with a particular emphasis on the use of wastewater and reduced waters and their potential impact on human health and environment. This project highlights the importance of transmissions uniformity as a crucial tool for reducing water demand at the farm level and minimizing negative environmental effects, including salt overstretched areas [16]. The article also delves into the development of irrigation systems and their corresponding performances, citing recent study results. Additionally, the article stresses the need for appropriate methods to analyze the social, economic, and environmental benefits of improved irrigated areas and the use of advanced technology for water management [17]. With a focus on how they are utilized in irrigation management, this study covers the emergence of water deficiency in field crops and the techniques used to assess the deficiency [18]. Examples from the agricultural industries of cottons, sunflowers, lupin, pasture medic, and sugarcane are presented to show that modest water shortfalls do not always result in reduced yields and, in some cases, even increase yields. The creation, operational capabilities, and spatial modeling applications of an irrigation management system based on GIS for use

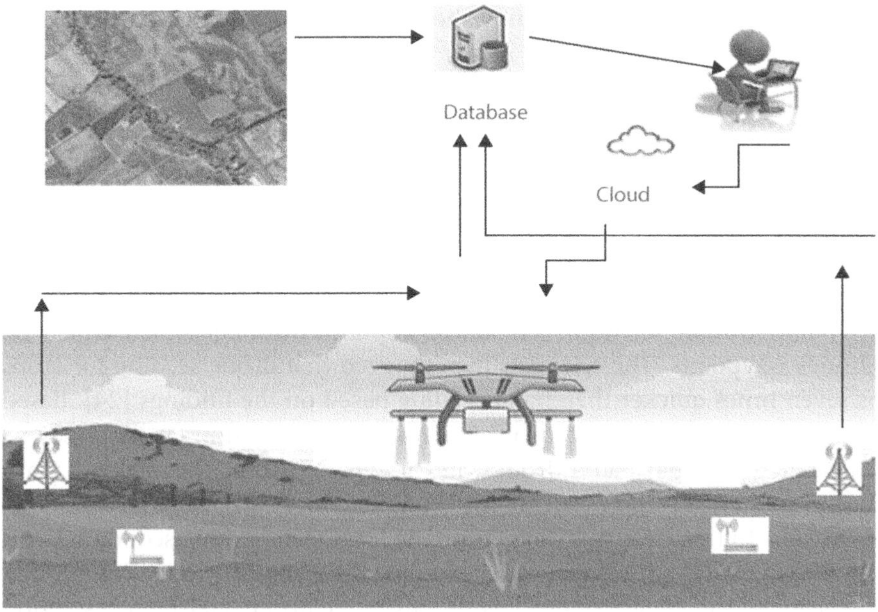

Figure 7.1 The representation of drone technology-based irrigation management.

by irrigation organizations and local governmental entities are described in this paper [19]. The representation of drone technology-based irrigation management is displayed in Figure 7.1.

7.3 Seed Dropping

A novel approach that has the possibility of lowering agricultural labor is drone seed distribution technology. Drones may be used for airborne seeding, especially for cover crops cultivated in woods and grasslands, and can deliver seed packages weight up to 25 kg [20]. In order to fulfill the agronomic needs of rapeseed planting, the novel approach may function effectively, consistently, and steadily. Inaccessible places may be used for drone planting, which is also a more effective technique than conventional sowing. Although a quadcopter drones with a load container works well for agricultural seeding, a hexacopter drone offers superior payload and performance for drone seeding. The plastic and metal seeding mechanism are driven by a servo motor [21, 22]. The dimensions of the rice seeds were determined by selecting 150 rice seeds at random, measuring their length, width, and thickness with vernier calipers, and calculating the arithmetic mean value. The study shows that drone seeding is seven times faster than manual seeding. A motor drives the canister door, which holds 40 *Tamarindus indica* seeds. A measurement of 10000 m^2 is the projected drone coverage area. The hexacopter drone is recommended in order to increase drone seeding performance and payload. The quad copter drone for agricultural seeding that uses a 0.17 kg load container dropped onto a disc with holes and is controlled by a servo motor, the seeding mechanism was constructed from plastic and metal [23]. A schematic representation of drone technology enabled seed dropping is shown in Figure 7.2.

While Arduino software is utilized for control, ArduPilot is used to plan the drone's waypoints. The scientists hypothesized that drone seeding for crops is seven times quicker than hand seeding based on the findings [24]. Based on the payload equipment, drone applications in precision farming may be researched. The payload a drone will carry is known as the payload. Crop health monitoring and pesticide application are the two primary topics covered here. In this essay, the many forms of UAV technology used for agricultural monitoring have been evaluated following a short introduction to their usage in the area of agriculture. It is also discussed how to take high-resolution

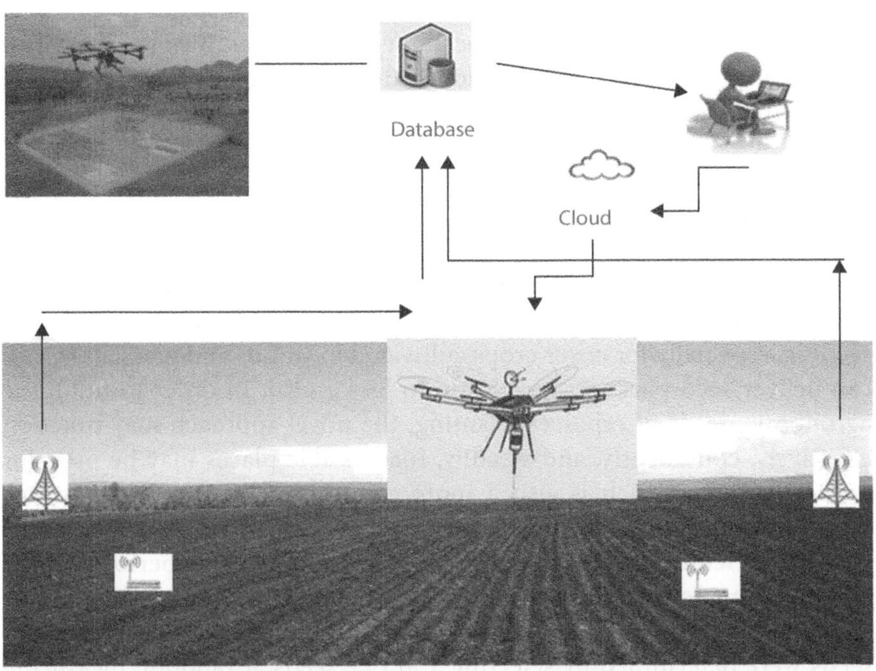

Figure 7.2 Schematic representation of drone technology enabled seed dropping.

pictures and how to analyze them to monitor crop health. A review of drone development for spot spraying of pesticides and advancements in pesticide spraying drones takes place in real world without human interference [26].

7.4 Pesticide and Fertilizer Spraying System

The combination of a UAV with a spraying system offers the possibility of creating a platform for vector and pest management. This site-specific treatment is suitable for big agricultural fields. For large-scale spraying, big lifts UAVs are needed for this purpose. Through the use of a pulse width modulation (PWM) controller, the spraying system affixed to the UAV operates more effectively when applying pesticides. A UAV is equipped with a sprayer system for applying pesticides. A platform for vector control and pest management might be created by integrating UAVs with sprayer systems [27]. The UAV's built-in sprayer has a universal sprayer that may be used to apply pesticides and fertilizers with the same sprayer. In order to control both liquid and solid contents, a universal nozzle is required. Instead of spraying fertilizers, pesticides are applied using a pressure pump. Images from remote sensing are taken using a multispectral camera, and these photos are then utilized to spot crop margins and green fields [28, 29]. In order to precisely apply pesticides and fertilizers to large agricultural fields, site-specific, and big lift UAVs are required. To achieve this, a hexacopter mounted sprayer was designed to examine the discharge and pressure rate, spray uniformity and liquid loss, droplet density and sizes, and field evolutions [30]. To decrease pesticide loss, an electrostatic sprayer based on electrostatic spray technology was invented. Additionally, quadcopters combined with sprayers are often used in open agriculture fields to spray pesticides and fertilizers [31]. The primary objective of this research is to minimize negative effects on humans. The GPS is used in the UAV's auto guiding system. Drones equipped with aerial remote sensing technology, which captures photos at various wavelengths and measures vegetation indices to identify different crop conditions, are critical for precision agriculture and smart farming [32]. Representation of drone technology enabled pesticide and fertilizer spraying in farmland is pointed in Figure 7.3.

7.5 Improving Soil Productivity

Even before the planting of crops over a field, drones might collect and develop an accurate map of the soil for soil analysis, which helps to direct

Figure 7.3 Representation of drone technology enabled pesticide and fertilizer spraying.

seed planting patterns. This information is useful for planning irrigation and nutrition after planting [33]. The use of subsurface drip irrigation (SDI) has several advantages, including reduced evaporation losses, lower runoff creation, and significant water savings. The dry soil surface resulting from SDI also leads to improved weed control, which reduces the need for herbicides and decreases pollution hazards. In addition, SDI improves traffic and maneuver conditions [34]. However, when using SDI, wastewater should be applied to the ground surface only when soil moisture levels are low, and caution should be taken to avoid contact with plant leaves and fruits. The depth of roots plays a critical role in determining the extent of groundwater accessible to crops, and limitations due to genetics or soil characteristics can severely restrict a crop's potential for growth [35]. Due to the fact that the majority of the alleged advantages only manifest while crop survival is in jeopardy, it is probable that significant evidence for the positive effects of the effects on production of osmotic adjustment (OA) was not found. In the majority of agricultural situations, crop preservation is not a significant concern since yields under such extreme stress are so minimal that any potential advantages are not advantageous to farmers [36]. Moreover, the potential advantages of maintaining turgor in reaction to drying soil often make things worse. After the soil has dried to its last stages before planting, conserving water is the best course of action for the majority of cropping circumstances. Due to OA's sustained water loss caused by increased stomatal permeability and leaf turgor, crop death is threatened as the crop quickly moves into the survival stage [37]. Plants may have access to a larger water reservoir that will allow them to grow and/or survive if osmolyte buildup in the roots enables ongoing, or even accelerated, root expansion into deeper, moist soil. When a water level is made up between both the plants and the ground water is maintained, water will continue to migrate into the plants. Reducing plants concentration of water by OA is one alternative whenever the soil dries up in order to maintain a freshwater gradient and extend ground water absorption through the plant [38].

7.6 Supporting Crop Growth

The main problem the agricultural industry is now experiencing is the poor crop monitoring performance brought on by extensive industrial farming. Drone technology might be used to continually monitor the progress of the agricultural sector, delivering information that is essential for comprehending crop growth. Smallholder farming is strongly reliant on rainfall in South Africa, which often causes crops to experience water stress and insufficient moisture [39]. In smallholder croplands, however, there is a shortage of precise and dependable techniques to detect crop water stress, particularly in areas where maize is the main subsistence crop. Since leaf pigments like chlorophyll absorb water, the optical (blue, green, and red) with relatively close (NIR) wavelengths are very interesting for water prediction [40, 41]. Although there are limits to satellite imagery, the deployment use UAVs outfitted with (VHR) cameras may get beyond these restrictions by offering sophisticated crop image processing analytics. Recent research has shown that deep learning-based strategies, which make utilize both spectral and spatial data, provide a complete solution and outperform conventional techniques. These methods also show promise in the identification of vegetation and the categorization of land cover [42, 43]. Finally, using the Normalized Difference Vegetation Index (NDVI), images were processed and evaluated. The outcomes are an accurate reflection of the crop's circumstances. The UAVs' potential is also increased with the addition of sensors and visual systems [44]. By combining geophysical data with multispectral aerial data on soil and crops, it is possible to customize the administration of fertilizers to each field zone. Monitoring the amount of nutrients in the soil is therefore crucial for determining crop development and enables the application of suitable variable-rate fertilization to prevent unneeded harm to the capacity for sustainable output [45]. The drone technology-enabled crop growth monitoring is pointed in Figure 7.4.

7.7 Crop Management Strategies

It is extremely difficult to predict emerging outcomes for agricultural system while taking complex and connected socioeconomic and environmental

Figure 7.4 Drone technology-enabled crop growth monitoring.

elements into account [46]. It is challenging to achieve long-term global ecological systems due to agro-ecological limitations, which is presented with the task of maintaining global foods security under a rapidly evolving climate. So, it is essential to have a proper grasp of the detailed dynamic of the occurrence of various abiotic stresses, as well as the prevalent mechanisms by which plants survive, avoids, or escape such severe environmental circumstances [47]. No one solution can totally alleviate this threat, thus substantial attention must be made by all facets of the community to combat such harsh climatic circumstances and the problems of feeding the quickly expanding population. In order to ensure a solid seed-soil contact, an automated planter would often open a small hole in the able to prepare soil seed bed, plant the seed inside the exposed, moist soil, cover the seed, and then frequently pack the soil down. No-till techniques include drilling the crop into the ground directly using crop leftovers as a drilling medium [48]. In the Mediterranean area, crop productivity is influenced by water usage both overall and during specific growth phases. From the stage of stem elongation through the stage of grain filling, wheat is more vulnerable to water stress than it is at other stages. The curved water-use-efficiency (WUE)-yield connection and the linear link between yield and seasonal evapotranspiration (ET) emphasize how crucial it is to guarantee a sufficient supply of water, as well as reasonably high yields, in order to accomplish effective water usage [49]. When there is a vapor pressure differential between the plants and the soil water, water continues to migrate into the plants. In case a plant dies out, reducing the plant water potential through OA is an option to maintain a water concentration difference and increase plant soil water absorption. However, once the soil water has drained to a certain point, there will not be much water left to be taken up. The extent to which OA can collect additional water is limited and can be determined by analyzing the amount of water that can be collected from the soil due to a decrease in its water content caused by OA [50].

7.8 Increasing Crop Yield

Typically, many different types are used in breeding, each of which is planted in a small, individual plot of a few meters. Obtaining and evaluating attributes from the numerous diverse experimental plots planted across tiny areas is currently a problem in plant research and calls for the development of new methodologies. One of the most valuable qualities that may be predicted from spectral data is grain yield [51]. We take into account the possibilities of plant sciences to solve post-Green Revolution difficulties in agriculture and investigate cutting-edge methods for boosting resilient sustainable agricultural output [52]. In order to produce the robust production systems required to assure future harvests, accelerated crop development must make use of naturally evolved features and revolutionary innovation guided by mechanistic knowledge. Using a lot of inorganic fertilizer and high-yield crop varieties considerably raised agricultural output, which improved food supply [53, 54]. Breeding in plants involves using different types, each of which is grown in small individual plots. However, obtaining and analyzing data from the diverse experimental plots is a challenge in plant research, and new methodologies are needed. Spectral data can help predict valuable qualities, such as grain yield, which is associated with vegetation indices acquired from aerial cameras during the blooming stage of maize [55, 56]. The UAVs can provide high-resolution imagery to bridge the gap between remote sensing and ground-based surveys. The use of remote sensing technologies offers several benefits over traditional field sampling procedures, including improving agricultural production forecasts. While inorganic fertilizers and high-yield crop varieties have increased agricultural output, their usage contributes to greenhouse gas emissions that cause climate change [57–59]. Osmotic adjustment has been suggested as a useful strategy for crops to adapt to water scarcity, but studies have found inconsistent advantages in agricultural output. The potential benefit of OA in agricultural plants includes enhancing soil water absorption, osmoprotection of cells, preservation of tissue and cell turgor, and maintenance of tissue mortality at higher stress levels [60]. To increase crop productivity with OA, a correlation must exist between rising transpiration water losses and rising crop yields, and greater soil water recovery is required. Two hypotheses have been proposed to enable increased soil water absorption: widening the gradient of water content between the plant and the soil and expanding the area of soil used to extract water [61].

7.9 Preventing Crop Disease

For the purpose of identifying plants that are contaminated by bacteria or fungus, drones using infrared and visible (IR) light may scan crops and capture images of the affected plants [62]. As a result, multi-spectral disease cannot spread to other crops. The research is discussed in the article, which assesses the effects of manure, compost, non-compost plant wastes, and liquid solutions on the frequency and extent of diseases and pests in both agricultural and landscape crop production. Although past studies have mostly concentrated on aesthetic plants grown in containers, there is growing interest in the possibility of employing manures to control and prevent diseases in field crops used in moderate horticulture and agriculture [63, 64]. This article examines how pests and illnesses are impacted by non-compost plants wastes, organic fertilizer, organic manures, and compost extract and teas in productive temperate fields and covered crop production. The research also looks at how successfully these organic manures manage and prevent diseases and pests, as well as the techniques employed to do so [65]. The integrative strategy of biomedical agriculture, that seeks to identify genotypes of food crops that might reduce the risk of chronic illnesses including cardiovascular disease and obesity, is highlighted in the essay. In order to find bioactive components that may lower the risk of food-mediated disease, the genotypes of both conventional and specialty crops are studied. This multi-stage process is first evaluated in human volunteers before crop genotypes and dietary combinations are studied structurally, cellularly, and in animals [66, 67]. Furthermore, covered in the paper is how polymorphisms might be utilized to pinpoint cohorts that may be receptive to or resistant to a certain class of plant-derived bioactive compounds. The vegetable and fruit-cancer conundrum is being approached in novel ways thanks to the availability of high-dimensional sets of data from new genomic, proteomic, and energy metabolism evaluations. In light of the findings of this work, databases may be useful for identifying promising leads and creating follow-up cohort studies and food-based projects that are classified by community and scenario as well as metabolic pathway [68].

7.10 Predicting Crop Yield

Having a model of how crops react to weather is essential for predicting the possible impact that climate change will have on agricultural productivity. Understanding the causes of the frequent discrepancies between projections from different models is necessary to developing a thorough grasp of the predicted implications of climate change [69]. Using statistical models that are based on previous yields and specific climatic factors, such as the typical planting season temperature and precipitation, is a standard practice [70]. To use a crop model approach, this research investigated the accuracy with which statistical models can forecast how crop yields would respond to variations in mean precipitation and temperature. Initially, the Perseus model was used to simulate past variations in maize output, as well as the consequences of speculative possible futures, including such 2°C heating and a 20% reduction in precipitation. The research indicated that when it came to forecasting how crops would respond to temperature fluctuations, panel and cross-sectional mathematical analysis performed better than time series models. Additionally, statistical models were more effective at larger geographic scales and when specific locations were combined to national averages [71–73]. Factors such as weed densities and environmental conditions also impact competition and crop yield, and a strategy based on the responsible factors is necessary to minimize the variability in forecasting methods [74]. While statistical models offer a valuable tool for forecasting future production responses, they are still considered to be unreliable [75]. The following factors influence competition: 1) the effects of density on the important growth factor competed for; 2) variations in the competitive strength of weed species for the essential signaling pathways; 3) the different impact of environmental condition on the competitiveness of weeds species; and 4) the effect of time of beginnings on competition. A strategy based on the factors(s) responsible for yields decrease must be developed in order to minimize the variability in forecast approaches [76, 77].

7.11 Conclusion

Now, we use drone technology in agriculture for seed dropping and pesticide spraying. Many problems can now be solved using information technologies. In recent years, drones for agriculture have made remarkable technical advances. Farmers need drones because they will no longer need to walk around their farms to inspect the soils, crops, and structures. The best surveillance and inspection technology is now available to farmers for their farms. For commercial farmers with vast tracts of land, it is true that farming has advanced technologically. Perhaps one day these drones will be fully automated and give farmers precise information on how much fertilizer, pesticide, or fungicide to apply after surveillance on particular areas or crops. Farmers can expect a new day to dawn. On the other hand, drones have drawbacks. Drones, for instance, raise privacy issues, are not entirely safe, and pose a serious security risk to nations. Drones are revolutionizing precision agriculture by offering farmers an efficient and cost-effective way to manage their crops. Drone technologies can be used for seed dropping, fertilizer and pesticide spraying, and field monitoring. Seed dropping using drones can help farmers plant crops more accurately and efficiently, reducing the time and labor needed for manual planting. Drones can also help with the distribution of fertilizers and pesticides, ensuring that they are applied evenly and in the correct amount. This can reduce waste and save farmers money while also improving crop yields. Field monitoring using drones can provide farmers with detailed data on crop health and growth, helping them to identify and address potential issues before they become serious problems. Drone technologies have tremendous potential for precision agriculture. By using drones for seed dropping, fertilizer and pesticide spraying, and field monitoring, farmers can save time and money while also improving crop yields and reducing waste. As the technology continues to improve and become more widely available, we can expect to see even greater benefits for farmers and the environment.

References

1. Anderson, G.W.; Batini, F.E. Pasture (1983), sheep and timber production from agro-forestry systems with subterranean clover sown under 15-year-old Pinusradiata by a method simulating aerial seeding. *Aust. J. Exp. Agric.* 23, 123–130.
2. Khalkheili, T. A., & Zamani, G. H. (2009). Farmer participation in irrigation management: the case of Doroodzan Dam Irrigation Network, Iran. *Agric. Water Manage.*, 96(5), 859-865.
3. Todorovic, M., & Steduto, P. (2003). A GIS for irrigation management. *Phys. Chem. Earth. A/B/C/*, 28(4-5), 163-174.
4. Dry, P. R., Loveys, B. R., McCarthy, M. G., & Stoll, M. (2001). Strategic irrigation management in Australian vineyards. *OENO One*, 35(3), 129-139.
5. Valipour, M. (2017). Global experience on irrigation management under different scenarios. *J. Water Land Dev.*, 32(1), 95.

6. van Schilfgaarde, J., Bernstein, L., Rhoades, J. D., & Rawlins, S. L. (1974). Irrigation management for salt control. *J. Irrig. Drain. Div.*, 100(3), 321-338.
7. Simonne, E. H., Dukes, M. D., & Haman, D. Z. (2010). Principles and practices of irrigation management for vegetables. *Vegetable production handbook for Florida*, 2011, 17-26.
8. Seckler, D., Sampath, R. K., & Raheja, S. K. (1988). An Index For Measuring The Performance Of Irrigation Management Systems With An Application 1. *J. Am. Water Resour. Assoc. (JAWRA)*, 24(4), 855-860.
9. Andrio, A. (2019). Development of UAV technology in seed dropping for aerial revegetation practices in Indonesia. *IOP Conf. Ser.: Earth Environ. Sci.*, 308(1).
10. Berner, B. (2020). Fertilization and Sowing From Unmanned Aerial Vehicles. *Pol. Tech. Rev.*, 2, 18–22.
11. Farrell, M., Gili, A., & Noellemeyer, E. (2018). Spectral indices from aerial images and their relationship with properties of a corn crop. *Precis. Agric.*, 19(6), 1127-1137.
12. Novikov, A.I. Ersson, B.T. (2019). Aerial seeding of forests in Russia: A selected literature analysis. *IOP Conf. Ser. Earth Environ. Sci.*, 226, 12051.
13. Sharma, L.; Pandey, P.C.; Nathawat, M. (2017). Assessment of land consumption rate with urban dynamics change using geospatial techniques. *J. Land Use Sci.*, 7, 135–148.
14. Pandey, P.C.; Mandal, V.; Katiyar, S.; Kumar, P.; Tomar, V.; Patairiya, S.; Ravisankar, N.; Gangwar, B. (2015). Geospatial Approach to Assess the Impact of Nutrients on Rice Equivalent Yield Using MODIS Sensors'-Based MOD13Q1-NDVI Data. *IEEE Sens. J.*, 15, 6108–6115.
15. Thomas, D.E.; Weyerhaeuser, H.; Saipathong, P.;Onpraphai, T. (2000). Negotiated land use patterns to meet local and societal needs. In *Proceedings of the Cultures and Biodiversity Congress 2000*, Yunnan, China, Yunnan Science and Technology Press: Yunnan, China, 2000; pp. 414–433.
16. Tokekar, P.; Hook, J.V.; Mulla, D.; Isler, V (2016). Sensor Planning for a Symbiotic UAV and UGV System for Precision Agriculture. *IEEE Trans. Robot.*, 32, 1498–1511
17. Alsalam, B.H.Y.; Morton, K.; Campbell, D.; Gonzalez, F. (2017). Autonomous UAV with vision based on-board decision making for remote sensing and precision agriculture. In *Proceedings of the IEEE Aerospace Conference*, Big Sky, MT, USA, 4–11 March 2017; pp. 1–12.
18. Pathak, S. V., Mohod, A. G., & Sawant, A. A. (2020). Review on effective role of UAV in precision farming. *J. Pharmacogn. Phytochem.*, 9(4), 463-467.
19. Laliberte, A. S., & Rango, A. (2008). Incorporation of texture, intensity, hue, and saturation for rangeland monitoring with unmanned aircraft imagery. *The International Archives of the Photogrammetry, Remote Sensing, and Spatial Information Sciences*. Calgary, AB, Canada, 5–8.

20. Zhang, R. Wen, Y. Li, L. Chen, L. Xu, G. Huang, Y. Tang, Q. Yi, T. (2020). Method for UAV spraying pattern measurement with PLS model-based spectrum analysis. *Int. J. Agric. Biol. Eng.,* 13, 22–28.
21. S.R. Naqvi, M. Kamran. (2017). Towards real-time crops surveillance for disease classification: exploiting parallelism in computer vision *Comput. Electr. Eng.*, 59, pp. 15-26
22. Yang, S. Yang, X. Mo, J. The application of unmanned aircraft systems to plant protection in China. *Precis. Agric.*, 2018, 19, 278–292.
23. Huang, X., Zhang, S., Luo, C., Li, W., & Liao, Y. (2020). Design and experimentation of an aerial seeding system for rapeseed based on an air-assisted centralized metering device and a multi-rotor crop protection UAV. *Appl. Sci.,* 10(24), 8854.
24. Song, C., Zhou, Z., Luo, X., Lan, Y., He, X., Ming, R., Li, K. & Hassan, S. G. (2018). Design and test of centrifugal disc type sowing device for unmanned helicopter. *Int. J. Agric. Biol. Eng.,* 11(2): 55–61.
25. ME, S. M., Maguteeswaran, R., BE, N. G., & Srinivasan, G. (2016). Quadcopter UAV based fertilizer and pesticide spraying system. *Int. Acad. Res. J. Eng. Sci.,* 1, 8-12.
26. Jubair, M. A., Hossain, S., Masud, M. A. Al, Hasan, K. M., Newaz, S. H. S. & Ahsan, M. S. (2018). Design and development of an autonomous agricultural drone for sowing seeds. *Proceedings of the 7th Brunei International Conference on Engineering and Technology*. pp. 1-4.
27. Song, S., Dongxing, Z., Li, Y., Tao, C., Kehong, L., & Xiaowei, Y. (2015). Simulation and verification of seed-filling performance of pneumatic-combined holes maize precision seed-metering device based on EDEM. *Trans. CSAE,* 31(3), 62-69.
28. Kirchgessner, N., Liebisch, F., Yu, K., Pfeifer, J., Friedli, M., Hund, A., *et al.* (2016). The ETH field phenotyping platform FIP: a cable-suspended multi-sensor system. *Funct. Plant Biol.,* 44(1), 154–168.
29. Han, L., Yang, G., Dai, H., Xu, B., Yang, H., Feng, H., ... & Yang, X. (2019). Modeling maize above-ground biomass based on machine learning approaches using UAV remote-sensing data. *Plant Methods,* 15(1), 1-19.
30. Niu Y, Zhang L, Zhang H, Han W, Peng X (2019). Estimating Above-Ground Biomass of Maize Using Features Derived from UAV-Based RGB Imagery. *Remote Sens.* 11(11):1261.
31. Wani, A. A., Joshi, P. K., & Singh, O. (2015). Estimating biomass and carbon mitigation of temperate coniferous forests using spectral modeling and field inventory data. *Ecol. Inf.* 25, 63-70.
32. Lu, H. D., Xue, J. Q., & Guo, D. W. (2017). Efficacy of planting date adjustment as a cultivation strategy to cope with drought stress and increase rainfed maize yield and water-use efficiency. *Agric. Water Manage.,* 179, 227-235.
33. Pasqualotto, N., Delegido, J., Van Wittenberghe, S., Verrelst, J., Rivera, J. P., & Moreno, J. (2018). Retrieval of canopy water content of different crop

types with two new hyperspectral indices: Water Absorption Area Index and Depth Water Index. *Int. J. Appl. Earth Obs. Geoinf.*, 67, 69-78.

34. Nhamo, L., Magidi, J., Nyamugama, A., Clulow, A. D., Sibanda, M., Chimonyo, V. G., & Mabhaudhi, T. (2020). Prospects of improving agricultural and water productivity through unmanned aerial vehicles. *Agriculture*, 10(7), 256.
35. Maes, W. H., & Steppe, K. (2019). Perspectives for remote sensing with unmanned aerial vehicles in precision agriculture. *Trends Plant Sci.*, 24(2), 152-164.
36. C. Kwan, D. Gribben, B. Ayhan, S. Bernabe, A. Plaza, M. Selva. (2020). Improving land cover classification using extended multi-attribute profiles (EMAP) enhanced color, near infrared, and LiDAR data, *Rem. Sens.*, 12, 1392.
37. Papageorgiou, E. I., Markinos, A. T., & Gemtos, T. A. (2011). Fuzzy cognitive map based approach for predicting yield in cotton crop production as a basis for decision support system in precision agriculture application. *Appl. Soft Comput.*, 11(4), 3643-3657.
38. Gupte, S., Mohandas, P. I. T., & Conrad, J. M. (2012). A survey of quadrotor unmanned aerial vehicles. In *2012 Proceedings of IEEE Southeastcon* (pp. 1-6). IEEE.
39. Sarghini F., De Vivo A. (2017) "Analysis of preliminary design requirements of a heavy lift multi rotor drone for agricultural use" *Chem. Eng. Trans.*, 58, pp.625-630.
40. Yanliang, Z., Qi, L., & Wei, Z. (2017). Design and test of a six-rotor unmanned aerial vehicle (UAV) electrostatic spraying system for crop protection. *Int. J. Agric. Biol. Eng.*, 10(6), 68-76.
41. Brady, N. C., Weil, R. R., & Weil, R. R. (2008). *The nature and properties of soils* (Vol. 13, pp. 662-710). Upper Saddle River, NJ: Prentice Hall.
42. Sinclair, T. R., Purcell, L. C., & Sneller, C. H. (2004). Crop transformation and the challenge to increase yield potential. *Trends Plant Sci.*, 9(2), 70-75.
43. Kang, Y., Khan, S., & Ma, X. (2009). Climate change impacts on crop yield, crop water productivity and food security–A review. *Prog. Nat. Sci.* 19(12), 1665-1674.
44. Luedeling, E., Kindt, R., Huth, N. I., & Koenig, K. (2014). Agroforestry systems in a changing climate—challenges in projecting future performance. *Curr. Opin. Environ. Sustain.*, 6, 1-7.
45. Kadam, N. N., Xiao, G., Melgar, R. J., Bahuguna, R. N., Quinones, C., Tamilselvan, A., ... & Jagadish, K. S. (2014). Agronomic and physiological responses to high temperature, drought, and elevated CO2 interactions in cereals. *Adv. Agron.*, 127, 111-156.
46. Cooper, M., Tang, T., Gho, C., Hart, T., Hammer, G., & Messina, C. (2020). Integrating genetic gain and gap analysis to predict improvements in crop productivity. *Crop Sci.*, 60(2), 582-604.

47. Duvick, D. N. (2005). Genetic progress in yield of United States maize (Zea mays L.). *Maydica*, 50(3/4), 193.
48. Meng, J., & Cheng, Z. (2020). Improving the estimation of soil-available nutrients at the sub-field scale using time-series UAV observations. *Remote Sens. Lett.*, 11(8), 739-747.
49. Boldt, A. L., Eisenhauer, D. E., Martin, D. L., & Wilmes, G. J. (1999). Water conservation practices for a river valley irrigated with groundwater. *Agric. Water Manage.*, 38(3), 235-256.
50. Pereira, L. S., Oweis, T., & Zairi, A. (2002). Irrigation management under water scarcity. *Agric. Water Manage.*, 57(3), 175-206.
51. T. Akram, M. Kamran,Towards real-time crops surveillance for disease classification: exploiting parallelism in computer vision *Comput. Electr. Eng.*, 59 (2017), pp. 15-26
52. Hafeez, A., Husain, M. A., Singh, S. P., Chauhan, A., Khan, M. T., Kumar, N., ... & Soni, S. K. (2022). Implementation of drone technology for farm monitoring & pesticide spraying: A review. *Inf. Process. Agric.*
53. Devi, G.,Sowmiya,N.,Yasoda K.,Muthulakshmi, K.,& Balasubramanian, K. (2020). Review on application of drones for crop health monitoring and spraying pesticides and fertilizer. *J. Crit. Rev.*, 7(6), 667-672.
54. Sarghini, F., & De Vivo, A. (2017). Interference analysis of an heavy lift multirotor drone flow field and transported spraying system. *Chem. Eng. Trans.*, 58, 631-636.
55. Saseendran, S. A., Ma, L., Malone, R., Heilman, P., Ahuja, L. R., Kanwar, R. S., ... & Hoogenboom, G. (2007). Simulating management effects on crop production, tile drainage, and water quality using RZWQM–DSSAT. *Geoderma*, 140(3), 297-309.
56. Oron, G., Campos, C., Gillerman, L., & Salgot, M. (1999). Wastewater treatment, renovation and reuse for agricultural irrigation in small communities. *Agric. Water Manage.*, 38(3), 223-234.
57. Cedergreen, N., Felby, C., Porter, J. R., & Streibig, J. C. (2009). Chemical stress can increase crop yield. *Field Crops Res.*, 114(1), 54-57.
58. Sinclair, T. R. (1994). Limits to crop yield?. *Physiology and determination of crop yield*, 509-532.
59. Ray, D. K., Ramankutty, N., Mueller, N. D., West, P. C., & Foley, J. A. (2012). Recent patterns of crop yield growth and stagnation. *Nat. Commun.*, 3(1), 1293.
60. Zelitch, I. (1982). The close relationship between net photosynthesis and crop yield. *Bioscience*, 32(10), 796-802.
61. Khairunniza-Bejo, S., Mustaffha, S., & Ismail, W. I. W. (2014). Application of artificial neural network in predicting crop yield: A review. *J. Food Sci. Eng.*, 4(1), 1.
62. Day, P. R. (1977). Plant genetics: Increasing crop yield. *Science*, 197(4311), 1334-1339.

63. Long, S. P., ZHU, X. G., Naidu, S. L., & Ort, D. R. (2006). Can improvement in photosynthesis increase crop yields?. *Plant Cell Environ.*, 29(3), 315-330.
64. Serraj, R. A. C. H. I. D., & Sinclair, T. R. (2002). Osmolyte accumulation: can it really help increase crop yield under drought conditions?. *Plant Cell Environ.*, 25(2), 333-341.
65. Zörb, C., Geilfus, C. M., & Dietz, K. J. (2019). Salinity and crop yield. *Plant Biol.*, 21, 31-38.
66. Thompson, M. D., & Thompson, H. J. (2009). Biomedical agriculture: a systematic approach to food crop improvement for chronic disease prevention. *Adv. Agron.*, 102, 1-54.
67. Newman, R., Waterland, N., Moon, Y., & Tou, J. C. (2019). Selenium biofortification of agricultural crops and effects on plant nutrients and bioactive compounds important for human health and disease prevention–a review. *Plant Foods Hum. Nutr.*, 74, 449-460.
68. Litterick, A. M., Harrier, L., Wallace, P., Watson, C. A., & Wood, M. (2004). The role of uncomposted materials, composts, manures, and compost extracts in reducing pest and disease incidence and severity in sustainable temperate agricultural and horticultural crop production—a review. *Crit. Rev. Plant Sci.*, 23(6), 453-479.
69. Chen, J., Sun, L., Cheng, Y., Lu, Z., Shao, K., Li, T., ... & Han, H. (2016). Graphene oxide-silver nanocomposite: novel agricultural antifungal agent against Fusarium graminearum for crop disease prevention. *ACS Appl. Mater. Interfaces* 8(36), 24057-24070.
70. McComas, K. A., Besley, J. C., & Steinhardt, J. (2014). Factors influencing US consumer support for genetic modification to prevent crop disease. *Appetite*, 78, 8-14.
71. Legrève, A., & Duveiller, E. (2010). Preventing potential diseases and pest epidemics under a changing climate. *Clim. Change Crop Prod.*, 1, 50-70.
72. Oerke, E. C. (2006). Crop losses to pests. *J. Agric. Sci.*, 144(1), 31-43.
73. Lobell, D. B., & Burke, M. B. (2010). On the use of statistical models to predict crop yield responses to climate change. *Agric. For. Meteorol.*, 150(11), 1443-1452.
74. Van Klompenburg, T., Kassahun, A., & Catal, C. (2020). Crop yield prediction using machine learning: A systematic literature review. *Comput. Electron. Agric.*, 177, 105709.
75. Aldrich, R. J. (1987). Predicting crop yield reductions from weeds. *Weed Technol.*, 1(3), 199-206.
76. Sellam, V., & Poovammal, E. (2016). Prediction of crop yield using regression analysis. *Indian J. Sci. Technol.*, 9(38), 1-5.
77. Maestrini, B., & Basso, B. (2018). Predicting spatial patterns of within-field crop yield variability. *Field Crops Res.*, 219, 106-112.

8

Cognitive Intelligence and Distributed Computing Systems Applications in Smart Farming

Sangeetha Radhakrishnan[1]* and A. Prasanth[2]

[1]Department of BCA & IT, School of Computing Sciences, Vels Institute of Science and Technology and Advanced Studies (VISTAS), Chennai, India
[2]Department of Electronics & Communication Engineering, Sri Venkateswara College of Engineering Sriperumbudur, Chennai, Tamil Nadu, India

Abstract
Agriculture is the dominant sector of the economy in at least nine countries of the world. Farming provides fuel for billions of populations, as well as job opportunities to plenty of people. Traditionally, no automated approach is equipped. Hence, farmers have to do all the work manually. So, automated techniques are required to enhance the farming process. Employing artificial intelligence in the field of agriculture commonly known as agriculture intelligence is a surging form of technological revolution. In this chapter, we are throwing light on to the applications of cognitive intelligence and the distributing computing in smart farming techniques. The various applications for the smart farming like the disease prediction, precision farming and the crop analyzing using cognitive intelligence tools are explained in detail.

Keywords: Artificial intelligence, cognitive intelligence , distributed computing, smart farming

8.1 Introduction

Farming is considered as the predominant sector of the world's economy, which contributes to approximately 6.4% of economic production.

**Corresponding author*: sangeetha0988@gmail.com

Rajesh Kumar Dhanaraj, Balamurugan Balusamy, Prithi Samuel, Malathy Sathyamoorthy and Ali Kashif Bashir (eds.) Computer Vision in Smart Agriculture and Crop Management, (159–186) © 2025 Scrivener Publishing LLC

Farming opens the door to many employment opportunities along with feeding millions of people. The farming industries are searching for a creative approach to increasing the yield of crops due to the unpredictable climatic conditions.

Cognitive intelligence is one such approach that is referred to as the ability to generate knowledge by using the existing information. The mental ability of humans and their understanding can be developed through the thought process, senses, and the experiences. Cognitive computing is a self-learning system that performs human-like tasks intelligently using the data mining algorithms and machine learning (ML) techniques [1]. It focuses on solving the problems by imitating the behavior of humans and the logic of reasoning. Cognitive computing applications are in almost every field such as the retail industry, logistics, health care, farming, and cyber security.

8.1.1 Smart Farming

Artificial intelligence (AI) is the science and technology for developing smart machines, particularly computer programming systems. Computer algorithms for addressing real-time issues are intricate, so the developers find it hard to code. Modern AI techniques ensure that the complexity of the programmers in making predictions, optimization, and the feature selection will be performed by the computer programs. The employment of AI in the field of agriculture can be vital to tackle various issues in irrigation. Smart farming is considered as the encompassing of technology into machines to benefit in the farming production system. The smart farming techniques can be effective to resolve various issues like disease infestation, application of chemicals, drainage issues, and the prediction of yield.

Machine learning along with deep learning techniques can be deployed in agriculture for precision farming. The tedious labor and rigorous role of agriculture can be replaced by smart farming with more precise and effective techniques. Precision smart farming is based on satellite remote sensing, time management of yield, global positioning systems and so on [2]. Precision smart farming gathers assorted data, incorporate various technologies, and analyze properly to increase the quality of the production while decreasing the cost of production.

8.1.2 Challenges in Traditional Farming

In developing countries, the finite knowledge and expertise along with the higher maintenance to acquire the technology become the major hurdles for the individual farms. Hence, acquiring the new technology became restricted to the large farms [3]. The profit of the ICT is finite to the industrialized countries, and it focuses on the development of extensively grown yield namely rice, wheat, and corn. It heightens threats of unfeasible intensification practices. Disease outbreaks may be detained by fungicides in an industrialized scenario. But once they outpower the impediment measures, it comes with a higher threat of producing resistant fungal strains, which can be further acts more devastatingly.

8.1.3 Techniques Used in Smart Farming

8.1.3.1 Machine Learning

Machine learning is a division of AI in which the major issues are hard to solve by the humans. Thus, the decision-making tasks and the problem-resolving tasks will be deployed to the algorithm. Machine learning is an area of study where machines get the capability of studying with no prior programming specifically [4]. An objective of ML is to build a predictive engine, which is for a specific test case programming for each object variant, which is required to identify.

8.1.3.2 Deep Learning

Deep learning (DL) is a sub section of ML where the programmer's time and the effort taken to perform the feature selection and the optimization is saved. Deep learning transformed the AI world drastically [5]. The lead in deep learning computer vision paved the way for smart device-imposed disease identification.

8.1.3.3 Image Processing

As the plants are highly unconcealed to the outer world, it is sensitive to various diseases. Hence, precaution and blockage of diseases are mandatory. The symptom of the plant becoming diseased is when colored marks and dots are visible on the seeds, stems, and the leaf. The detection of the disease traits in the early stage and looking after the crops in the huge farming areas are intense. Hence automated tools and techniques are useful. So, for the automatic detection of disease and the leaf disease classification, image processing is used.

8.1.3.4 Internet of Things

The Internet of Things (IoT) uses analytical tools combined with ML techniques. To save money, time and effort deploy drones, robots, remote sensors, and imaging along with the analytical tools for surveying the farming fields and for monitoring crops. It also provides data to farmers in a rational form [6]. The deployment of the IoT in agriculture enhanced the farmer's decisions with accurate data along with decreasing the costs and improving the crop yield.

8.1.3.5 Artificial Neural Networks

Artificial neural network (ANN) is an effective and flexible DL technique with three layers is called deep neural network. The three layers are the input layer, output, and multiple layers. In order to process the information, the nervous system, like the brain, inspired the ANN as an information processing system to execute information.

8.1.3.6 Wireless Sensor Network Technology

The wireless sensor network (WSN) technology and its applications are in the various agricultural aspects such as the requirement of sensors, which are wireless in the field of farming. The major purpose is the usage of networks and sensors for fetching various benefits to resolve agricultural domain issues [7]. With respect to the review, the conclusion is that the main concerns to the solution is very complex and expensive, and that the proper solution was lacking for the different issues.

WSNs are normally used for the data collection, sharing and the storage of the sensed data. The target is to get the perfect choice for the reason of the original data sensing of environmental properties and climatology. The main purpose was the remote monitoring system and grid topology for design reporting, testing, and construction of infield-distributed WSN. The result was a huge decrease in the expense and the quality improvement production in agriculture and precision agriculture by concatenating the applications of precision irrigation and the WSN.

8.1.4 Benefits of Smart Farming

The rise of digital irrigation and the relevant technology unlocked a path for fresh data opportunities. Over an entire field, the remote sensors, unmanned aerial vehicles (UAVs), and the satellites can gather all the information 24/7. Also, the loam condition, plant health, humidity and the temperature can be monitored [8]. The importance of the number is negligible in data avalanche, and the data quantity generated by sensors are huge.

The goal is to help the farmers to understand the situation of the farm ground using the latest technology like remote sensing, which can be used to analyze the situation by the naked eye. Traditionally, the farmers will walk or drive through the field to understand the situation, but it is not accurate. But using the remote sensing like the latest technologies gives accurate results. These sensors use algorithms to convert a farming land circumstance as statistical data, which is easily understandable and helpful for the farmers for decision-making. The goal is to make better decisions in the field by making use of AI for a better yield.

8.1.5 Applications of Smart Farming

8.1.5.1 Disease Detection

Plants are exposed to the exterior conditions; hence, it is extremely prone to diseases. So, the control and the prevention is mandatory. The rate at which the disease spread depends on certain factors such as the yield conditions and the infection susceptibility. The identification of the diseases in the plant is the initial step to prevent the loss of crops and the agricultural products quantity. When the plant is diseased, colored streaks and spots can occur on the plant leaves, seeds, and stems [9]. The quick identification of plant diseases becomes a major issue in the world. The development in the computer vision by DL techniques make the road to the smartphone-assisted disease detection.

The automated techniques are highly beneficial when monitoring the huge crop farms and early-stage disease symptom identification becomes a tedious task. In case of developing countries, solely relying on naked eye monitoring to classify and identify the disease is costly. So, the goal is to employ the image processing-based software solution for the identification, as well as the classification of the diseases in the plant.

Image processing is beneficial to identify the diseased stem, fruit or leaf, to identify the shape of the affected area, size, and the shape of fruits. Through the computer vision, tools, methods, and algorithms are employed for the classification and identification of the disease. A higher success rate of 99.53 % achieved in detecting of corresponding diseases in the plant using the deep convolutional neural networks. The detection of diseases in a crop such as rice are also detected using neural networks. The coefficient of the variation Vosper Vector Machines SVM and K-means algorithm principal component analysis are few of the substitutions and on another instance more effective basis model [10]. The latest tools for crop disease detection comprise hybrid spectroscopic and image tools with an independent farming machine, which can give information on detection of plant disease at the very prior stage for controlling extent of disease in plants.

8.1.5.2 Crop Phenotyping

Crop phenotyping is defined as the noticeable behavior of a being, which resulted by its genetic inheritance type interaction along with the environment. The properties may include the behavioral, biochemical, shape, size, and shape. The phenotype represents the huge number of processes, structures, and the functions that are rapidly changing during growth and development. The structured assessment of phenotyping for the breeding, cultivar, and the phenomics study is necessary. The primary goal is the yield enhancement and issues in the plant breeding.

In crop phenotypic, it is mandatory for having a strong sensor since the collection of data in a systematic means with respect to the time and the space is required. In plant breeding, a system consists of five sensors such as infrared thermal radiometers, ultrasonic distance sensors, portable sensors, and the RGB webcam for the maximum throughput phenotyping [11]. From the farming plot field, crop canopy is measured using those multiple sensors. For geo referencing the sensor calculations, a GPS was used. Two environmental sensors were integrated to collect the simultaneous environment details.

The outcomes retrieved from the wheat and soya bean land with the performance of the sensor system are robust, as well as satisfactory in yield test. The properties of the temporal dynamics of the traits acquired from sensor plotting attribute as a function of time. The plant breeders may find the field-based throughput plant phenotyping data sensor system to be an effective tool.

Phenotyping is scrutinized under the three criteria. The first criterion is met with using the high throughput phenotyping in controlled conditions. Few of the methods used for sensing with a higher throughput phenotyping are 3D laser scanning, hyper and multi spectral imaging and thermal IR cameras, and fluorescence sensing. To attain the 3D data precisely, light detection and ranging (LiDAR) is a substitutional remote sensing technology. The LiDAR has its capability in crop phenotyping applications and is equipped for crop phenotyping with 3D high throughput.

The other criterion is determined by conducting a test on crop phenotypic enhancement in a semi-controlled environment with elements like drought, resistance and lodge. The last strategy involves managing agricultural plants' many environmental attributes in uncontrolled conditions in accordance with the farmer's practices.

8.2 Cognitive Intelligence

The capability of the human brain to absorb the information and form meaning and intelligence is cognitive intelligence. It uses the existing knowledge attained from practices and various experiences. Cognitive intelligence is defined as the ability and understanding of humans, which is developed based on the experiences, sensing, and the thought process [12]. It is the capability to build knowledge with the real time data and information. It comprises intellectual functions like learning, memory, judgement, attention, and logical reasoning.

8.2.1 Applications of Smart Farming

Applications of cognitive computing are in various fields like the logistics, healthcare, retail and in the cyber security field.

8.2.1.1 Retail Industry

The application of cognitive computing in the retail industry assists the marketing team in collecting more relevant data and further analyze to make retailers more adaptive and efficient. It can be extremely helpful for the electronic commerce and for on floor shopping. It helps retailers to show their customers the personalized products by collecting basic data and information from the customers.

8.2.1.2 Logistics

Cognition is the new horizon in the logistics, supply chain, and transportation. It is used in all stages of logistics such as designing decisions in the warehouse, management and automation of the warehouse, networking, and the IoT. In the process of warehousing, cognition helps in the storage code compilation, automated picking along with the automatically guided machine and the usage of the robots helps in the improvisation of work effectively. It uses cognition to design a way for improving the recognition rate that saves a lot of human labor.

8.2.1.3 *Banking and Finance*

Cognitive computing in the banking sector improves the operational efficiency, customer experience and engagement and grow revenues. It will completely revamp the banking and other financial institutions on 3D, which involves a deeper contextual engagement, new analytics perception, and enterprise conversion. Such transformations include digitally performing different banking transactions, claims, and loans processing in minutes. Cognitive technology is said to be beneficial in product management and the customer care support.

Cognitive banking will help the customers to decide the personalized investment plans according to the customer being risk averse or a risk taker. In addition to this, it provides personalized engagement between the customer and the banking firm by understanding the personal fashion of every customer and focusing on the requirements. Hence, the system will understand intelligently the customer's personality based on the various content available online by the customer.

8.2.1.4 *Cyber Security*

Cognitive technologies are used to provide the end-to-end security, threat detection, access, and amend threats. The cognitive algorithm helps to avert cognitive hacking; hence, making the customer least vulnerable to attacks and provide a technical resolution for the detection of any inconsistent information and data.

8.2.1.5 *Power and Energy*

The technologies process the huge data to support the decisions taken and learn from the outcomes using the cognitive computing. Cognitive technology helps to make prominent decisions of future such as commercial oil wells, making existing power stations more accurate and providing a higher hand to the existing power companies.

The vital security players in the industry introduced the cognitive services for the detection of security threats and analysis. It detects the cyber threats along with scanning the security system for vulnerabilities and provide a solution [13]. To get better use of cognitive computing, there is a need of building a huge database and to maintain the privacy and the user confidentiality along with the prevention of the data leakage.

8.2.1.6 Healthcare

In the health care, cognitive computing paves the way for the better decision making while treating patients along with improving the medical professional efficiency. It is an algorithm, which uses the ML tools, data mining techniques, language processing, and the visual recognition depends on the patient information, medical reports, and the other related information.

A huge amount of data is processed by the system to solve some specific questions and to make relevant recommendation. Cognitive computing in the medical field connects the human and machine functions in which the machine and the human brain overlap each other to make better decision. Hence, it helps the doctors and other medical practitioners to diagnose their patient disease better and treat accordingly. It empowers the doctors in planning customized disease treatment.

8.2.1.7 Education

With the help of the cognitive technology assistance, personal lessons can be provided to the students and mentor them through the coursework and assist learners to understand tough concepts well. It helps to guide the learners to select the courses according to their wish. It is also like a career counsellor that helps the students to take decision to choose their career. To deliver the better service such as developing the student report and getting the feedback.

8.2.1.8 Agriculture

Cognitive computing application in farming paves the way for the green revolution. It uses the cognitive technology, which helps to learn, understand from the previous experience, environment, interaction and hence provide higher efficiency [14]. Plenty of cognitive techniques are used in farming with respect to the usage of the fertilizers, pesticides, and for increasing the crop production in farming.

8.2.2 Agricultural Robotics

Cognitive technology is used in farming with extreme use of robots. It results in another branch of study called agricultural robotics. Throughout the world, these existing robotics are used effectively in different farming tasks.

A robot named See and Spray was developed by the Blue River Technology commercializing computer vision for monitoring and accurately weed spray on the cotton plant. For helping the strawberry farmers to pluck and pack the crops, a robot named harvest CROO has been developed. A deep learning app called Plantix has been developed by a Berlin-based technology start-up PEAT, which identifies the potential problems and deficiencies of soil. Software algorithms corelate specific foliage pattern with some plant pests and diseases. Like the Plantix app, to provide soil analysis services for framers, Trace Genomics gives soil analysis service.

Sky Squirrel Technology is one among the company, which equips the drone technology to the vineyards. A Colorado-based firm uses ML techniques linked with the satellites for climate prediction, analyzing the sustainability of the crop and monitor farming field to find the existence of pests and diseases [15]. Farm shots are another start-up based in North Carolina, focused on the analysis of farming data retrieved through the images taken by the drones and the satellites. Particularly, the goal of the company is the detection of disrespects and poor plant nutrition on the farming field.

8.2.3 Benefits of Cognitive Computing

A human capability to make better decisions for a sustainable environment can be improved using cognitive computing. Due to its capability for the comprehension, understanding from the previous experiences and the environment, interaction, and the reasoning, which in turn increases the efficiency. Using ML, we can determine the better performing model on certain conditions. Cognitive technology can assist people to learn the standard and identify irregularities or the troublesome areas with the help of ML and the video analytics [16]. It helps in decreasing de-forestation, monitoring urbanization, eliminating diseases, and better comprehension, along with the management of ecosystem.

8.2.4 Scope of Cognitive Computing

Cognitive technology provides a new way of computable issues. Machines can reply to the complicated situations identified by obscurity and has a wide-ranging influence on our personal life, medical field, and business.

8.2.4.1 Engagement

A huge amount of unstructured and organized data are stored in the cognitive system repository. It has the capability of gaining profound insights of the domain and provide the professional advice. The contextual links between the various items in a system's reality, which allows to develop hypotheses and the arguments included in the built model by the systems. These can bring together the contradictory or inconsistent facts. As the output, the systems can have an insightful communication with people. A best example of the engagement model is the chatbot technology. Some AI chatbots are already trained in commercial applications and are equipped with the domain knowledge.

8.2.4.2 Decision

The cognitive system is rapidly evolving in response to the fresh data, actions, and the outputs. The prerequisites for decision-making include the capacity to determine the reasoning behind a particular choice and to change the degree of trust in a system response. One example of this idea being put to the test is IBM Watson's use in the medical industry. The computer that is capable of compiling and analyzing the patient data, his medical history, and the diagnosis. The solution-based recommendations on the capability to decode the explanation and analyzing the questions in the context of complex healthcare data and natural language, including physician's notes, patient records, medical reports, and clinical assessment, as well as annotations on the texts in the healthcare records. The result improves in accuracy over time using learning. Clinicians can spend much time with their patients by providing decision support tools and minimizing the paperwork.

8.2.4.3 Discovery

Discovery comprises finding insights, comprehending a large quantity of the data and information, and gaining skills. These models are based on ML, which is unsupervised and deep learning. There is a definite need for the systems, which assist in using the information much effectively than people could do on their own increasing amounts of data. One among the cognitive solutions is the cognitive management information shell at Louisiana State University (LSU). In order to build an inspection, visualization, and the interactive sensing, which offers real-time analysis and monitoring, A distributed intelligent system in the model gathers streaming information and data including the text and the video. Hence, to identify a significant occurrence and resolving the issues, the Cognitive Information Management (CIM) shell sends an alert and drastically changes its configuration.

8.2.5 Limitations of Cognitive Computing

8.2.5.1 Limited Analysis of Risk

The risk analysis capacity of the cognitive system falls short when it comes to the unstructured data. This encompasses the cultural, political, and social-economic issues. As an instance, a predictive model can identify a potential site for the oil development. But the cognitive model should take into account this factor if the country is undergoing a change in the atmosphere. Hence, human engagement is mandatory for finishing the tasks like risk analysis and final decision making.

8.2.5.2 Meticulous Training Process

The cognitive system requires data for training to fully comprehending the procedure and improve. The lethargic adoption of the cognitive system is probably due to the tough training procedure. IBM Watson is holding Wellpoint's finance management in a similar position. The process of educating Watson for the usage by insurer entails having the IBM experts go over the content of each healthcare insurance policy. The nursing staff continues to maintain feeding instances until the system is fully comprehended a certain healthcare situation. Moreover, using a cognitive system, which is a difficult and expensive process, makes the situation worse.

8.2.5.3 More Intelligence Augmentation Rather than Artificial Intelligence

The most effective roles of the cognitive computing system are of assistants, which are more comparable to intelligence enhancement than AI. Although cognitive technology complements human reasoning and analysis, important decisions may still be taken by humans. Chatbots and intelligent assistants are two such examples. These focused projects are an efficient way for the firms to begin utilizing the cognitive technology as opposed to the enterprise-wide adoption.

The advancement in computing, which automation sparked, is undoubtfully cognitive computing. It establishes a standard for computing systems to meet in order to compete with the human brain. Yet, because of various problems, AI is challenging to use when there is huge ambiguity or a rapid change or even when creativity is required. The quantity of information sources increase the complexity of the problem. It is difficult to combine, analyze, and integrate unstructured data. A compound cognitive solution may include lots of technologies, which co-exist to provide deep domain insights.

Along with AI, ML, and natural language processing technologies, the cognitive system should include other technologies like NoSQL, Hadoop, Elasticsearch, Kafka, and Spark. Both static historical data and dynamic real time information or data might be handled by the comprehensive system. The business who wants to implement the cognitive solutions ought to start with a certain specific industry. Strong business rules, which serve as the algorithms guides and a huge amount of data for the machines to learn, should be included in these sections.

8.3 Distributed Computing

A distributed computing system is predominantly a group of processors connected by a network in which each has their local memory space and other peripherals. Any two processors in the system can interact with one another by sending the information through the network. Typically, the distributed computing system refers to a processor and its resources as a site, node, or machine.

8.3.1 Evolution of Distributed Computing Systems

It started with the development of potent microprocessors. They were initially 8-bit devices, but 16-bit, 32-bit, and 64-bit processors rapidly gained popularity. Several of them offered mainframe computer performance at a fraction of the cost. From a 10-million-dollar machine that carried out one instruction each second. Now, computers with a price performance increase of 1013, which cost $1,000 but can carry out one billion commands per second.

Highspeed computer networks are created, which was the second advancement. Local area networks (LAN) enable the connection of several devices so that tiny bits of information may be exchanged between them in just a few microseconds. Due to the challenges of centralized processing in the main frame applications, distributed computing has gained popularity.

All the components of a mainframe software architecture's are within a central host computer. Using a terminal that records keystrokes and transmits that data to the host, users communicate with the host. In the previous 10 years, mainframes have discovered a fresh application as a server in the distributed client-server systems. The file sharing systems, in which the server sends the files from a common location to a desktop computer, formed the foundation of the first PC networks, which essentially replaced the main frame computer [17]. The requested task is subsequently executed on the desktop environment, complete with the logic and data. When shared, utilization, update contention, and data transfer volume are minimal, file-sharing design platforms. As the number of online users increased and graphical user interfaces (GUI) gained popularity in the 1990s, PC LAN computing underwent changes.

The next significant development in distributed computing is the division of software architecture into two or three levels. The GUI is often found on the user's desktop in the two-tier client server systems whereas the database management services are typically found on a server, which is a much powerful machine that serves several clients. The database management (DBM) server and the user interface environment are responsible for different aspects of processing management. When work people are defined as a dozen to a hundred people working on a LAN simultaneously, the two-layer client server architecture is a solid choice for the locally distributed computing. Nevertheless, performance starts to decline if there are more than 100 users, and scaling the system becomes challenging. To address the limitations of the two-layer design, the three-tier architecture also known as the multi-tier architecture was created. In between the user interface client and DBM server environment, a 3rd tier was added. The logical design of systems was successfully separated by the three-layer designs, but the complexity of the cooperating interfaces was remained challenging because of the technological interdependence between interconnecting operations. Afterwards, remote procedure call (RPC) standards are employed to try and standardize how the processes interacted with each other.

8.3.2 Features of Distributed Computing System

8.3.2.1 No Common Physical Clock

It is very crucial in introducing the concept of distribution into a system and assumes responsibility for supplying built-in asynchrony among the processors. The nodes in a dispersed network do not share a single physical clock.

8.3.2.2 No Shared Memory

It is a vital component of information passing communication between the nodes that built up the network. With the memory architecture, there is no idea of a shared physical clock. Nonetheless, the distributed shared memory abstraction may still use to offer the abstraction of a common address space.

8.3.2.3 Geographical Separation

The processors in the distributed system are dispersed all over the world. Nevertheless, having the processors interconnected to a wide area network is not needed. A network or a collection of workstations are interconnected to an LAN can be viewed as tiny distributed system. The advent of less cost and rapid speed of shelf processors has led to a rise in the popularity for the NOW configuration. The NOW architecture is the foundation of the google search engine.

8.3.2.4 Autonomy and Heterogeneity

Since they have separate memory varied compositions and are often not a part of specific system interconnected by a network, processors are autonomous in nature. Instead, they work together to provide services or solve issues.

8.3.2.5 Distributed Computing System Model

Systems for distributed computing are constructed using a variety of paradigms. Workstation server, processor pool, workstation server and the minicomputer workstation, and hybrid are five major categories into which these machines may be divided.

8.3.2.6 *Minicomputer Model*

A distributed computing system based on minicomputers comprise of a few minicomputers (it can be large computers too) connected by a network. It is an expansion of the time-sharing centralized system, as illustrated in the Figure 8.1. Each minicomputer often has several users signed on at once. Each minicomputer has a number of interactive terminals linked to it for this purpose [18]. Each user has remote access to one particular minicomputer while signed on to another minicomputer. Using the network, a user can access distant resources, which are located on the computer systems other than the one they are presently logged into. When resource sharing with the remote users is needed like sharing database information of various sorts, with each database housed on different system, the minicomputer model may be employed. A distributed computing system built on the minicomputer architecture is the early ARPANET.

8.3.2.7 *Workstation Model*

A communication network connects a number of workstations to form a distributed computer system based on workstation concept. Each station in an office or department of a university may have its own disc and function as a single computer user, and they may be dispersed throughout

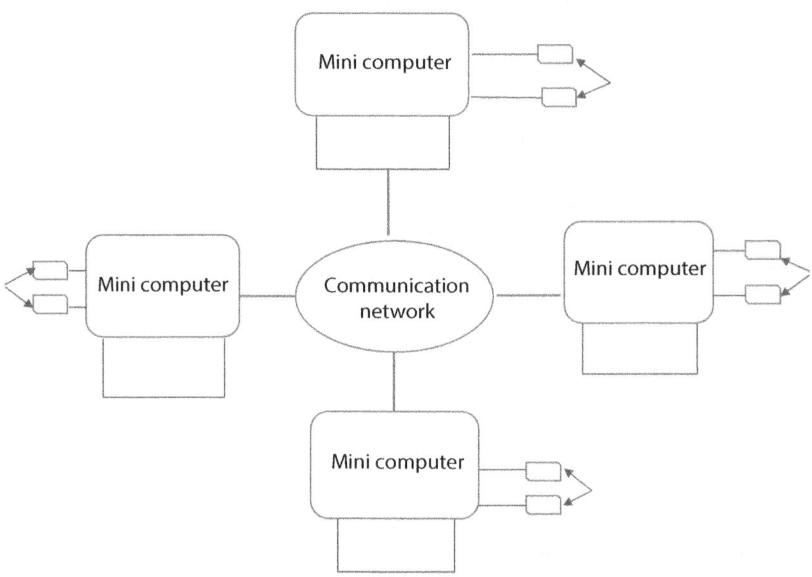

Figure 8.1 Distributed computing system based on minicomputer model.

a building or campus. It has been observed that in such a setting, a sizable fraction of the workstations remains idle at any one time wasting a considerable amount of CPU time. Figure 8.2 illustrates the distributed system based on workstation. The purpose of the workstation is to connect all these workstations via a fast LAN in order to allow idle workstations to process tasks but not enough power at their workstation to process those tasks effectively.

In this architecture, users submit jobs for execution by logging in to one of the workstations designated as their home workstation. When the system determines that the user's workstation lacks the processing power necessary to execute the processes of the submitted jobs efficiently, it moves one or more of the processes from the user's workstation to another idle workstation and has those processes executed there before returning the execution results to the user's workstation.

8.3.2.8 Processor Pool Model

The processor pool approach is based on a user needing any processing power for a short period of time; for instance, when recompiling a program made up of a lot of files after altering a fundamental shared declaration. As a result, under the processor pool model, rather than the workstation

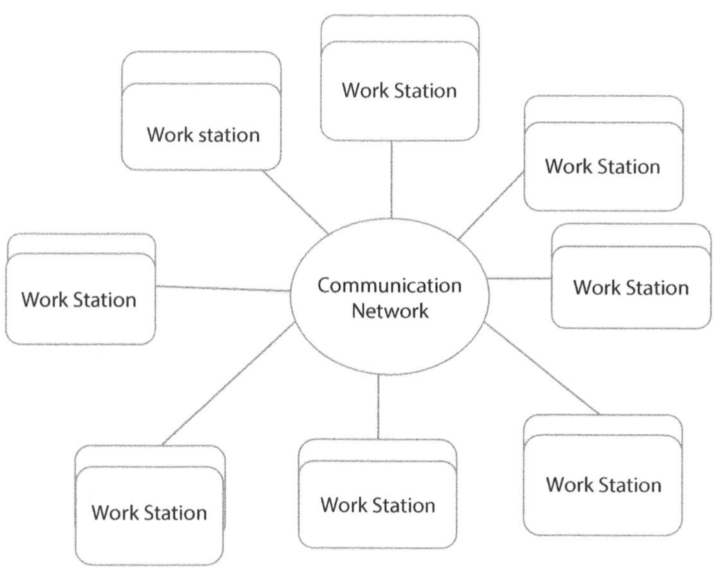

Figure 8.2 Workstation model.

server approach, where a processor is assigned to every user, processor is pooled so as to be used by people when in need [19]. The majority of microcomputers and minicomputers connected to the network make up the pool of processors. Every processor in the processor pool has their memory for loading and also running a program either system or distributed system application.

8.3.2.9 Hybrid Model

The most popular approach for creating distributed computing systems is the workstation server architecture. This is due to the fact that a sizable portion of computer users merely carry out basic interactive operations such as editing documents, sending emails and running small applications. For such basic usage, the workstation server paradigm is perfect. The processor pool model is more appealing and appropriate in working current environments where user groups frequently perform tasks requiring massive computation.

A hybrid paradigm can be used to create a distributed system that maintains benefits of both the workstation server and the processor pool architectures. The workstation server paradigm serves as the foundation for the hybrid architecture, which also includes a processing pool. For calculations, which are massive for work station or which need numerous systems running at once for the effective execution, the pool's processors can be dynamically assigned [20]. The hybrid approach provides assured interactive jobs response by enabling it to be handled on the workstations of the users, in addition to efficient execution of computation-intensive operations. Nevertheless, compared to the workstation server paradigm or the processor pool approach, the hybrid model is more expensive to install.

8.3.3 Advantages of Distributed Computing System

8.3.3.1 Inherently Distributed Computation

Applications that involve consensus among participants that are used globally such as financial transactions like money transfers and travel reservations such as flight travel are inherently distributed by nature.

8.3.3.2 Resource Sharing

Resources are spread throughout the system since it is not practicable nor cost effective to replicate them across all sites in order to increase performance. Placing all the resources at one location is also not feasible because it might result in noticeable performance degradation. Partition the data sets over several servers, with replication taking place at several places, for rapid access and improved dependability in distributed databases like DB2.

8.3.3.3 *Access to the Geographically Remote Data and Resources*

Due to its huge size, data is frequently impossible to copy at every site, and maintaining the vital data may be risky. For instance, due to its sensitivity, banking system data cannot be reproduced everywhere. Hence, it is instead kept on a central server that branch offices may access remotely, improvements in mobile communication that allow access to the central server, which requires distributed protocols and middleware.

8.3.3.4 *Enhanced Reliability*

A distributed system, which has intrinsic efficiency by resource replication, offers increased reliability. Moreover, the dispersed resources often do not fail or crash. Reliability incorporates a number of factors:

8.3.3.4.1 Availability

The resources can be accessed at any time and are always available.

8.3.3.4.2 Integrity

As the resources can be acquired by the multi processors, it should always be in correct state.

8.3.3.4.3 Fault-Tolerance

Even if distributed systems work efficiently when some of its resources do not work, it is fully tolerant.

8.3.3.5 *Increased Performance/Cost Ratio*

By pooling resources and using geographically distant data and resources, the performance cost ratio may be enhanced. In a distributed system, each work can really be partitioned and dispersed among a number of computers rather than being assigned as a whole to parallel machines.

8.3.3.6 *Scalability*

There may be more nodes linked to the wide area network, which will not have an impact on the communication performance.

8.3.3.7 *Modularity and Incremental Expandability*

Heterogeneous processors can integrate into the system, which is executing the same middleware algorithm without affecting the performance, and it is currently easily swapped out with new ones.

8.3.4 Scope of Distributed Computing

Every field of computation uses distributed computing. Distributed computing always has a greater cost benefit ratio than other dedicated computing. Scientific applications, business applications, applications in the computer systems, engineering and design and other sectors all heavily rely on distributed computing.

In engineering and design, distributed computing is frequently used to create products like air foils, rapid circuits, internal combustion engine, and nano-micro electromechanical system. Mostly optimization is required for these kinds of applications [21]. The majority of optimization algorithms, such as genetic programming for the discrete optimization and simplex, branch and bound, and interior point way for linear optimization were parallelized and calculated via distributed computing.

Scientific applications involving the human genome sequencing, the study of biological sequence to create drugs and cure for illness, the analysis of huge datasets in astrophysics and bioinformatics and the comprehension of quantum phenomena and macro molecular structure in computation physics and chemistry all heavily rely on the high-performance computing.

Commercial applications make heavy use of distributed computing. While web and database servers are regularly used by the applications, optimization is essential for querying and making rapid choices for the improved business operations. The benefit of efficient distributed and parallel algorithms is required for the issues such as time series analysis, classification, association rule mining and clustering due to the enormous volume of data and geographical distribution of the data.

Since the computer systems have proliferated throughout all the applications of computer science, parallel distributed computing has been in corporate into a variety of fields, including analyzing system security, detecting network intrusion, crypto analysis, and ad hoc mobile computations.

8.4 Cognitive Intelligence and Distributed Computing in Smart Farming

8.4.1 Need for Digitization of Agriculture

The IoT plays a vital role to meet the growing food demand in farming, which results to the improvement of efficiency by the use of inputs such as soil, fertilizers and pesticides, predicting the pest and the diseases, and monitoring crops are watered properly. With less expenses, productive farming improves output. A rapidly gaining notion in the agriculture industry is smart farming. Farmers are utilizing a variety of technology such as drones, smart agriculture, sensors, terrain contour mapping, self-driven and GPS-equipped tractors, and robots to produce more food effectively and sustainably. Existing agricultural land will be under a lot of pressure to increase productivity to a sufficient level for the next decades.

The IoT has played a big part in the enormous changes that have occurred in agriculture over the past several decades. Agriculture has greatly benefited from the IoT, which has taken on a prominent role in the industry. Farmers can increase output while reducing costs, and they can also monitor their fields from a distance. In addition to crop monitoring, smart irrigation and the other agricultural tasks are also conveniently accessible in smart farming while utilizing the IoT. Smart farming uses a platform that incorporates sensors, microcontrollers, transmitters, drones, and other technologies to enable farmland to perceive, analyze and accurately measure environmental data.

8.4.2 Existing System

In order to map the hazards associated with the pests and diseases, the crucial metric 'accumulated degree-days' is derived using land surface data acquired by a meteorological satellite. In order to cope with missing data, logistic regression is utilized and findings of land surface, monthly temperature degree and days data are analyzed using linear regression [22]. A significant association is demonstrated between the accumulated values derived from meteorological stations and the accumulated values received from the satellite.

An article by Lottes *et al.* provided a method that recognizes beetroot sugar plants, extracting features to provide an accurate weed and crop estimation by integrating Random Forest classification and utilizing nearby information through a Random Markov Field (RMF) [23]. On a real farm robot in three separate sugar beetroot fields, the suggested strategy was put into practice and assessed. A robot can carry out essential spraying or weed removal by autonomously segregating the two plant types.

To extract information on farmed land, Lu suggested deep convolutional neural networks (DCNN) and transfer learning. With the exception of integrity and consistency, a suggested technique outperforms the latter in studies when the two methods are compared to each other [24]. High altitude UAV photographs are employed.

Convolutional neural network and crops lines method were proposed by Bah to identify weeds in beetroot, spinach, and bean fields. Drones are utilized to obtain high vegetable photos, which are around 20 m. The outcomes show that the beetroot field achieves the greatest accuracy. The article discusses various challenges with weed detection, particularly when they are just appearing or when they are growing close to crops [25].

Robust vision systems are the foundational technology for constructing autonomous robots for precision farming. The systems that are created must be able to keep an eye on the crops and target just those plants that need to be treated. Based on the morphology of properties, the publication [26] offered a unique technique for the identification and categorization of crops and weeds. The use of the support vector machine (SVM) method was made for crop/weed discrimination. The discovered regions are divided into three groups using SVM: crop, weed, and mixed. Using two different types of crops – onion and beetroot sugar – the results demonstrate an efficient categorization rate of weed and crop discrimination.

Several DCNN models are employed to [27] identify weeds in turfgrass. These are based on digital camera photographs of weeds. These models include Detect Net, Google Net, and VGG Net. The outcomes demonstrated that weed detection is a very good application for DCNN.

Research has been conducted about the challenges that researchers face when using IoT in the green agricultural sector and roughly 22 UAV applications, which are dedicated to airborne crop observation procedures or spraying duties [28]. Villa-Henriksen *et al.* recognized numerous difficulties experienced while using IoT in several applications, hence suggested various remedies to deal with them [29].

Studies by Friha *et al.* hypothesize the use, categorization, application, and comparison of the most advanced technologies, including Cloud/Fog computing, UAVs, wireless technologies, open source IoT platforms, software defined networking (SDN), and middleware platforms [30].

8.4.3 Technologies Used in Cognitive Intelligence and Distributed Computing in Smart Farming

8.4.3.1 Genetic Algorithm, Fuzzy Logic, ANN Technology

In order to reliably categorise crops and weeds during the whole growing season, a system was created using the Genetic Algorithm (GA), Fuzzy Logic technique (FL), and ANN technologies. The use of the constraints allowed for the identification of each crop centre. Although 300 training samples were employed, more training samples may be used to improve accuracy.

8.4.3.2 Remote Sensing, Decision Tree, Hyper Spectral Data Mining

Hyper spectral data was categorized using the classification and regression trees (CART) decision tree algorithm into classifications like water stress, weed presence, and nitrogen application rates. This categorization accuracy ranged from 75% to 100%, and the validation output was attained with a 96% accuracy for irrigation, 83% for nitrogen factor, and 100% for weed control tactics at early growth stage and single factor analysis. Instead of focusing primarily on the early phases, excellent precision should be achieved at all levels.

8.4.3.3 WSN Technology

In order for a beginner user to quickly choose the right sensor whenever necessary, evaluations of the need for wireless sensors and how they may be applied using WSN technology in many parts of agriculture are made. If sensor nodes are placed in an open, hostile environment, they are vulnerable to physical harm, obstruction, and interference. The network's overall goal of maintaining system stability should not be impacted by a failed node.

8.4.3.4 Image Processing, K-Mean Clustering

The color, as well as the texture traits, may be utilized to identify and categorize various agricultural and horticultural products produced in the healthy and diseased areas, which have been proven to be extremely successful for detection of disease in *Malus Domestica* using the various ways. In future, Bayes can be used for classification along with K-means clustering and principal component analysis.

8.4.3.5 Machine Vision, Image Processing Techniques

In order to distinguish between healthy and sick citrus leaves in a lab setting, the hue, the saturation, and the intensity color properties of the texture were combined with statistical classification algorithms using the color co-occurrence approach. In the future, the algorithms would be employed in outdoor settings and a merged dataset created for leaf front and rear. It would be difficult to go from evaluating a single leaf to a leaf canopy due to the inherent color variations in natural illumination.

8.4.3.6 Hyperspectral Imaging, Artificial Neural Networks

The obtained hyperspectral signatures were analyzed with artificial neural networks. In order to achieve the desired outcome, all spatial pictures were rebuilt and computed in accordance with the membership of each pixel's color cluster. In reference to the characteristics, the outcomes were compared. As there were just a few genotypes and a small sample of each plant, the output could not be evaluated with accuracy. According to the findings, ANN is a good method for phenotyping crops.

8.5 Conclusion and Summary

8.5.1 Concluding Thoughts

Using AI-driven technology, industries in the agricultural sector may solve issues including crop yields, plant and soil health, weeds, and disease. The effectiveness of work may be greatly enhanced with the aid of the tools at hand. Using scientific methods like remote sensing, GPS, data analytics and other techniques, more practical farming may be done with the aid of precision farming, which helps to increase agricultural productivity and lower possible environmental danger. In addition, early disease detection in plants is possible by using the ANN, picture recognition software, and several techniques. The health of the crop can be observed and high-quality productivity can be acquired with little or no loss because of early disease detection.

Artificial intelligence in agriculture has a significant potential to address issues like labor and resource shortages. For agricultural features including plant leaf color, leaf area index, plant health, biomass, yield and chlorophyll, traditional procedures involve laborers, which takes a lot of time.

Nondestructive and rapid throughput phenotyping with the advantages of convenient and flexible operations with on-demand access to spatial and data resolution would be possible with the aid of the many methodologies presented. This chapter aims to discuss the use of automation in agriculture using cognitive intelligence and the distributive computing systems to increase the crop quality while maximizing the output and using the least amount of labor and time.

8.5.2 Future Enhancement

Artificial intelligence equips agronomists with a weapon against insects that feed on the grain and offers a fix for several issues, including foliar diseases and nutritional deficits to mention a few. The most common uses of AI in agriculture, according to the study assessments, seem to fall under categories like agricultural robotics, which businesses are creating, and autonomous programming robotics handle crucial tasks like soil-crop monitoring, where businesses use DL algorithms for data processing collected by the software or drone-based technology to observe soil and yield health. Crops can be harvested at a huge volume and rapid rate than human laborers.

The prediction analysis, depending on the images where ML models are created, analyze the massive amounts of generated data each day on historical climate patterns, fresh research, soil reports, pest infestation, rainfall, and drone and camera images that provide significant insights to increase crop yield. To guarantee that leaf pictures are divided into sections such as backdrop, diseased and non-diseased part, paratope processing of the image is done before disease detection. The stimulated region is cut out after and transferred to various labs for further examination. Moreover, it helps in the identification of any nutritional deficiencies and pests. Crop readiness identification: To assess how mature the green fruits are, pictures of different yields are taken in white/UVA light. Before placing the products on the market, the cultivators can build several phases of readiness based on the type of crop or fruit and place them into different stacks. By establishing a field map and identifying places estimations may be created during the cultivation phase using high-definition photos from drones. This has a remarkable positive impact on the optimization of resource. Artificial intelligence can assist humans in overcoming its greatest challenges, which is feeding an additional 2 billion people by 2053, irrespective of climatic changes and overpowering fertile deltas with sea water. It can be done by AI using pests' detection and predicting the crop, which yield the best returns.

References

1. Misbah Pathan *et al.*, "Artificial cognition for applications in smart agriculture: A Comprehensive review", *Agriculture*, Volume: 04, 2020, pp. 81-95.
2. T. Manjula and T. Sudha, "Cognitive Computing For Sustainable Agriculture", *Asian J. Comput. Sci. Inf. Technol.* ISSN: 2249-0701 Volume: 08 No.S3, 2019, pp. 159-161.
3. Andale Kalila Tesfaye *et al.*, "Combining machine learning, space-time cloud restoration and phenology for farm-level wheat yield prediction", *Artif. Intell. Agric.* Volume: 05, 2021, pp. 208–222 .
4. X. Hu and R. Zhang, "Text classification based on machine learning," 2022 *IEEE International Conference on Artificial Intelligence and Computer Applications (ICAICA)*, Dalian, China, 2022, pp. 911-916.
5. Balasubramaniam, C.Vijesh Joe, TA.Sivakumar, A.Prasanth, "Optimization Enabled Deep Learning-Based DDoS Attack Detection in Cloud Computing", *Int. J. Intell. Syst.*, 2023, pp.1-14.
6. P. Paruthi Ilam Vazhuthi, A. Prasanth, SP. Manikandan, KK Devi Sowndarya, "A hybrid ANFIS reptile optimization algorithm for energy-efficient inter-cluster routing in Internet of Things-enabled Wireless Sensor Networks", *Peer-to-Peer Networking Appl.*, 2023,pp. 1-29.
7. P. Vazhuthi, S.P. Manikandan, A. Prasanth, An Energy-Efficient Auto Clustering Framework for Enlarging Quality of Service in Internet of Things-enabled Wireless Sensor Networks using Fuzzy Logic System, *Concurr. Comput.*, 34, 2022, 1-28.
8. V.K. Jyothi *et al.*, "Retrieval of flower videos based on a query with multiple species of flowers", *Artif. Intell. Agric.*,Volume: 05 ,2021, pp.262–277.
9. S. Jain, I. Verma and S. Sharma, "E-Agriculture Integration with Cloud Computing," 2022 Seventh International Conference on Parallel, *Distributed and Grid Computing (PDGC)*, Solan, Himachal Pradesh, India, 2022, pp. 298-303.
10. Kirtan Jha, Aalap Doshi, Poojan Patel and Manan Shah, "A comprehensive review on automation in agriculture using artificial intelligence", *Artif. Intell. Agric.*, Volume: 2, 2019, pp. 1-12.
11. Hasina Ait Issa *et al.*, "A comprehensive review of Data Mining techniques in smart agriculture", *Food*, Volume: 12, Issue 4, October 2019, pp. 511-525.
12. Neha Gupta, Sachin Sharma, Pradeep Juneja and Umang Garg, "A Network Data Analytic Technique in a 5G-IoT-Based Smart Healthcare System Using

Machine Learning" in *5G Internet of Things and Changing Standards for Computing and Electronic Systems*, IGI Global, 2022, pp. 81-93.
13. Faisal Jamil, Muhammad Ibrahim, Israr Ullah, Suyeon Kim, Hyun Kook Kahng and Do-Hyeun Kim, "Optimal smart contract for autonomous greenhouse environment based on IoT blockchain network in agriculture", *Comput. Electron. Agric.*, vol. 192, 2022, pp. 106-573.
14. Lijin Santhosh and Ankitha Philip, "The Importance of Cloud Computing in Agriculture", *Proceedings of the National Conference on Emerging Computer Applications (NCECA)*, vol. 4, no. 1, 2022, pp. 168.
15. Yiannis Amperozides, "ArcView: Cloud-based application to process, analyse and visualize UAV-collected data for precision agriculture applications utilizing artificial intelligence", *Agriculture*, Volume: 174, July 2020, 105457, pp.1-24.
16. Nabila Husna Shabrina *et al.*, "Deep learning models for automatic identification of plant-parasitic nematode", *Artif. Intell. Agric.*, Volume: 7, 2023, pp. 1–12.
17. Md. Firoj Ali and Rafiqul Zaman Khan, "Distributed Computing: An Overview", *Int. J. Adv. Netw. Appl.*, Volume: 07, Issue: 01, 2015, pp. 2630-2635 ISSN: 0975-0290.
18. Zijue Chen,Semantic segmentation for partially occluded apple trees based on deep learning, *Comput. Electron. Agric.*, Volume: 181, February 2021, pp.105952.
19. Dan Wei *et al.*, "Effects of intelligent feeding method on the growth, immunity and stress of juvenile *Micropterus salmonids*", *Artif. Intell. Agric.* Volume: 05, 2021, pp. 118–124 .
20. Ma, X., Jiang, L., Huang, H. *et al.* "Imbalanced gradients: a subtle cause of overestimated adversarial robustness", *Mach. Learn.*, 2023. https://doi.org/10.1007/s10994-023-06328-7.
21. Saini Vandana *et al.*, "Weed identification using machine learning models, i-manager's *J. Pattern Recognit.*, Volume: 9, Issue: 2, January 2022.
22. Da Silva, J.R.M., Damasio, C.V., Sousa, A.M.O., Bugalho, L., Pessanha, L., Quaresma, P, "Agriculture pest and disease risk maps considering MSG satellite data and land surface temperature", *Int. J. Appl. Earth Obs. Geoinf.*, Volume: 38, 2015, pp. 40–50.
23. Lottes, P., Hoeferlin, M., Sander, S., Müter, M., Schulze, P., Stachniss, L.C., 2016. An effective classification system for separating sugar beets and weeds for precision farming applications. In: *2016 IEEE International Conference on Robotics and Automation (ICRA)*, pp. 5157–5163.
24. Lu, H., Fu, X., Liu, C., Li, L., He, Y., Li, N, "Cultivated land information extraction in UAV imagery based on deep convolutional neural network and transfer learning", *J. Mt. Sci.* Volume: 14, 2017, pp. 731–741.
25. Bah, M.D., Dericquebourg, E., Hafiane, A., Canals, R., "Deep learning based classification system for identifying weeds using high-resolution UAV imagery", *International Conference on Science and Information*, pp. 176–187, 2018.

26. Bosilj, P., Duckett, T., Cielniak, G., "Connected attribute morphology for unified vegetation segmentation and classification in precision agriculture", *Comput. Ind.* Volume: 98, 2018, 226–240.
27. Yu, J., Sharpe, S.M., Schumann, A.W., Boyd, N.S., "Deep learning for image-based weed detection in turfgrass", *Eur. J. Agron*, Volume: 104, 2019, pp. 78–84.
28. Ricardo S. Alonso *et al.*, "An intelligent Edge-IoT platform for monitoring livestock and crops in a dairy farming scenario", *Ad. Hoc. Netw.*, Volume 98, March 2020, 102047.
29. A. Villa-Henriksen, G.T. Edwards, L.A. Pesonen, O. Green, C.A.G. Sørensen, "Internet of things in arable farming: implementation, applications, challenges and potential", *Biosyst. Eng.*, Volume: 191, 2020, pp. 60-84.
30. O. Friha, M.A. Ferrag, L. Shu, L.A. Maglaras, X. Wang, "Internet of things for the future of smart agriculture: a comprehensive survey of emerging technologies", *IEEE CAA J. Autom. Sinica*, Volume: 8, 2021, pp. 718-752.

9

Blockchain-Based Smart Agriculture with the Internet of Things: A Revolutionary Approach in Agriculture and Food Supply Chain

Vasanth R. and Pandian A.*

Department of Computing Technologies, School of Computing, College of Engineering and Technology, SRM Institute of Science & Technology, Kattankulathur, Chengalpattu, Tamil Nadu, India

Abstract

Globalization trends have caused the food supply chain to lengthen and become more complex than ever. The food supply chain, however, has a few difficulties, including inefficiencies in the supply chain and issues with food safety, quality, trust, and traceability. These elements harm society and the economy while endangering the health of the public. By making it easier for producers and customers to create confidence, blockchain technology aids in the resolution of many of these problems. Offering detailed product information on the blockchain can greatly improve process transparency. Supply chain traceability and security are enhanced by blockchain technology. An intelligent model protocol for many Internet of Things (IoT) sensors that monitor crop development and the agricultural environment is suggested in this paper, and it also offers a transparent, decentralized blockchain tracking solution. All transactions in decentralized supply networks were managed by smart contracts. The usage of blockchain technology enhances the quality of transactions, and clients can use the system to confirm an item's validity and legality. The proposed model combines AI, IoT, and blockchain to create a smart and futuristic agricultural system that gives farmers access to a rich, new, and effective decision support that can specifically create the scope of greater agricultural process productivity and assist farmers in maximizing their benefits.

**Corresponding author*: pandiana@srmist.edu.in

Vasanth R.: ORCID: https://orcid.org/0000-0002-0295-8015

Rajesh Kumar Dhanaraj, Balamurugan Balusamy, Prithi Samuel, Malathy Sathyamoorthy and Ali Kashif Bashir (eds.) Computer Vision in Smart Agriculture and Crop Management, (187–220) © 2025 Scrivener Publishing LLC

***Keywords*:** Blockchain technology, agricultural applications, food supply chains management, data integrity, traceability

9.1 Introduction

Due to the confluence of cutting-edge technologies, a new era in agriculture has recently begun; this era offers increased efficiency, transparency, and sustainability throughout the food supply chain. Among these revolutionary technologies, blockchain and the Internet of Things (IoT) have emerged as essential components in defining the future of agriculture. This study aims to look at the advantages of smart agriculture based on blockchain and how it may help the food supply chain.

9.1.1 Food Supply Chain in Agriculture's Context

Agriculture, the backbone of our global economy, is essential to preserving food security for the expanding global population. But there are a number of problems that traditional agricultural practices have had to deal with, including a shortage of resources, climate change, food fraud, inefficient supply systems, and concerns about food safety. These concerns have highlighted the need for innovative solutions that may address these issues while ensuring the sustainability and resilience of our food systems.

The rise of precision agriculture, or "smart agriculture," uses cutting-edge technology to improve many areas of farming. Real-time data on the weather, soil quality, crop health, and animal conditions are gathered via IoT devices like sensors and drones under this paradigm shift. This data-driven approach helps farmers increase agricultural output, save resources, and make wise decisions [1].

9.1.2 The Purpose of Blockchain Technology

Originally designed as a distributed ledger system for crypto currencies, blockchain has evolved into a revolutionary tool for many industries, including agriculture. Its main strength is its ability to increase trust, security, and transparency in data transfers. To provide tamper-proof, immutable records of every step of the supply chain, from farm to fork, blockchain technology may be employed in agriculture. This raises the quality of food safety, reduces the likelihood of fraud, and ensures the traceability of food products [2].

9.1.3 The Synergy Between Blockchain and IoT

The true potential of blockchain in agriculture is realized when blockchain and IoT collaborate. By connecting IoT devices with blockchain networks, data generated from farms, warehouses, transportation, and retail can be safely and publicly recorded and shared. This combination enables access to real-time, verifiable information on the location of origin, the quality of the items, and their journey by farmers, distributors, regulators, and consumers.

The value of blockchain-based smart agriculture: In today's networked society, blockchain-powered smart agriculture is crucial. It might change the way that food is supplied in a number of ways, including:

Food traceability: Consumers want to know the exact origins of the food they eat. Blockchain offers a traceable path from farm to table, increasing customer safety and confidence.

Food waste reduction will be aided by improved supply chain logistics, real-time monitoring, and data analytics made possible by blockchain and IoT devices.

Efficiency of the Supply Chain: Automation and data-driven decision-making improve the efficiency of the supply chain, reducing costs and generally increasing production.

Sustainability: Blockchain-based smart agriculture promotes environmentally friendly farming practices by reducing resource waste and enhancing farming techniques.

As the world battles to feed a rising population, combat the consequences of climate change, and ensure the integrity of our food supply, blockchain-based smart agriculture is emerging as a ray of hope. By examining the procedures, case studies, challenges, and potential benefits of this unique technique in further detail, this research study hopes to shed light on its transformative potential for agriculture's future and the world's food supply chain [3].

9.1.4 An Overview of IoT and Its Role in Agriculture

The IoT is a ground-breaking technological concept that is quickly catching on in a number of industries, including agriculture. The IoT is essentially a network of physically networked objects, sensors, and devices that have connections, software, and sensors to automatically collect, exchange, and act on data. Agriculture has been significantly impacted by IoT, which has revolutionized farming practices and enhanced resource management. An overview of the IoT and how it relates to agriculture is provided below:

9.1.4.1 *Sensing and Data Collection*

IoT devices used in agriculture include a wide range of sensors and instruments, such as weather stations, soil moisture monitors, drones, and GPS trackers. These sensors collect data in real-time on a range of crucial agricultural factors, including weather patterns, soil moisture content, temperature, humidity, and crop health [4].

9.1.4.2 *Data Sensing and Gathering*

Many sensors and pieces of equipment are used in agriculture, including weather stations, soil moisture monitors, drones, and GPS trackers. Real-time data is collected by these sensors on a variety of crucial agricultural factors, including weather patterns, soil moisture content, temperature, humidity, and crop health. Thanks to the data produced by IoT devices, farmers are now able to make educated decisions regarding the state of their crops [5].

9.1.4.3 *Automation and Control*

Numerous agricultural processes can be automated thanks to the IoT. For instance, irrigation systems may be automated based on data regarding soil moisture to make sure that crops are appropriately hydrated. It is feasible to remotely control machinery and equipment like tractors and harvesters with IoT devices and applications.

9.1.4.4 *Animal Observation*

In addition to crop farming, the IoT is used in animal husbandry thanks to tools like wearable sensors and smart tags. By keeping an eye on factors like body temperature, feeding patterns, and location, these devices maintain tabs on the health and behavior of cattle. With the use of livestock monitoring, farmers can optimize feeding regimens, detect illnesses early, and improve breeding practices.

9.1.4.5 *Supply Chain Management*

The IoT helps with supply chain management in agriculture. Sensors are used to track agricultural products and keep an eye on their location and condition throughout storage and transportation. Produce freshness and quality are ensured, spoiling is reduced, and traceability is enhanced.

9.1.4.6 *Environmental Monitoring*

The IoT promotes environmental sustainability by monitoring factors such as water usage, air quality, and soil quality. The data acquired is used to assist environmental standards, sustainable land management, and conservation projects.

9.1.4.7 *Technologies for Supporting Decisions*

IoT device data is handled and analysed using sophisticated analytics and machine learning techniques. Through simplified dashboards and mobile applications, farmers have access to this data, which provides useful insights for improving farming operations [6].

9.1.4.8 *Scalability and Honesty*

Farmers may start with a few sensors and gradually expand their IoT infrastructure thanks to the tremendous scalability of IoT solutions for agriculture. The IoT systems may potentially be linked with other technologies, like blockchain, to enhance traceability and data security.

In conclusion, the IoT, a game-changer in the agriculture business, enables data-driven, efficient, and sustainable farming practices. It enhances resource management, provides farmers with real-time information, and increases the overall productivity and resilience of agricultural systems. Future advancements in intelligent and sustainable agriculture have even greater promise when combined with IoT integration and other cutting-edge technologies like blockchain.

The following are the primary research aims of the study "Blockchain-based Smart Agriculture with the Internet of Things: A Revolutionary Approach in Agriculture and Food Supply Chain": Analyze the current situation in agriculture; analyze the risks, inefficiencies, and issues that the supply chain is now experiencing; analyze the use of blockchain technology in agriculture; learn about the fundamental concepts behind blockchain technology and how it may be applied to agriculture to enhance traceability, transparency, and security; look at the application of the IoT in agriculture; consider the IoT and its various applications in monitoring, data collection, automation, and decision support systems; examine how blockchain and IoT technology may be used to build a smart agriculture ecosystem that overcomes the difficulties and inefficiencies found; case studies and examples are provided; present case studies and examples of successful blockchain-based smart agricultural implementations from real-world settings; paying particular attention to the effects on the food supply chain; determine obstacles and future directions; talk about the difficulties and barriers to IoT and blockchain adoption in agriculture and provide possible solutions; and highlight current trends and potential future paths for this field of study [7].

9.2 Literature Review

9.2.1 Blockchain Technology and Its Applications in Agriculture

9.2.1.1 Enhancing Traceability and Transparency

Numerous studies have stressed the potential of blockchain to increase traceability and transparency in the agricultural supply chain. Research has shown that blockchain can provide customers with a transparent record of a product's journey from the farm to the consumer in order to guarantee food safety and authenticity. When there are foodborne outbreaks, it might be particularly crucial to swiftly identify the cause.

9.2.1.2 Supply Chain Efficiency

Blockchain has won accolades for its ability to improve the effectiveness of agricultural supply systems. Research suggests that it might reduce errors, simplify logistics, and do away with paperwork—all of which could result in cost savings. It has been investigated how well blockchain can be used for tasks including inventory management, transportation, and agricultural harvests. By automating and digitizing these operations, agricultural enterprises may improve their overall performance.

9.2.1.3 Reduction of Food Fraud

Food fraud and counterfeit commodities are top concerns in agriculture. Blockchain, according to studies, can help with these issues by providing a tamper-proof record of a product's journey. Customers may, for example, scan a QR code on a food item to access a blockchain-based database that verifies its authenticity and reduces the likelihood of purchasing phony goods.

9.2.1.4 *Data Sharing and Cooperation*

Cooperation between several entities is essential in the agriculture business. Thanks to blockchain technology, data can be shared between farmers, suppliers, distributors, and authorities in a more secure and efficient manner. Studies have looked at the possibility for blockchain-based networks that would let stakeholders share real-time information on crop status, weather forecasts, and market prices. The interchange of real-time data can lead to more effective decision-making [8, 9].

9.2.1.5 *Sustainability and Certification*

The verification and preservation of sustainability certificates, such as organic or fair-trade labels, is made possible by blockchain technology. Customers can more easily make informed judgements as a result of the increased trustworthiness. Research examines how blockchain may support sustainable agriculture practices by recording data on the condition of the soil, the usage of water, and the use of pesticides. Using this information, sustainable farming practises may be confirmed and approved.

9.2.1.6 *Adoption Problems and Roadblocks*

Even while blockchain has a lot of potential, there are still a lot of barriers preventing its widespread use in the agriculture industry. Data privacy, scalability, interoperability with existing systems, and the cost of implementing blockchain solutions are all discussed. Researchers underline the need to tackle these difficulties in order to fully grasp the potential of blockchain in agriculture.

9.2.1.7 *Future Perspectives*

Many academics stress the importance of conducting further study and pilot projects to fully comprehend the agricultural blockchain's potential applications. A future study will focus on how to use blockchain to improve agricultural solutions by combining it with other cutting-edge technologies like the IoT, artificial intelligence (AI), and machine learning. Additionally, academics investigate how blockchain may be used to address universal issues like food security and climate change adaptation.

In conclusion, the research that is now available on blockchain technology in agriculture shows how this field may be transformed by solving issues with traceability, transparency, supply chain effectiveness, and sustainability. While several pilot projects and case studies show that blockchain applications are feasible, continuing research is still looking for ways to get beyond acceptance obstacles [10, 11].

9.2.2 Exploration on IoT in Agriculture and Its Benefits

Precision agriculture: Precision farming is made possible by the IoT, which offers real-time information on crop health, weather, and soil conditions. Precision irrigation, fertilization, and pesticide application allow farmers to increase crop yields while using less resource.

Intelligent Irrigation: To optimize water use, IoT-based smart irrigation systems monitor soil moisture levels and weather forecasts. By doing this, water waste is decreased, irrigation expenses are decreased, and the proper quantity of moisture is provided to plants.

Crop Monitoring: By early illness, nutrient deficiency, or pest infestation detection, IoT sensors and drones can monitor crop health. As a result, early actions may be taken to prevent crop losses and the overuse of pesticides.

Management of livestock: IoT tools like smart collars or tags can monitor the condition and whereabouts of cattle. When an animal gets ill or leaves its authorized area, farmers are notified, allowing for speedier reactions and better animal welfare.

Supply Chain Optimization: By monitoring the movement and state of agricultural products from the farm to customers, IoT helps the supply chain function better. Transparency prevents fraud, ensures food safety, and lowers food spoilage. The IoT weather stations capture real-time data, which helps with more precise weather predictions. By using forecasted weather conditions as a guide, farmers can safeguard crops from bad weather events.

Enhanced Decision-Making: The IoT generates large amounts of data that may be evaluated to learn more. Making data-driven decisions about planting dates, crop rotations, and resource allocation may help farmers improve farm operations [12].

Environmental Sustainability: The IoT encourages sustainable agriculture by lowering resource waste. Precision farming methods, less pesticide usage, and effective water and fertilizer use are all part of a more ecologically friendly approach.

The IoT has made it possible for farmers to remotely monitor and control their equipment, doing away with the requirement for continual on-site presence. This saves time and resources while enhancing comfort and security.

Greater Productivity and Profitability: By maximizing agricultural practices, reducing losses, and increasing yields, the IoT ultimately leads to better productivity and profitability for farmers.

Data-Driven Research: By using the data collected by IoT devices, academics may evaluate trends, develop fresh ideas, and expand their understanding of agriculture.

Risk mitigation: The IoT helps farmers immediately detect and mitigate risks. For example, early insect detection helps prevent big infestations, reducing the overall risk to crops.

The IoT solutions are very scalable and versatile since they can be customized to the unique requirements of farms of any size and adapted to different crops and climatic circumstances. The IoT in agriculture generally provides several advantages, including improved decision-making, resource optimization, enhanced production, and sustainability. The agriculture sector can meet the difficulties of feeding a growing global population while reducing environmental consequences if it continues to adopt IoT technology [13].

9.2.3 Challenges in the Agriculture and Food Supply Chain

Blockchain and IoT technologies offer innovative solutions to address challenges in the agriculture and food supply chain, including traceability, transparency, resource management, data security, supply chain efficiency, and sustainability listed in Table 9.1. By combining these technologies, the agriculture industry can create a more resilient, efficient, and transparent supply chain that benefits both producers and consumers.

Table 9.1 Challenges in the agriculture and food supply chain.

S. no.	Challenges in the agriculture and food supply chain	Challenge	Blockchain solution	IoT solution
1	Traceability and Transparency:	Due to a lack of transparency and traceability in the supply chain, it may be difficult to identify the source of food contamination during an outbreak.	Blockchain offers complete product traceability for food goods by offering an immutable record of every transaction. From the farm to the client, every stage of the supply chain is watched over to ensure accountability and transparency.	The IoT sensors can monitor the weather to ensure that temperature, humidity, and other factors are optimal for ensuring the safety of food. The IoT device data may be linked together via the blockchain, allowing for real-time monitoring and transit condition verification.
2	Food Fraud and Counterfeiting	Food fraud is a significant issue since phony commodities routinely enter the supply chain.	The trustworthiness of products is protected by block chain's tamper-proof feature. Customers may examine a product's blockchain history by scanning a QR code on it to verify its origins and legitimacy.	The IoT devices may monitor products for any unusual situations or deviations from the norm, warning to potential product tampering or phony goods.

(*Continued*)

Table 9.1 Challenges in the agriculture and food supply chain. (*Continued*)

S. no.	Challenges in the agriculture and food supply chain	Challenge	Blockchain solution	IoT solution
3	Resource Management	Inefficient resource management may squander resources like water and fertilizers, which can also be harmful to the ecology.	Blockchain enables efficient resource allocation by monitoring resource usage at every stage of manufacturing. By autonomously beginning processes, like irrigation, based on real-time data, smart contracts may optimize resource utilization.	The IoT sensors in the field can monitor the soil's moisture content, allowing for precise irrigation. Using this information, one may make informed decisions about water usage and waste reduction.
4	Data Security and Privacy	In the digital age, privacy problems and data breaches are frequent, particularly with highly sensitive consumer and agricultural data.	Block chain's better data security is enabled through encryption and decentralized storage. It guarantees that confidential information is kept intact and is only accessible to those with authorization.	The IoT data may be encrypted and delivered securely to the blockchain in order to guarantee the security and integrity of the information acquired.
5	Supply Chain Inefficiencies	Inefficient supply chains can lead to delays, food spoilage, and increased costs.	Blockchain streamlines supply chain operations by providing real-time data visibility to all stakeholders. As a result, less paper is produced, and the supply chain as a whole is more effective.	The IoT devices like GPS trackers and temperature sensors enable real-time tracking of items in transit. This information may be included on the blockchain, allowing for preemptive action in the event of delays or adverse conditions.

(*Continued*)

Table 9.1 Challenges in the agriculture and food supply chain. (*Continued*)

S. no.	Challenges in the agriculture and food supply chain	Challenge	Blockchain solution	IoT solution
6	Sustainability and Environmental Impact	By using excessive amounts of pesticides, destroying forests, and contaminating water supplies, unsustainable farming practices can harm the ecosystem.	Blockchain helps buyers to make informed judgements about environmentally friendly products by tracking sustainability practices and certifications.	The IoT can monitor environmental conditions and provide information for sustainable resource and land management, aiding farmers in adopting green practises.

9.3 Methodology

9.3.1 Data Collections

Blockchain-based smart agriculture depends heavily on the IoT, which serves as the basis for automation, decision-making, and process optimization. The IoT devices furnished with a range of sensors are first strategically positioned across agricultural regions in order to capture data from numerous parts of the farm. This category of sensors also includes cameras, GPS tracking, humidity, soil moisture, and temperature. These IoT sensors continually monitor the agricultural environment, collecting real-time information on weather patterns, soil quality, crop health, and equipment performance before transmitting it to a central system for analysis. There are many different types of information collected, including information about the environment (such as temperature, humidity, and rainfall), measurements of the soil (such as moisture content and pH levels), information about crops (such as growth stage and yield estimates), information about livestock (such as location and health information), and information about machinery (such as equipment status and upkeep requirements). Many different communication channels are used for data delivery, including Wi-Fi, cellular networks, LPWAN, and satellite connections. Using a standard platform, which might be Cloud-based or on-premises, the data gathered from various IoT devices is aggregated. A crucial step in maintaining transparency and accountability is the recording of certain data points onto the blockchain, which are generally associated with crucial operations like supply chain monitoring and quality control. These data are safeguarded using encryption and access controls. Data quality is maintained at a high level by frequent sensor calibration and validation testing. The fact that storage and archiving solutions are selected in accordance with data volume and regulatory restrictions enables long-term analysis. Data analytics, incorporating machine learning algorithms, enables anomaly identification, resource optimization, and crop production estimates. Real-time data may help farmers decide when to apply fertilizer and irrigation in the best possible way. Efforts are being made to create user-friendly dashboards and visualization tools that will enhance data comprehension through charts, maps, and graphs. Particularly while tracking cattle, respecting personal data privacy necessitates abiding by data privacy regulations and obtaining consent [14].

9.3.2 Data Analysis

For blockchain-based smart agriculture employing the IoT, data analysis, which involves the processing and interpretation of enormous datasets received from IoT sensors and, in certain circumstances, data preserved on the blockchain, is crucial. The initial phase in this process is data preparation, which entails cleaning, structuring, and modification of the data for analysis to tackle issues like missing values and outliers. Using visual exploration techniques like scatter plots and heatmaps, exploratory data analysis (EDA), the following phase enables the identification of patterns and correlations. Using descriptive statistics, which provide information on central tendency and variability, the characteristics of the data are then summarized. Moving averages and other time-series analysis techniques are utilized to spot temporal patterns since agricultural data are time-series data. Predictive modeling, which typically makes use of machine learning methods like regression and decision trees, is used to forecast agricultural yields, disease outbreaks, and the optimal periods to produce crops. Anomaly detection techniques, such isolation forests, assist in locating unusual data points suggestive of crop diseases or equipment issues. In order to maximize resource allocation, decrease waste, and increase yields, statistics on water consumption, fertilization, and pesticide application are also evaluated. Transparency and traceability are guaranteed when the supply chain is monitored using blockchain analysis. Risk assessment works to lessen the possibility of unfavorable weather, pest, disease, or market circumstances through data analysis. By providing them with timely guidance based on recent environmental and agricultural data, decision support systems assist farmers in making decisions like irrigation and harvesting timing. Complicated data insights are made simpler for farmers and decision-makers by user-friendly data visualization through interactive dashboards. Continuous monitoring systems keep analysis results up to date and provide judgements based on the most recent information. The results of data analysis help guide activities in research and development for smart agriculture, promoting innovative solutions and improved practices. In general, effective data analysis supports efficient agriculture in line with overarching goals, equips farmers with useful information, and boosts output and sustainability [15].

9.3.3 Experiments and Case Studies

Research on blockchain-based smart agriculture utilizing the IoT usually combines experiments and case studies to evaluate the practical applications and benefits of this technology in actual agricultural settings. Here are a few examples of tests and case studies that have been conducted in this area:

9.3.3.1 Monitoring the Supply Chain Trial

Researchers created a supply chain tracking system using blockchain and IoT devices. It tracked how agricultural products got from the farm to the consumer. This experiment was created to demonstrate how blockchain ensures supply chain traceability and transparency, allowing buyers to verify the authenticity and quality of goods.

9.3.3.2 Case Research on Crop Monitoring

A case study was conducted to deploy IoT sensors to monitor crop health and growth on a large-scale farm. We obtained data on the state of the crops, the temperature of the soil, and the moisture content. The study investigated how real-time data and blockchain technology may be used to improve irrigation plans and detect crop ailments early in order to increase yields [16].

9.3.3.3 The Experiment on Intelligent Irrigation

In this study, the IoT sensors were installed in fields to collect data on the quantity of soil moisture and the weather. The block chain's smart contracts were used to automate the decision-making process for watering. The research assessed how effectively this method conserved water and maintained crop health.

9.3.3.4 Livestock Tracking Case Study

Researchers conducted a case study on a cow farm where IoT sensors were put on animals to track their whereabouts and health. The data was stored on a blockchain to ensure data integrity. The study looked at how blockchain and IoT may improve livestock management, reduce theft, and improve animal welfare.

9.3.3.5 Quality Control Test

An experiment was designed to monitor the condition of perishable goods while they were being transported. The IoT sensors tracked temperature and humidity levels, and this data was recorded on a blockchain. The research examined how immutable data on the blockchain will make it simpler to identify quality issues and stop spoilage.

9.3.3.6 A Decentralized Market Case Study

A case study looked at how blockchain technology may be used to create a decentralized farm market. Farmers may sell their products directly to consumers and other stakeholders via smart contracts. The study examined the benefits of eliminating intermediaries and raising price transparency.

9.3.3.7 The Carbon Footprint-Reducing Test

Scientists conducted an experiment on a sustainable farm where IoT sensors were used to monitor pollution, water consumption, and power use. Data stored on a blockchain was used to calculate and minimize the farm's carbon impact. The goal of the study was to demonstrate how blockchain technology may promote ecologically responsible conduct [17].

9.3.3.8 Case Study for Prediction of Crop Yield

For a case study, it was necessary to compile historical data on crop growth, weather trends, and soil quality. Machine learning techniques were used to estimate crop yield. The research assessed the precision of these predictions as well as their usefulness to farmers' crop planning.

9.3.3.9 *Farmer Cooperative Experiment*

In an effort to facilitate the exchange of tools, information, and resources among farmer cooperatives, a blockchain-based platform has been created. The study examined the potential benefits of such a platform for small-scale farmers, including increased resource access, improved teamwork, and cost savings.

9.3.3.10 *Case Study on Pest and Disease Monitoring*

To monitor pest and disease infestations in crops, researchers used IoT cameras and sensors. Data was stored on a blockchain, enabling early discovery and focused action. The case study evaluated the impact on crop health and reduced pesticide use.

The numerous applications of blockchain and IoT in smart agriculture are highlighted in these experiments and case studies, as well as how these developments may improve the efficiency, sustainability, and transparency of agricultural systems. They also provide beneficial vantage points for future agricultural research and practical uses.

9.3.4 Selection Criteria for Blockchain and IoT Technologies in Agriculture

The finest blockchain and IoT technologies for agriculture must be chosen carefully after carefully weighing a number of crucial factors. Scalability is a crucial selection factor that ensures the chosen solutions can effectively handle the massive scale of agricultural operations without performance degrading. This fits well with the aims and purposes of agricultural activities. Effective data transfer, the eradication of data silos, and seamless interaction with existing systems and devices all depend on interoperability. Security, which includes robust features like encryption and access restrictions, is of the highest significance when managing sensitive agricultural data. Given that agricultural activities are continuous, reliability and uptime are essential for avoiding significant losses brought on by downtime. Low-power consumption IoT devices offer longer battery lifetimes and require less upkeep [18].

Farmers need access to their data, and network connectivity needs to be adaptable enough to work in remote areas with little bandwidth or no internet. The advantages and disadvantages of different technologies should be compared, and contemporary data analytics tools enable data-driven decision-making and operational optimization. The chosen technology must be flexible enough to be used for a range of agricultural use cases and farming operations. Respecting relevant rules pertaining to data privacy, food safety, and environmental sustainability is crucial.

User-friendliness, appropriate training, and support resources are crucial for users with varying degrees of technical competency. It is crucial to have continual assistance and updates from reliable providers or companies in order to manage shifting needs and security risks. Sustainable agricultural practices are congruent with considering the consequences of the environment, such as energy utilization. Proof of concept (PoC) or pilot programs enable the examination of the viability, efficacy, and impact of these technologies on specific agricultural operations prior to their widespread adoption. Agricultural stakeholders may improve agricultural practices' efficiency, sustainability, and transparency by carefully weighing these factors and coming to informed decisions that are tailored to the demands of the industry [19].

9.3.5 Data Sources and Tools

Blockchain-based smart agriculture leveraging the IoT employs a variety of data sources and technologies to collect and analyze agricultural data. These data sources and technologies are essential for enhancing farming operations' efficiency, transparency, and decision-making. Here is a detailed breakdown:

9.3.5.1 Data Sources

The IoT sensors are crucial in modern agriculture since they gather information about many aspects of farming activities in real-time. These sensors can be divided into several groups. For instance, environmental sensors track parameters like temperature, humidity, air pressure, wind speed, and rainfall to provide critical weather information. Using soil sensors, such as moisture, pH, and nutrient content sensors, farmers can keep an eye on the condition of their soils. Crop sensors are still another essential component since they monitor crop development, health, and production forecasts, as well as look for potential issues like pests and diseases. Livestock sensors are used to keep an eye on the location and wellbeing of animals.

High-resolution photography of fields is made possible using satellite and drone technologies in agriculture, allowing for detailed monitoring, disease identification, and yield calculation. By keeping track of information regarding financial transactions, quality control, and supply chain management, blockchain technology provides immutability, transparency, and traceability. It is necessary to have access to weather information from meteorological services in order to make weather-aware decisions about when to plant, water, and harvest. Market data, which contains specifics on prices, demand, and trends, has a significant impact on the selection of crops and pricing strategies. Last but not least, historical data, such as records of crop performance, weather patterns, and soil conditions, serves as the foundation for predictive modelling and long-term planning in current agriculture. Together, these components form a large network of data-driven technologies that provide farmers the chance to boost sustainability and output [20].

9.3.5.2 Tools and Technologies

On top of blockchain technologies like Hyperledger Fabric, Corda, and Ethereum, distributed ledgers in agriculture are constructed. These technologies offer transparency and immutability for supply chain tracking and data integrity and utilize systems like AWS IoT, Azure IoT, and Google Cloud IoT Core to manage and analyze the data generated by IoT sensors. They offer resources for controlling hardware, absorbing data, and doing in-the-moment analytics. Machine learning libraries like TensorFlow and scikit-learn enable predictive modelling while data analytics tools and frameworks like Python, R, and Jupyter notebooks are used to handle and analyze agricultural data. To make data easier for farmers and decision-makers to grasp, create dynamic dashboards and visualizations using data visualization platforms like Tableau, Power BI, and D3.js. The use of GIS software, such as ArcGIS, to combine geographic data with agricultural data enables precision farming. Platforms for smart contracts, such as Solidity (for Ethereum), make it feasible to automate and enforce agreements relevant to agricultural activities, such as payment upon delivery. Blockchain wallets are used to manage crypto currency transactions on blockchain networks in terms of financial transactions in agriculture. By processing data locally on IoT sensors or gateways utilizing edge computing devices and platforms, one may decrease latency and improve real-time decision-making. Security measures including encryption, access controls, and cybersecurity processes are essential to protect IoT data and blockchain networks from threats.

Making sure that agricultural operations adhere to all applicable laws and regulations, particularly those relating to data security and food safety, is made possible through compliance and regulation tools. Thanks to remote monitoring and control solutions, farmers may remotely manage IoT equipment and devices, increasing operational effectiveness. Mobile applications allow farmers to access data in real-time and make decisions while in the field. By combining various data sources and technologies, blockchain-based smart agriculture and IoT create an integrated ecosystem that optimizes agricultural operations, enhances traceability, and boosts overall sustainability and productivity in farming [21].

9.4 Blockchain Technology in Agriculture

Despite being a decentralized and distributed ledger system, the blockchain technology that underpins cryptocurrencies like Bitcoin also has applications well beyond digital currency. It provides a secure and uncrackable way to confirm and record transactions or any kind of data. One of the guiding concepts of blockchain technology is decentralization, which ensures that no single entity has total power by distributing a copy of the whole blockchain among a network of computers. In order to preserve the integrity of the ledger, this digital ledger records transactions in immutable blocks, each of which includes a cryptographic hash of the block before it.

A new block must be agreed upon by all network users in order to be added to the blockchain. Several techniques, such as Proof of Work (PoW) and Proof of Stake (PoS), may be used to achieve this; PoW relies on miners completing difficult puzzles to create a block, whereas PoS relies on validators holding and staking bitcoin to add new blocks. The immutability of blockchain technology, which makes it nearly impossible to edit or remove information after it has been entered, is a fundamental feature. The security of the blockchain is strong against fraud, hacking, and data manipulation since altering one block implies changing all future blocks on every copy of the blockchain.

Because everyone can view the contents of a blockchain, which is famed for its openness, it is simpler to verify and account for transactions. Smart contracts, which are computer programs that automatically and directly enforce contractual requirements, are another feature that many blockchains provide. The versatility of blockchain extends to a multitude of applications, boosting efficiency and transparency across industries. These applications include real estate transactions, voting systems, healthcare records, identity verification, and supply chain management.

Scalability, energy consumption (in PoW systems), and regulatory difficulties are still problems for blockchain. Additionally, it distinguishes between public blockchains, which are open to all users and maintained by a decentralized network (like Bitcoin and Ethereum), and private blockchains, which are restricted to a small number of participants and are typically utilized for commercial purposes. In conclusion, the potential for blockchain technology to upend industries through trustworthy and transparent record-keeping continues to grow as new use cases are researched and implemented, impacting the future of numerous sectors [22].

9.4.1 Several Ways of Blockchain Technology for Agriculture

Blockchain technology has the potential to transform the agriculture industry, particularly in the areas of traceability, transparency, and security. There are several ways blockchain technology may be applied to agriculture:

9.4.1.1 Traceability of the Supply Chain

Blockchain enables full product tracking from the farm to the consumer's plate. Every stage of the production, processing, packaging, and distribution of agricultural products may be precisely recorded on the blockchain, resulting in an immutable and transparent trail that traces the product's entire journey. Nowadays, it is easy to verify provenance, making it possible for customers and enterprises to do so. This lessens the possibility of fraud and false labelling while preserving the integrity of locally produced, fairly traded, or organic goods. To help clients make informed decisions and improve quality control practices along the supply chain, blockchain also contains data on environmental conditions, quality ratings, and certifications.

9.4.1.2 Responsibility and Openness

The monitoring of fair-trade practices is made possible by blockchain, guaranteeing that farmers and laborers are paid fairly and have respectable working conditions. This recently revealed openness encourages moral behavior and the practice of religion. Additionally, by offering a reliable record of compliance with agricultural regulations and safety criteria, blockchain aids farmers and distributors in avoiding costly fines and recalls. This transparency also facilitates quicker responses to food safety issues.

9.4.1.3 Security and Confidence

The blockchain's immutable ledger guarantees that data is safe and irreversible, reducing the risks associated with fraud or tampering with records.

Smart contracts further increase efficiency by automating and enforcing agreements along the whole agricultural supply chain. For instance, payments to farmers may be automatically provided when specific quality standards are reached, doing away with the need for middlemen and reducing the possibility of disputes.

9.4.1.4 *Efficiency and Reduced Waste*

Blockchain enhances inventory management by giving real-time visibility into the availability of agricultural products, which can ultimately reduce waste brought on by overproduction or spoilage. Administrative processes like paperwork, inspections, and customs clearance may be sped up because to blockchain's automation and transparency, cutting costs and delays.

9.4.1.5 *Data Sharing and Cooperation*

A few of the many participants in the agricultural supply chain that can work together and share data safely and successfully with blockchain technology are farmers, distributors, retailers, and regulators. This improves decision-making and industry-wide collaboration.

9.4.1.6 *Consumer Involvement*

Customers may obtain comprehensive information about the food they purchase by using their cellphones to scan QR codes on items. By giving them knowledge about the food's origin, production methods, and safety, this helps them make better judgements.

9.4.1.7 *Reduction in Food Fraud*

Blockchain functions as a strong deterrent to the introduction of counterfeit products into the market because it makes it simple for buyers to easily validate the authenticity of the goods they buy.

9.4.1.8 *Catastrophe or Pollution*

Blockchain enables targeted recalls and minimizes consumer harm by swiftly locating the origin of contamination outbreaks or recalls connected to food safety.

9.4.1.9 *Sustainable Environmental Practices*

Blockchain technology may be used to track the carbon footprint of agricultural goods, allowing customers to choose more environmentally friendly choices, and promoting green business practices.

In conclusion, by enhancing traceability, transparency, and security throughout the supply chain, blockchain technology has the potential to revolutionize the agriculture business. Blockchain fosters confidence among stakeholders, reduces inefficiencies, and guarantees the safety and quality of agricultural goods thanks to its trustworthy and tamper-resistant ledger. A more responsible, environmentally friendly, and consumer-friendly agriculture sector is what this breakthrough promises [23].

9.4.2 Case Studies Blockchain Implementation in Agriculture

Numerous case studies and examples explain the genuine advantages of blockchain technology in the agriculture industry and demonstrate how it has been implemented successfully in this industry. Here are a few notable examples:

9.4.2.1 *IBM Food Trust*

The blockchain-based IBM Food Trust technology aims to improve the traceability and transparency of the food supply chain. This platform has been utilized by significant food and retail corporations including Walmart, Nestlé, and Unilever. For instance, Walmart used IBM Food Trust to quickly determine the country of origin of mangoes, as opposed to the days it would have taken to do so using conventional techniques. Here is an example of how blockchain technology may speed up and improve food safety and traceability. Traceability Enhanced (TE)-Food is a blockchain-based solution for agricultural supply chain food safety and traceability. It has been utilized in a number of countries, including Vietnam and Hungary, to track food products from farm to table. TE-Food successfully established the Vietnamese origin of pig products, reducing the danger of disease outbreaks and raising the bar for food safety.

AgriDigital is an Australian-based blockchain platform that offers services to the agriculture and commodities industries. Farmers, traders, and buyers are now able to follow the flow of grain from the field to storage and beyond thanks to the simplification and digitization of grain trading procedures. Less paperwork, greater openness, and speedier payouts are the results. Previously known as Chronicled, Viant is an industry leader in supply chain management and product tracing. It collaborated with American agricultural cooperative Organic Valley to keep tabs on the distribution and manufacturing of organic milk. This raised consumer trust and boosted sales by enabling customers to confirm the veracity of organic claims.

Provenance: Provenance emphasizes supply chain sustainability and transparency. To follow the path of fresh produce, it worked with the British supermarket Sainsbury's. In order to promote transparency and ethical consumerism, consumers could use their cellphones to get detailed information about the things they purchased, including origin, quality, and environmental effect. Bext360, a blockchain and AI platform, can trace coffee beans from the farm to the consumer. It worked with coffee cooperatives in regions like Africa and South America to ensure fair payment for producers and openness for buyers. By tracking every step of the supply chain, from coffee bean harvest to roasting, Bext360 increases transparency and fairness in the coffee industry. These case studies demonstrate how blockchain technology can be applied to agriculture to increase traceability, transparency, and security. These implementations benefit a wide range of stakeholders, including farmers, consumers, and businesses along the supply chain. They might increase food safety, reduce fraud, promote moral behavior, and increase consumer confidence in agricultural products. Blockchain is proving to be a disruptive force [24].

9.5 Internet of Things in Agriculture

The IoT applications have significantly changed the landscape of smart agriculture over the past few years. These ground-breaking ideas are helping to improve crop management, farming productivity, and overall agricultural sustainability. In this section, we examine a few illustrative IoT applications in smart agriculture. Precision farming is one such field where IoT is having a huge impact. The IoT sensors are essential in precision farming, especially when determining the amount of soil moisture present. Farmers can successfully adjust their watering schedules thanks to these sensors. Utilizing this information, cutting-edge systems like the John Deere Field Connect system decide when and where to irrigate. Additionally, IoT-enabled weather monitoring stations provide real-time meteorological data, which is important for farmers to plan crucial tasks like planting, harvesting, and pest control. This reliance on precise weather predictions boosts total agriculture productivity.

Furthermore, crop health monitoring has found the use for drones and IoT-enabled cameras. They take pictures of crops to help farmers spot diseases, pest infestations, or nutrient deficits early on. These innovations are essential tools for maintaining crop health and increasing yields. Leading companies like Arable are experts at providing specialized IoT devices made just for tracking and enhancing crop health in precision farming. An additional application for IoT is in livestock monitoring. Livestock can be constantly monitored for health, behavior, and location using sensors attached to them. To monitor cattle and ensure early disease detection, IoT-enabled ear tags are one example. Additionally, IoT-enabled automated feeders and water dispensers ensure accurate and effective hydration and feeding of animals, minimizing waste, and increasing overall effectiveness. The IoT in agriculture also includes smart irrigation systems. The IoT-based drip irrigation systems can dynamically adjust water flow based on real-time soil moisture levels and weather conditions, thereby optimizing water usage and minimizing wastage. Similar to smart irrigation systems, which can be operated from a distance and are responsive to current weather information, smart sprinkler systems help conserve water. These examples demonstrate the versatility of IoT applications in agriculture, allowing data-driven decisions, resource conservation, yield improvement, and general sustainability of farming practices. IoT technology is still essential for transforming conventional agriculture into effective smart agriculture that is also ecologically benign [25].

9.5.1 Examples of IoT Applications in Smart Agriculture

Thanks to IoT technology, the agricultural industry has seen a significant revolution. By offering real-time data and automation capabilities, it has opened prospects for raising agricultural yields, decreasing resource wastage, and enhancing overall farm management. The topic of smart agriculture might benefit from a variety of IoT applications.

One of the key applications for IoT sensors in this industry is precision farming. Farmers may adjust irrigation and fertilization procedures with the aid of these sensors, ultimately utilizing less water and chemicals. Important properties like as soil moisture, temperature, and nutrient levels are assessed by these sensors. In addition, weather stations connected to the IoT provide real-time meteorological data that helps farmers plan their planting and crop management operations.

Smart irrigation is yet another crucial use. With IoT-enabled irrigation systems that automatically change water distribution based on real-time data, water waste may be reduced, and crop health can be improved.

To further improve water use, field-based soil moisture monitors that turn on irrigation when necessary.

In the discipline of supply chain management, IoT sensors are utilized for cold chain monitoring. These sensors keep an eye on the temperature and humidity of perishable goods while they are in transportation and storage, ensuring product quality and minimizing spoiling. By enabling farmers and distributors to monitor stock levels, storage conditions, and automate replenishment processes, IoT technologies also make inventory management simpler.

The IoT also helps to enhance pricing and market data. The IoT-enabled real-time market data aids farmers in choosing when and where to sell their products for the maximum prices. The IoT data is integrated into farm management software to give farmers a comprehensive dashboard and analytics platform. This connection allows them to examine and assess data, which enables them to draw more defensible and fact-based decisions. To sum up, the multiple IoT applications for smart agriculture are crucial for supporting farmers in making decisions regarding their farming operations that are sustainable, economical, and yield-improving [26].

9.6 Integration of Blockchain and IoT in Agriculture

The application of blockchain and IoT technologies in agriculture may result in a disruptive smart agricultural ecosystem with better efficiency, transparency, and sustainability along the whole agricultural value chain. The following is a thorough description of how these two technologies may be used to develop such a system: This smart agriculture ecosystem is centered on IoT devices that have been strategically placed on farms. These sensors may gather a wide range of data, including essential indicators like soil moisture content, temperature, and humidity, as well as information on crop growth, livestock health, and machinery operation. This real-time data collection is the cornerstone of data-driven decision-making in agriculture.

The automation of this system depends on smart contracts, which are self-executing contracts with predetermined rules and circumstances. A smart contract may, for instance, be used to automate an irrigation system to activate automatically when soil moisture levels drop below a specified level. Similar to this, a smart contract can begin delivering signals during goods delivery if the consignment's temperature exceeds the permitted range. Additionally, smart contracts may automate payment processes, guaranteeing that farmers are paid immediately after successfully delivering their harvest. The blockchain ledger contains the unalterable history of every agricultural product, including information about its country of origin, manufacturing processes, method of transit, and handling. Consumers and stakeholders can access this data through mobile applications or QR codes to monitor. The security and transparency of blockchain technology may be used to the advantage of financial transactions occurring within the ecosystem. It ensures secure payments and transparent financial transactions, reducing fraud and ensuring equitable compensation for all parties.

The IoT sensors are necessary for environmental monitoring and sustainable development. By maximizing the use of water and fertilizer, reducing the usage of energy, and minimizing their environmental impact, they help farmers make sustainable decisions. They maintain track of resource consumption and environmental conditions. In conclusion, the application of blockchain and IoT in agriculture paves the way for a smart agricultural ecosystem that significantly raises productivity, sustainability, and transparency throughout the industry. This transformation is ushering in a new era of agricultural innovation and efficiency, which ultimately benefits farmers, consumers, and environment.

9.6.1 Advantages of Combining Technologies

A vast range of advantages that might radically alter the agricultural industry are offered by the application of blockchain and IoT technologies. The assurance of data security and integrity is crucial among these benefits. With blockchain, records become immutable, safeguarding crucial knowledge about agricultural conditions, animal health, and environmental factors. The IoT sensors are essential for preventing unauthorized access and manipulation since they safely collect and distribute data to the blockchain. This strengthening of data security provides a solid foundation for modern farming methods. Real-time observation and decision-making abilities are significantly enhanced by this combination. Farmers now have instant access to data from IoT sensors on important farming aspects including soil quality, weather predictions, and equipment status. Such real-time monitoring helps farmers manage their crops and resources more efficiently by permitting swift decisions. Predictive analytics is another benefit of combining historical blockchain records with IoT data, since it helps farmers forecast weather changes, disease outbreaks, and agricultural yields. This enhances planning and strategy [27].

The combination of blockchain and IoT dramatically enhances supply chain management. The transparency that blockchain records offer the supply chain benefits from stakeholders having access to comprehensive information on the handling, origin, and transit of agricultural products. Smart contracts' automation streamlines the supply chain by triggering automated payments upon delivery, eliminating the need for additional administrative effort, and preventing payment delays. Furthermore, the blockchain's traceability features enable swift and accurate product recalls in the event of quality issues, limiting customer harm and safeguarding the reputation of agricultural firms. Another advantage is provided by the areas of resource efficiency and waste reduction. With the aid of real-time IoT data analysis, farmers can make the most of resources like water, fertilizer, and power while minimizing waste, expenses, and the environmental effect of agriculture. Precision agriculture is made possible by IoT sensors and blockchain data, which allows for the precise application of resources when and where they are needed, resulting in higher crop yields and less resource consumption.

These technologies are combined to aid farmers in using more sustainable practices. The IoT sensors monitor the weather, assisting farming operations in implementing sustainable practices like using fewer chemicals and less water. These sustainability activities are documented on the blockchain, providing verifiable proof of compliance with sustainable farming standards and certifications. The combination of blockchain and IoT also enhances trust and accountability. Consumers benefit from the transparency and traceability that these technologies offer, allowing them to verify the authenticity and quality of agricultural items. Every interaction is meticulously recorded on the blockchain, making all parties responsible for their actions and putting a halt to unethical or dishonest activity. Finally, efficiency and cost-cutting are significant benefits. The requirement for manual labor and administrative overhead is reduced by the automation of numerous procedures, notably in payment processing and supply chain management. Because of the transparency of blockchain, disputes and discrepancies may be resolved more quickly and for less money [28].

9.7 Case Studies

Table 9.2 provides a concise overview of each case study, followed by the specific outcomes and benefits of implementing blockchain and IoT technologies in agriculture and food supply chains.

Table 9.2 Case studies on blockchain and IoT solutions in agriculture.

Case study	Overview	Outcome and benefits
Walmart and IBM's Food Safety Blockchain	To improve food safety and traceability, Walmart and IBM worked together on a groundbreaking effort to incorporate blockchain technology into their food supply chain.	- Traceability: The time required to track food contamination was reduced from weeks to seconds. - Transparency: Customers may read product routes in great detail by scanning QR codes. Recall management costs are reduced via automation.
TE-Food (Vietnam)	TE-Food, a blockchain-based traceability solution, is used in Vietnam's agriculture sector, primarily to track pig products.	- Provenance tracking: Counterfeit dangers are reduced by giving source data through QR codes. Digitization reduces administrative expenses while increasing supply chain efficiency. - Customer confidence: Increased customer confidence leads to higher sales.
AgriDigital (Australia)	The blockchain technology AgriDigital is used to manage the networks that deliver grain to Australia.	Smart contracts facilitate efficient transactions by streamlining contract management and settlements. - Transparency: Building trust through a shared ledger. - Fewer conflicts: Fewer arguments result in cost savings.
FruitChain (Honduras)	Blockchain technology is used by the Honduran project FruitChain to track the organic banana supply chain.	- Provenance and Quality: Customers are reassured of a product's veracity and perfection thanks to QR codes. By collecting information on pay and working conditions, fair trade upholds moral integrity.

(Continued)

Table 9.2 Case studies on blockchain and IoT solutions in agriculture. (*Continued*)

Case study	Overview	Outcome and benefits
Provenance (UK)	The Provenance platform, which is built on blockchain technology, is used globally to give transparency into the food supply chain.	- Sustainable Seafood: It is guaranteed that seafood products are sustainable. - Decreased Fraud: The seafood business has seen a decrease in labeling and fraud.

9.8 Challenges and Future Directions

Using blockchain and IoT technology in agriculture has a number of benefits, but there are also a number of problems that need to be fixed. For smaller farms or areas with limited resources, connecting IoT sensors with blockchain systems presents a significant barrier due to the hefty initial costs. Governments, agricultural groups, and technology suppliers may team up to offer financial incentives like subsidies or low-interest loans in order to reduce this. Improved access to power and the internet in rural areas can also aid in overcoming challenges in isolated areas. The need to safeguard private agricultural data makes data security and privacy a challenge. Using strong encryption and access control techniques for IoT data, as well as creating private or consortium blockchains with limited access can improve security. The variety of IoT devices and blockchain platforms may cause interoperability problems, but middleware solutions and standardization initiatives can make integration simple. Using scalable blockchain platforms and consensus methods can help address scalability issues in addition to assisting farmers and agricultural employees in effectively utilizing these technologies [29].

Resistance to change in long-established agricultural practices and supply chain structures may make adoption difficult, therefore it is important to emphasize the benefits of efficiency, cost savings, and improved product quality. Clear data ownership rules, governance structures, and frameworks for regulatory compliance are critical to resolving ownership and control challenges. The agricultural community may be able to overcome its poor adoption awareness by raising adoption knowledge through campaigns and demonstration initiatives. By addressing these concerns jointly and putting the offered solutions into action, the agriculture industry can fully leverage the advantages of blockchain and IoT technologies for a more transparent and sustainable future [30].

9.9 Conclusion

In conclusion, the adoption of blockchain-powered smart agriculture is a game-changing force that will have substantial effects on the food supply chain. It is significant because it can make all elements of agriculture, from production to consumption, more efficient, sustainable, and transparent. Due to blockchain's potential to enhance traceability, consumers can safely track the origin and safety of their food, reducing the danger of fraud and foodborne illnesses. Additionally, it improves operational efficiency and reduces costs by automating processes with smart contracts in the supply chain.

The data-driven insights provided by IoT sensors and safely saved on the blockchain may be used by farmers and politicians to make informed decisions, maximize resource utilization, and engage in sustainable farming practices. The biggest advantage of blockchain-based smart agriculture is that it boosts consumer confidence by ensuring complete transparency across the food supply chain. This recently discovered openness encourages client loyalty and supports businesses committed to excellence and transparency by assuring them of the authenticity and quality of the items. To effectively benefit from these benefits, officials must establish clear legal frameworks and allocate resources. Technology, safe data management, and peer cooperation should be embraced by farmers. Stakeholders must continue supporting research, promote interoperability, and inform users about the advantages of this revolutionary technology. In this approach, blockchain-based smart agriculture has the potential to change how food is produced and delivered in the future, bringing about a food supply chain that is safer, more efficient, and transparent while also being advantageous to society.

References

1. Biswas M, Mahi M, Nayeen J, Hossen R, Acharjee UK, Md W. BUVOTS: A Blockchain based Unmanipulated Voting Scheme. Rakib and Acharjee, Uzzal Kumar and Md, Whaiduzzaman, *BUVOTS: A Blockchain Based Unmanipulated Voting Scheme* (November 23, 2020). 2020 Nov 23.
2. Mukherjee PP, Boshra AA, Ashraf MM, Biswas M. A Hyper-ledger Fabric Framework as a Service for Improved Quality E-voting System. In *2020 IEEE Region 10 Symposium (TENSYMP)* 2020 Jun 5 (pp. 394- 397). IEEE.
3. Al-Amin S, Sharkar SR, Kaiser MS, Biswas M. Towards a BlockchainBased Supply Chain Management for E-Agro Business System. In *Proceedings*

of International Conference on Trends in Computational and Cognitive Engineering 2021 (pp. 329-339). Springer, Singapore.

4. Akib AA, Ferdous MF, Biswas M, Khondokar HM. Artificial Intelligence Humanoid BONGO Robot in Bangladesh. In *2019 1st International Conference on Advances in Science, Engineering and Robotics Technology (ICASERT)* 2019 May 3 (pp. 1-6). IEEE.
5. Akhund *et al.*, 2020, June. IoT Waiter Bot: A Low Cost IoT based Multi Functioned Robot for Restaurants. In *2020 8th International Conference on Reliability, Infocom Technologies and Optimization (Trends and Future Directions)(ICRITO)* (pp. 1174-1178). IEEE.
6. Hossain, M.P., Khaled, M., Saju, S.A., Roy, S., Biswas, M.: Vehicle registra-tion and information management using blockchain based distributed ledger frombangladesh perspective. In: *2020 IEEE Region 10 Symposium (TENSYMP)*. IEEE
7. Datta P, Bhowmik A, Shome A, Biswas M. A Secured Smart National Identity Card Management Design using Blockchain. In *2020 2nd International Conference on Advanced Information and Communication Technology (ICAICT)* 2020 Nov 28 (pp. 291-296). IEEE.
8. Mahi MJ, Hossain KM, Biswas M, Whaiduzzaman M. SENTRAC: A Novel Real Time Sentiment Analysis Approach Through Twitter Cloud Environment. In *Advances in Electrical and Computer Technologies 2020* (pp. 21-32). Springer, Singapore.
9. Akhund *et al.*, 2018, December. ADEPTNESS: Alzheimer's disease patient management system using pervasive sensors-early prototype and preliminary results. In *International Conference on Brain Informatics* (pp. 413-422). Springer, Cham.
10. Ahamed, S., Siddika, M., Islam, S., Anika, S., Anjum, A. and Biswas, M., 2021. BPS: Blockchain Based Decentralized Secure and Versatile Light Payment System. *Asian J. Res. Comput. Sci.*, pp. 12-20.
11. V. Matyushok, V. Krasavina, A. Berezin and J. S. García, "The global economy in technological transformation conditions: A review of modern trends," *Econ. Res.-Ekon. Istraz.*, vol. 2021, pp. 1–41, 2021.
12. L. Movilla-Pateiro, X. M. Mahou-Lago, M. I. Doval and J. Simal-Gandara, "Toward a sustainable metric and indicators for the goal of sustainability in agricultural and food production," *Crit. Rev. Food Sci. Nutr.*, vol. 61, no. 7, pp. 1108–1129, 2021.
13. H. C. Morang and D. Laskar, *Problems and prospects of agricultural development in the tribal areas of Golaghat and Sivasagar Districts of Assam*, 2020.
14. L. Lipper and D. Zilberman, "A short history of the evolution of the climate smart agriculture approach and its links to climate change and sustainable agriculture debates," in *Climate Smart Agriculture*, pp. 13–30, Springer, Cham, 2018.

15. V. Saiz-Rubio and F. Rovira-Más, "From smart farming towards agriculture 5.0: a review on crop data management," *Agronomy*, vol. 10, no. 2, p. 207, 2020.
16. M. A. Bouras, F. Farha and H. Ning,"Convergence of computing, communication, and caching in Internet of Things," *Intell. Converged Netw.*, vol. 1, no. 1, pp. 18–36, 2020.
17. S. R. J. Ramson, S. Vishnu and M. Shanmugam, "Applications of Internet of Things (IoT)–an overview," in *2020 5th International Conference on Devices, Circuits and Systems (ICDCS)*, pp. 92–95, Coimbatore, India, March 2020.
18. S. Nižetić, P. Šolić, D. López-de-Ipiña González-de-Artaza and L. Patrono, "Internet of Things (IoT): opportunities, issues and challenges towards a smart and sustainable future," *J. Clean. Prod.*, vol. 274, p. 122877, 2020.
19. A. U. Mentsiev, A. R. Isaev, K. S. Supaeva, S. M. Yunaeva and U. A. Khatuev, "Advancement of mechanical automation in the agriculture sector and overview of IoT," *J. Phys. Conf. Ser.*, vol. 1399, no. 4, p. 044042, 2019.
20. J. Mahalakshmi, K. Kuppusamy, C. Kaleeswari and P. Maheswari, "IoT sensor-based smart agricultural system," in *Emerging Technologies for Agriculture and Environment*, pp. 39–52, Springer, Singapore, 2020.
21. M. Nazari Jahantigh, A. Masoud Rahmani, N. Jafari Navimirour and A. Rezaee, "Integration of internet of things and cloud computing: a systematic survey," *IET Commun.*, vol. 14, no. 2, pp. 165–176, 2020.
22. A. Panarello, N. Tapas, G. Merlino, F. Longo and A. Puliafito, "Blockchain and IoT integration: a systematic survey," *Sensors*, vol. 18, no. 8, p. 2575, 2018.
23. R. Haas, D. Imami, I. Miftari, P. Ymeri, K. Grunert and O. Meixner, "Consumer perception of food quality and safety in Western Balkan Countries: evidence from Albania and Kosovo," *Foods*, vol. 10, no. 1, p. 160, 2021.
24. K. Pal and A.-U.-H. Yasar, "Internet of things and blockchain technology in apparel manufacturing supply chain data management," *Procedia Comput. Sci.*, vol. 170, pp. 450–457, 2020.
25. E. M. Torroglosa-Garcia, J. M. A. Calero, J. B. Bernabe and A. Skarmeta, "Enabling roaming across heterogeneous IoT wireless networks: LoRaWAN MEETS 5G," *IEEE Access*, vol. 8, pp. 103164–103180, 2020.
26. A. I. Badran and M. Y. Kashmoola, "Smart agriculture using Internet of Things: A Survey," in *Proceedings of the Proceedings of the 1st International Multi-Disciplinary Conference Theme: Sustainable Development and Smart Planning, IMDC-SDSP 2020*, Cyperspace, 28-30 June 2020, p. 10, 2020.
27. T.-h. Kim, V. S. Solanki, H. J. Baraiya, A. Mitra, H. Shah and S. Roy, "A smart, sensible agriculture system using the exponential moving average model," *Symmetry*, vol. 12, no. 3, p. 457, 2020.
28. H. S. Anupama, A. Durga Bhavani and A. B. A. Z. Fayaz, "Smart farming: IoT based water managing system," *IJITEE*, vol. 9, no. 4, pp. 2383–2385, 2020.

29. M. R. Ramli, P. T. Daely, D.-S. Kim and J. M. Lee, "IoT-based adaptive network mechanism for reliable smart farm system," *Comput. Electron. Agric.*, vol. 170, p. 105287, 2020.
30. K. Lova Raju and V. Vijayaraghavan, "IoT technologies in agricultural environment: a survey," *Wirel. Pers. Commun.*, vol. 113, no. 4, pp. 2415–2446, 2020

10

Computer Vision Systems in Livestock Farming, Poultry Farming, and Fish Farming: Applications, Use Cases, and Research Directions

Balasubramaniam S.[1]*, Vijesh Joe C.[2], A. Prasanth[3] and K. Satheesh Kumar[4]

[1]*School of Computer Science and Engineering, Kerala University of Digital Sciences, Innovation and Technology (Formerly IIITM-K), Digital University Kerala, Thiruvananthapuram, Kerala, India*
[2]*Computer Engineering, Sankar Polytechnic College, Tirunelveli, Tamilnadu, India*
[3]*Department of Computer Science and Engineering, Vel Tech Rangarajan Dr. Sagunthala R&D Institute of Science and Technology, Chennai, Tamilnadu, India*
[4]*School of Digital Sciences, Kerala University of Digital Sciences, Innovation and Technology (Formerly IIITM-K), Digital University Kerala, Thiruvananthapuram, Kerala, India*

Abstract

Agriculture provides human facilities through raising plants and livestock. Agriculture fuels a sedentary lifestyle. Smart agriculture is new, and a number of farmers do not know what it means. The field of artificial intelligence (AI) known as "computer vision" analyzes visual data such as digital pictures, movies, and other visual inputs to form conclusions and recommendations. Computer vision is revolutionizing agriculture. Livestock, poultry, and fish farms should supply foods like meat, milk, eggs, and offal as 30% of the daily protein requirement. Thus, we require a precise and timely animal count. This technology makes cattle tracking and counting easy. Assessing and managing poultry welfare requires monitoring. Computer vision's wide variety of information and being non-intrusive make it an exciting new method for automating online poultry tracking and health monitoring. Reproduction in the aquaculture industry will require intelligent feeding methods in the future; its implementation will also be a priority. Computer vision as an intelligent feeding system is also evolving

**Corresponding author*: baluttn@gmail.com

Rajesh Kumar Dhanaraj, Balamurugan Balusamy, Prithi Samuel, Malathy Sathyamoorthy and Ali Kashif Bashir (eds.) Computer Vision in Smart Agriculture and Crop Management, (221–258) © 2025 Scrivener Publishing LLC

rapidly. The merging of the two systems will improve the state of aquaculture, and its potential growth depends on technology innovation and execution.

Computer vision models can assess cow, pig, sheep, and hen closed-circuit television (CCTV) feeds to guarantee the best possible living conditions and social well-being. Insights can be gained via automating home management, analyzing behavior, predicting or detecting diseases, measuring weight, checking eggs for defects, counting fish, and so on. These AI-powered robots feed animals, collect eggs, and clean dung. After removing eggs that are near hatching, cracked, or unclean, they may count and package them automatically contain feathers and blood. Drones can observe environmental changes; offer fish stock statistics; and measure water oxygen, pH, salinity, and pollution. Farms and robots can use fish hunger sensors to better feed their fish. The ability for veterinarians to have continuous and potentially remote access to the data generated by the Internet of Things (IoT) embedded in the animals opens interesting possibilities for teleconsultation or tele-expertise capable of optimizing the health and well-being of the animals, especially in areas of medical deserts that are also used daily in non-desert areas. This chapter emphasizes the necessity of livestock, poultry, and fish farming systems based on computer vision technology. It identifies the hurdles or issues that still need to be resolved and also discusses the applications and use cases of this system.

***Keywords*:** Computer vision, smart farming, livestock, poultry, fish, management

10.1 Introduction

The latest buzzword in agriculture is "smart farming," which aims to replace antiquated practices with cutting-edge information communications and technology (ICT)-based alternatives. Imaging technology, artificial intelligence (AI), large amounts of data, the cloud, next- generation networks are all examples of emerging technologies that are predicted to make major advancements in this field [1]. Bringing information technology to agriculture is only part of what we mean when we talk about "smart agriculture," which is more concerned with the generation and application of knowledge than anything else. Information technology should be built into agricultural machinery and tools so data can be processed and analyzed, allowing judgments or at least enabling partially automating their preparation [2]. The pressure on crop and animal production, as well as environmental effects, stem from the need to feed a fast-expanding population. As a counterpoint, smart farming technology is becoming more widespread in today's agriculture to help maximize crop yields and livestock productivity while reducing expenses. Precision agriculture (PA) is a method of farm monitoring, which uses data collected from the careful observation, measurement, and analysis of each field and crop. Precision livestock farming (PLF) is a kind of farming that uses automated keeping tabs on specific animals in order to increase crop yields, milk yields, illness detection, animal behavior, and environmental conditions [3].

The modern era of technology and inventions is fast modifying our daily lives. We can barely fathom the infinite possibilities of such technologies. The development of cutting-edge technologies like computer vision is making significant contributions in many different fields. The agricultural sector is only one of many that have begun using computer vision to improve its practices. Agriculture is widely regarded as the economic backbone that sets each country apart in international trade. Countries that produce a lot have a huge advantage in the export market. Production costs are more than they could be because of factors such as excessive labor cost in some countries, inadequate research and development, and lack of machinery.

Humans make meaning of their environments by using visual information processed by the brain. The field of computer science known as computer vision tries to automate human visual processing using digital tools. Computer vision is an emerging field that aspires to give computers vision capabilities in an increasingly visual environment. By using computer vision, machines can examine a series of photos, recognize and detect items within them, and draw conclusions about what they show. Computer vision is expanding its roots in the farming industry. The agricultural industry has benefited from computer vision's increased efficiency, as seen by increased output and decreased production costs, thanks to automation. Computer vision's automation and detection skills are finding widespread use in the agricultural sector recently [4].

There will be a rise in demand for poultry around the world, more than double between 2005 and 2050, while the demand for chicken eggs is expected to increase by about 40% during the same period. Production, health, and welfare monitoring will become more challenging as a result of this practice, labor shortages, and other biosecurity measures. To maximize output while decreasing unit costs and resource consumption, the implementation of intelligent systems for managing poultry flocks is crucial. A smart poultry management system incorporates PLF technology including sensible sensors, agricultural process automation, and data-driven determination support platforms. There are many emerging technologies that promise to have far-reaching effects on the poultry industry, particularly in terms of poultry house layout, animal care, feeding, and early disease detection. Large amounts of data, generated by monitoring, need to be organized and made available to users using new sensory technologies. As data are being generated and made available, fast action is required depending on the results, especially if practical applications are not feasible. Accordingly, utilizing PLF technologies calls for the establishment of crucial data analysis systems that permit consistently data-driven judgments to be made [5].

The average age of farm workers is higher than the national average in several nations; fishery output will face the challenge of scarcity of staff, necessitating adjustments in aquaculture technology. As current information technology enters agriculture, intelligent fish farms are also emerging. The intelligent fish farm uses machines to replace human work, to increase oxygen, optimize feeding, reduce disease occurrences, and precisely harvest to free up personnel and achieve green and sustainable aquaculture [6]. Implementing AI-powered image recognition in farming has enabled the detection of flaws in products and the classification of crops according to weight, color, size, and ripeness. When combined with the right machinery, they can drastically cut down on costs and lead times [7]. Cameras are used by computer vision systems to keep an eye on livestock. Thus, cattle husbandry necessitates the application of AI to visual analysis such as data collecting, analysis, and decision-making, and poultry production systems have become widespread. Numerous fields can benefit from this technology, including the automated control of dwellings, the study of human and animal behavior, the protection of animals, the diagnosis of diseases, the measurement of body mass, and the inspection of eggs [8].

New technologies make it possible for machines to mimic human visual perception and comprehension. In agriculture, non-contact and scalable sensing technologies are made possible by combining computer vision algorithms and picture acquisition via remote cameras. Keeping an eye on animals, checking the quality of what is being produced, having robots inspect to ensure standards are met, and keeping tabs on certain animals are only some of the applications of AI. The discipline of image recognition has recently seen significant advances due to new deep learning technology [42]. Modern deep learning algorithms are far more resilient than older approaches to computer vision, enabling very accurate real-time image recognition [43]. As a result, footage captured by conventional surveillance cameras or webcams can be used in conjunction with deep learning techniques to do video analytics. Edge computing, also known as Edge AI, combines traditional computing with on-device machine learning. By offloading AI processing to edge devices, the Internet of Things (IoT) and AI can be used together to develop scalable computer vision applications.

10.2 Smart Agriculture

Agriculture is the backbone of human existence because it provides the bulk of the world's food grains and other essential raw materials. It is crucial to a nation's economic development. Moreover, it offers a wide variety of jobs to the public. The nation's economy cannot improve without agriculture seeing expansion. Traditional farming practices are still widely used, and as a result, crop and fruit yields are low. However, the output has increased in places where automation has been introduced and where machines have replaced human labor. That is why it is so important to use cutting-edge research and technology in farming to boost output [9]. The past decade has been marked by a period of climate change and rainfall variability. This is why many farmers in India have turned to climate-aware techniques advocated under the umbrella term "smart agriculture". The IoT enables "smart agriculture," an automated and steered information technology. The expansion of the IoT sector finds widespread use across all types of wireless settings [10]. To counteract the negative effects of global warming on agricultural output, climate-smart agriculture (CSA) proposes a systemic reorganization and refocus of agricultural practices. Farm productivity is threatened by widespread changes in rainfall and temperature patterns, making the world's poor even more vulnerable. When food markets are unstable, they can threaten the food supply for an entire community. Increasing farmers' ability to adapt, as well as the resilience and resource usage efficiency of agricultural production systems, can help mitigate risks [11].

10.2.1 Livestock Farming

Precision farming includes animal management, which is a difficult task in the industry at the moment. The inaugural PLF conference was held in 2003, introducing using a novel method for creating vital manufacturing systems. To entirely revolutionize cattle farming industry, there are a few promising current and future technologies and PLF may be one of the most significant of these. Understanding the grazing behavior adjustment mechanisms that allow adaptability to grazing settings is essential for effective grazing and pasture management. Continuous monitoring of each animal in the flock allows for the rapid identification and flagging of any little variation from normal behavior (for that specific animal) and so greatly improves animal health and well-being in addition to allowing the exact management of grazing [3].

Figure 10.1 Livestock farming.

Figure 10.1 shows the livestock farming illustration. The livestock sector relies heavily on the health of its animals because poor animal health can have negative effects on output, animal welfare, and even human health if antibiotics are misused or zoonoses spread. In fact, a high number of zoonosis diseases can be transmitted to people due to the high concentration of animals near human populations in various nations. One of the most significant concerns that has led to PLF is the failure to adequately monitor health problems in order to recognize the early clinical indications of diseases on the farm. Prevention is always preferable, as early detection of most ailments allows for simple treatment. Farmers can react to disease outbreaks as they become apparent or take preventive measures with the help of veterinary services, and crucial animal health factors like mobility, air quality, feed, and water intake can be monitored continuously with the help of sensors, huge amounts of data, AI, and machine learning. They may detect, anticipate, and avoid disease outbreaks by continually gathering these data and employing cutting-edge technology to predict deviations or irregularities. When compared to traditional methods of detection, this newer one is much more cost-effective.

Techniques dependent on sound have the ability to automate the monitoring of animals for industrial-scale agriculture. Automated pig cough identification is now possible with the help of a sound-dependent device, the Pig Cough Monitor™ (PCM). Mathematical analysis of all incoming sounds allows this gadget to detect and count the frequency of coughs in real time. Animals can both vocalize their sadness and exhibit it through abnormal behavior. Microphones could assess vocalization, and staff observations or surveillance cameras could record activities; together, they could be interpreted to provide useful data. It is now possible to consistently anticipate and then decrease the complexity of mastitis in cows using automated sensors and algorithms. Coccidiosis can be detected using air sensors in the poultry industry when the number of sick birds increases. These sensors measure levels of airborne vaporized chemical compounds. Air sensors could pick up on this shift much sooner than a farmer or veterinarian could. Non-invasive, low-cost sensor technology can detect a few widespread diseases in cattle. In addition, regular recording of feed and water in farms allows for an evaluation of the earliest stages of satiety. Temperature, humidity, and carbon monoxide (CO^2) sensors for climate control will enable autonomous evaluation of indoor thermal comfort.

Precision cattle feeding maximizes nutrient consumption without compromising performance. This highlights the significance of having a system in place to record the amount of feed consumed each day automatically and precisely by animals or groupings. A 25% decrease in protein intake, a 40% decrease in environmental nitrogen excretion, and a 10% rise in profitability are all attributed to precision feeding in the cattle industry. Automated feeding systems (AFS) offer low price replacement for traditional feeding methods. Cattle, sheep, and pigs are just a few of the livestock species for which specialized feeding equipment has been designed. These systems' user interfaces can be useful since they track not only when and how much feed is given to animals, but also their electronic identifiers, how long they are fed, and how much feed they ingest. To make people happier and healthier and for dairy farmers to decrease the amount of labor required to milk each cow, automatic milking systems (AMS) have become increasingly popular in recent decades, notably in western Europe. Initially, setups were linked to 'indoor' systems and the grasslands close by. More and more people are thinking about how to include robots into industrial-scale, pasture-based dairy operations. Now that milking may be done at any time during the day, the cow or goat can be selected when it is most convenient. The AMS is adaptable to both inside and outdoor feeding environments.

10.2.2 Poultry Farming

Smart farming employs cutting-edge systems that permit remote monitoring and management. Maintaining a chicken coop properly requires routine checks of temperature and humidity. In order to prevent incidents, it requires constant monitoring. Data storage and accessibility will become major concerns as new sensing technologies are implemented to monitor chicken production. It is important to act swiftly, if not in real time, based on the conclusions drawn from the data as they are being generated and made available. Accordingly, utilizing PLF technologies calls for the establishment of crucial data analysis systems that permit consistently data-driven judgments to be made [5].

Figure 10.2 shows the architecture of typical poultry farming.

This chapter's objective is to highlight the ways in which recent advancements in PLF sensor technology have the potential to improve poultry production systems, particularly in the areas of environmental optimization, feed precision, and bird welfare. Also discussed will be the challenges that may arise from implementing these technologies, as well as any unintended consequences that may result from their use. The quick detection and diagnosis of poultry diseases using sensor technology hold enormous promise, and their various mechanisms are critically explored in the context of the avian influenza virus. Data governance, which will be crucial to the systems that will be used to produce chickens in the future, is also discussed, as are the systems for storing, accessing, and analyzing data provided by sensors. Finally, the advantages that big data analytics and IoT technologies may offer for chicken production systems are discussed, along with their connections to emerging sensor technologies and data. It is possible to maximize output by carefully monitoring and adjusting the conditions within the chicken house. The quality of the litter, the level of humidity, the amount of carbon dioxide, and the presence of ammonia are all examples of environmental inputs. The PLF system can be used to keep tabs on and adjust the atmosphere

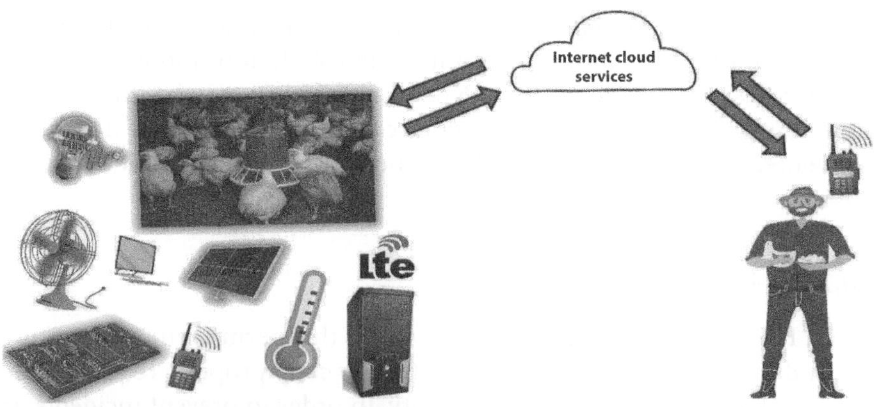

Figure 10.2 Poultry farming.

in a chicken house by adjusting the ventilation rate in response to variations in relative humidity detected by sensors. The levels of ammonia and carbon dioxide gas are known to correlate with the relative humidity in a chicken house, both of which can have negative effects on the health of the birds housed there. There may be a decline in output and avian health if these gas levels rise.

Robotics is another promising technology with use in the monitoring and upkeep of poultry homes' environments. Several companies are actively developing and manufacturing robots that can automatically monitor the conditions inside chicken homes. The robots are either small, self-driving cars that move around in the air, or lightweight devices that glide along rails on the floor, above the birds and carry out various jobs while they go about their business in the aviary. These robots have the potential to improve avian health and activity by encouraging the birds to roam about. In addition, litter in a poultry house can be aerated with the help of commercially available robots equipped with additional equipment. For example, experimental observations of reduced Salmonella infection in broilers in poultry rooms with aerated litter supports the hypothesis that aerated litter has a role in reducing the prevalence of infectious diseases in poultry. Disinfectant-applying litter-scrubbing robots are also in development for use in chicken farms. These modern robots can also be equipped with sensors to assess the coop's condition in real time. This kind of robot allows for remote monitoring of the environment, hence increasing biosecurity and decreasing the need for human intervention within the hen house.

As an essential part of chicken farming, PLF technology has the potential to greatly increase production in the feeding sector. If the goal is maximizing profit, achieving exact statistics on the transformation of feed, or maintaining the ideal weightage of the birds, modern technologies can help improve poultry feeding. The feeding practices of broiler breeder farms, of which the hatching of chicks is the sole function with a genotype that promotes rapid weight gain and maximum meat production, are one area under scrutiny. Although rapid growth genes are present in broiler breeders, rapid weight increase causes breeders to stop laying eggs, making it crucial to limit feed intake. Reduced efficiency in producing high-quality broiler chicks occurs as a result of tighter regulation of feed intake on farms, which causes significant variations in body weight. Producers of broiler breeds must routinely weigh chickens and adjust feed amounts based on the birds' average weight. Although doable, this is a time-consuming process that necessitates frequent sampling, and if done incorrectly, could result in suboptimal feeding schedules. For this reason, a precise feeding system is being tested at the moment, which allows the controlled dispensing of food to birds one by one, based on their respective weights. The contraption allows smaller birds to feed while keeping larger birds from doing so. The weight gap between different broiler breeds was reduced by half in trials demonstrating the efficacy of this precision feeding system. Multiple feedings per day are possible with this method, which has been found to improve feed conversion over once-daily feeding. The same team's subsequent efforts have yielded an improved precision feeding system that employs weighing individual birds. In addition, we monitor their diet using radio-frequency identification tags attached to their wings. Since then, research on precision feeding has shown improved feed conversion and marginally increased fertility in broiler breeders compared to conventional feeding methods.

10.2.3 Fish Farming

Current aquaculture practices have been overproducing due to the ever-increasing scale and density of aquaculture, which has led to an increased water environment rate imbalance, increased incidence of diseased fish, a general reduction in the quality of aquatic products. In addition, a big proportion of the world's agricultural workforce is getting up there in age, meaning that fishery productivity will struggle due to lack of workers, and aquaculture technologies will need to be updated soon. The concept of smart fish farms is also gaining traction, as modern information technology gradually permeates other areas of agriculture [6].

Figure 10.3 shows the fish farming process. Food security and nutrition initiatives of any scale must include fishery products. In 2018, it is estimated that aquaculture contributed 82 million tons to the global fish production total of 179 million tons. Between 1961 and 2017, the global population increased by a mean of 1.9% every annum, whereas the rate of expansion of the global economy per year fish usage of human consumption is 3.1%, nearly two times as fast as the growth rate of all proteins from other animals' meals, which was 2.2%. It is possible that fish may replace meat from animals and poultry in a big way. The higher quality, lower cost of protein found in fish is a big factor, and the absence of epidemic concerns like avian flu makes aquaculture an attractive alternative. Global fisheries have struggled to find migrant workers as a result of COVID-19-related travel and entry–exit restrictions. As the agricultural workforce in fisheries production ages, it becomes increasingly challenging to provide a steady global supply of aquatic products. Aquatic ecosystems are harmed, and aquatic product diseases are common due to overfishing, outdated aquaculture methods, and factors such as human actions,

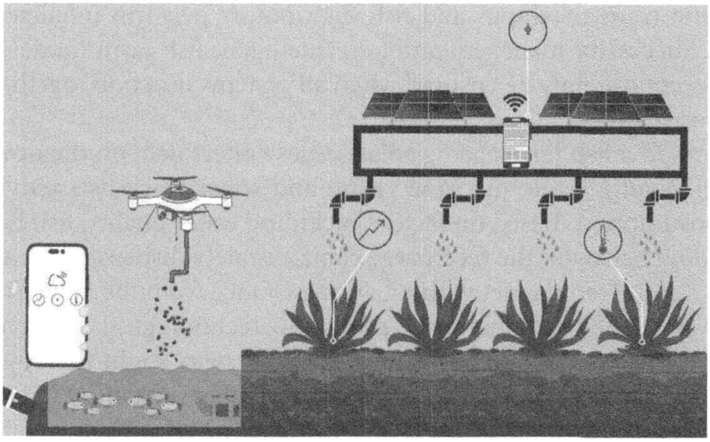

Figure 10.3 Fish farming.

industrial pollutants, and others. Because of increasing regulations meant to safeguard the environment at home and a decline in aquatic biological diversity, illnesses are concerns facing several of the world's major aquaculture production locations. Meanwhile, each year, there will be less land that can be used for aquaculture and less money made per square foot. All the way down, the fishing industry's supply chain lacks adequate funding for research and development and engineering technical support.

Technology advancements include IoT, Big Data, AI, 5G, Cloud Computing, and the IoT remote fish farm measurement and control robots are used to complete fish farm production and management. Finally, the problems of lack of available workers, water contamination, excessive danger, and low efficiency in the aquaculture industry are discussed by digital and intelligent technology-based intelligent fish farms. Future fisheries development will inevitably follow the path set by intelligent fish farms, which represent an industrial shift in the manner of fishery production. Intelligent fish farms can be broken down into four subcategories based on the sort of environment they provide the fish with: pond-type, land-based factory-type, cage-type, and marine ranch.

The water quality of pond-style intelligent fish farms is monitored in real time by sensors, and the fish's surface behaviors are gathered by unmanned aerial vehicle patrols. Bionic fish monitor fish health, nutrition, and development. Unmanned boats spray pesticides and fertilizer to maintain water quality. Robotic vehicles are used to transport bait. A smart aeration system allows for precise regulation of dissolved oxygen (DO). With the automatic fishing bait dispenser, precise automated giving is possible.

Automatic trawl machines and fish distributors perform intelligent harvesting. Successful management of an intelligent fish farm modeled on a pond system can only be ensured when all systems function together in an organized manner.

Sensors in a fish farms using smart cages collect data on the ocean and its water quality, while machine vision and sonar track fish activity and food consumption. Based on factors including water quality, fish biomass, and feeding behavior, the feed barge can accurately dispense bait. A washing robot can clean the net without any assistance from the crew. Fish may now be harvested automatically, thanks to technological advancements like the fish pump and automatic net lifting system. Light therapy with a backup oxygen supply achieves rapid water quality regulation in an emergency. There are two main categories of fish farm: floating and stationary. Depending on the water layer in which each aquaculture object resides, a three-dimensional aquaculture mode will be formed. Unmanned vessels will ferry bait, energy, and other production ingredients from a land-based storehouse to the breeding region. Space constraints on land are a major obstacle to conventional aquaculture operations, but intelligent fish cages offer a viable solution.

Superior cloud–IoT integration technologies characterize the conventional aquaculture IoT system. There are three distinct layers involved: the device, the network, and the Cloud. Sensing devices, controllers, and an information terminal make up the device layer. A variety of pieces of aquaculture machinery such as aerators (oxygen cones), feeders (food dispensing devices), and pump valves are controllable. Data from sensors are sent up to the data collection terminal, which also acts as a receiver for control commands. Platforms in the Cloud and apps for mobile devices that can store data create the Cloud service layer and analysis of aquaculture data and the provision of various aquaculture information services.

10.3 Computer Vision

When it comes to AI and machine learning, computer vision is the subfield that focuses on teaching machines to recognize and process images like photos, videos, and other media in order to draw conclusions and make decisions. It is like giving a computer program an eye. Computer vision technology allows for the realization of previously inconceivable endeavors, such as the operation of self-driving vehicles [12]. Cutting-edge methods, such as AI and linked technology, are bolstering agriculture's digital transition and making it more productive than ever before. In order to make "smart agriculture" a reality, it is necessary to automate a number of tasks in agriculture; deep learning-based computer vision, the "brain" of AI, makes this possible. Over the past few decades, scientists and engineers have dedicated a great deal of time and effort to the study and improvement of AI as people hope to create machines that can mimic human intelligence [13]. A visual inspection system consists of a light source, an image capture device, and a computing system for the extraction of characteristics and processing of the captured image. These kinds of setups are common in factories and other places where things are mass-produced quickly, and precise measurements are needed frequently to make sound decisions. As opposed to human visual examination, which can be flawed due to factors including subjectivity, fatigue, slowness, and high costs, these technologies provide accurate and repeatable measurements without physical contact [14]. This is a job made easier by computer vision. The goal of computer vision, a branch of computer science, is for programmatic development of human-level image/video analysis, interpretation, and comprehension. Teaching a computer to analyze and interpret an image down to the pixel level is the core idea behind computer vision. By employing specialized software algorithms, machines attempt to retrieve visual data, process it, and interpret the results [15].

In order to automate processes involving crops, cattle, or gardens, computer vision gathers the necessary visual data by identifying, detecting, and tracking certain items based on visual features. Computer-based, expert-level vision systems have discovered widespread use in agricultural operations over the past few decades [13]. Technology-driven solutions in agriculture can now be implemented without requiring human interaction, thanks to advances in computer vision and high-quality image acquisition via remote cameras [13].

Figure 10.4 shows the diagrammatic representation of human vision and computer vision systems.

10.3.1 Primary Operations of Computer Vision

The three primary operations of computer vision are the following:

10.3.1.1 Capturing an Image

In order to get an image, computer vision software or applications always employ a digital camera or CCTV system. First, it takes a snapshot of the scene and saves it as a binary file consisting of ones and zeros.

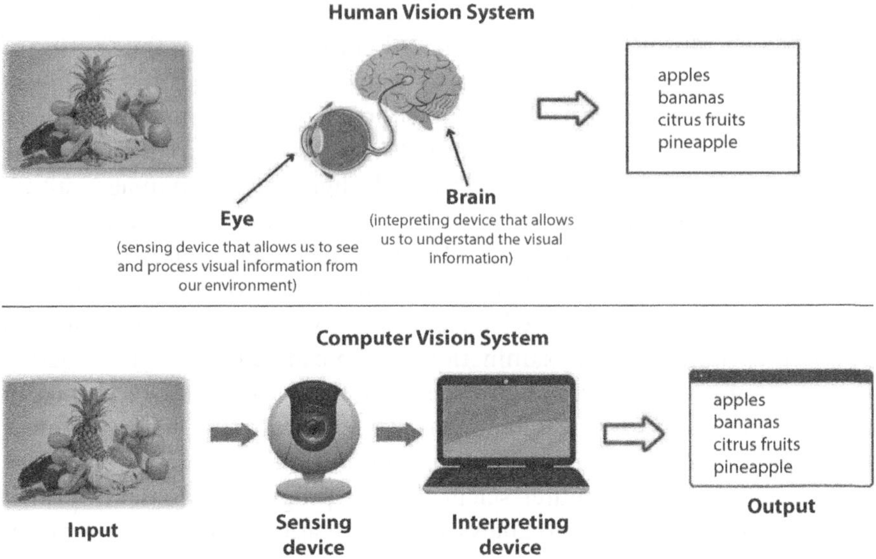

Figure 10.4 The human vision system vs. the computer vision system.

10.3.1.2 Image Editing Procedure

This phase involves applying several computer vision algorithms to the digital information included within a file. Input digital data are used to calculate the fundamental geometrical elements of an image and then the final image is generated utilizing these algorithms.

10.3.1.3 Analyzing and Taking Required Action

Lastly, based on the computer vision's analysis, the system performs the task for which it was intended.

10.4 Primary Computer Vision Techniques

10.4.1 Classification of Images

There are many obstacles to overcome while classifying images, such as different viewpoints, scales, intra-class variance, picture deformation, occlusion, illumination conditions, and background clutter. Researchers in the field of computer vision have developed a method that uses historical data to assign labels to images [16]. To aid in the development of learning algorithms, they supply the computer with a small sample of images representing each class. It examines the bars to get a feel for how various types seem. Simply put, they collect a labeled image dataset and then feed it to a computer for analysis. By far, the most well-known architecture for this task is using convolutional neural networks (CNNs). Most commonly, it is used to classify data presented in the form of images. Most CNNs have a preliminary "scanner" for incoming data that is not designed to process the entire training dataset all at once. One would not want, for instance, a layer with 10,000 nodes to input a 100 × 100 pixel image.

Figure 10.5 shows various primary computer vision techniques.

10.4.2 Detection of Objects

Another common computer vision task is object detection. It takes advantage of previously accomplished image classification to identify specific types of visual content. Its primary use is to identify the kind of objects contained inside a given set of bounds. When it comes to detecting objects, deep learning and machine learning technology are put to good use. Humans have an innate ability to recognize and localize familiar items within seconds of being exposed to a visual medium such as a photograph, film, or video. Thus, the purpose of object detection is to teach computers to recognize and find items with the same dexterity as a human. Tracking and retrieving objects, video surveillance, and image captioning are just few of the many uses for object detection.

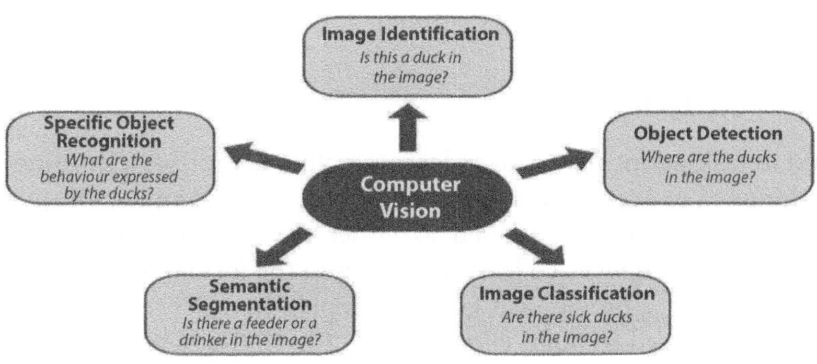

Figure 10.5 Primary computer vision techniques.

10.4.3 Tracking of Objects

The term "object tracking" refers to the action of keeping tabs on anything or something. Object detection is followed by additional observation in videos and the real world, and has long been a natural fit for this method. Based on this model of observation, we can classify it into two broad groups. One such technique is the generative one, which describes the outward features using a generative model. Additionally, the discriminative technique can be utilized to identify the object from the scene. Because of its superior efficiency, it is gradually replacing traditional tracking techniques as the gold standard.

10.4.4 Semantic Segmentation

Segmentation, a core component of computer vision, is the process by which a picture is broken down into smaller, more manageable groups of pixels that may then be assigned labels and placed into categories. The goal of semantic segmentation is to quickly determine the significance of each pixel. Picking a landscape with people, roads, cars, and trees all visible necessitates defining their respective bounds. This means that detailed pixel-wise predictions are required from the models, rather than just categorization.

10.4.5 Instance Segmentation

Segmentation, for example, can be thought of as assigning a color to each of five cars. When doing a classification, the goal is to identify the subject of a fixed-focus photograph. Multiple objects overlap and have distinct backgrounds, making for convoluted scenes. We not only identify their similarities and differences, but also their borders and connections to one another and the items we have already classified.

10.4.6 AI Techniques

To aid in task performance, deep learning often comprises a learning process designed to accumulate "experience" from data samples. Deep learning techniques can be classified as either supervised or unsupervised. Information for supervised learning comes in the form of inputs and their matching outputs, provided as labeled samples. In order to get from input to output, we need to build certain rules. The CNNs, which are modeled on the visual cortex of the human brain, are excellent at processing images, whereas recurrent neural networks are well-suited to handling sequential data. Unlabeled data are used in unsupervised learning, where a model looks for hidden patterns in a dataset with little to no human oversight. When it comes to unsupervised learning techniques, there is the generative adversarial network (GAN). In order for a GAN to generate high-quality results, it requires at least two modules to engage in mutual game learning: two models, one that generates and one that sorts long short-term memory and region-based CNN models are only two examples of the many variants and enhancements that have been generated from the basic deep learning models [17].

Figure 10.6 depicts the framework for classification using machine learning, while Figure 10.7 depicts the framework for classification using deep learning.

Image feature extraction and labeling those features with object categories are two of the main tasks of a computer vision model. Next, it will try to classify a new image by comparing its features to those already stored in its database. But how do we teach a computer to recognize patterns in data and pull-out relevant information? To achieve this, we employ a technique known as convolution.

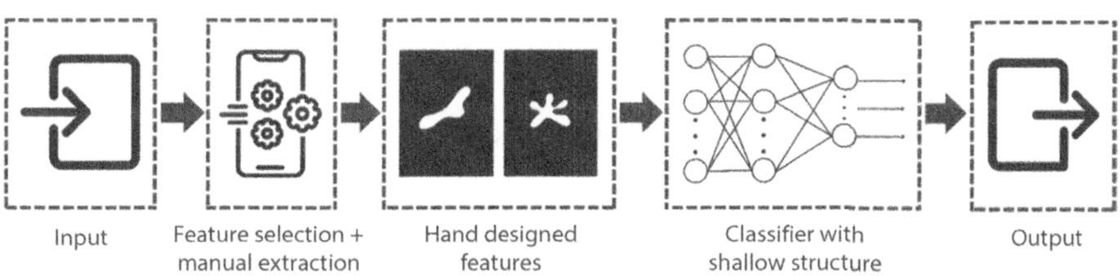

Figure 10.6 Machine learning-based classification.

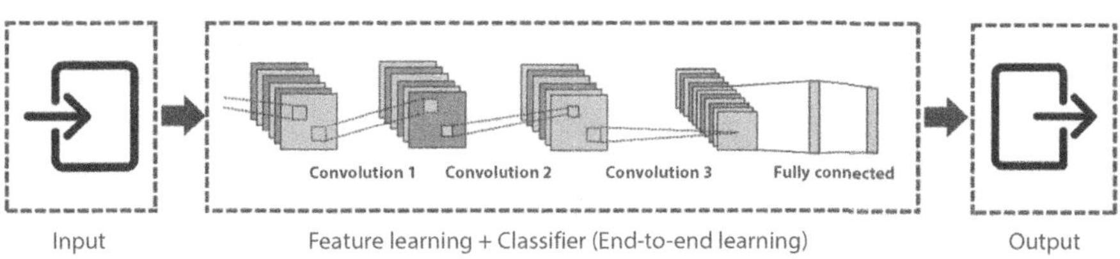

Figure 10.7 Deep learning-based classification.

10.4.6.1 Convolution and Feature Extraction

A convolution matrix is required to convert an unprocessed grid into a processed grid. Using this matrix, we may designate which "features" should be used for further analysis. Once we have this matrix, we can use it to perform operations on the original grid of numbers by multiplying its values by their corresponding values in the original grid and adding the resulting vector. In the illustration, we want to convolve a 9*9 grid using a 3*3 convolution matrix. Our original 9*9 grid was subdivided into nine 3*3 grids. If we visualize this process as being applied to the RGB values of each pixel, we can say that we have "extracted" certain characteristics from the original image. The underlying mechanism of every image filter is this incredibly basic procedure, but the effects can be rather impressive.

The following should be done in order while creating a computer vision model:

1. Transform visuals into data that computers can read (numbers).
2. Make use of convolutions to pull "features" from photos, then set them up.
3. A neural network is fed these data to "learn" which features are connected with classes of objects ("dogs" versus "cats," for example).
4. This model should be fitted and assessed.

Figure 10.8 shows the convolution process.

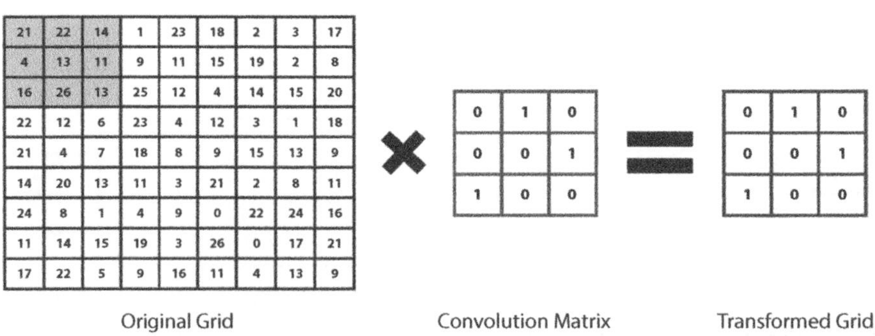

Figure 10.8 Convolution process.

10.4.6.2 Convolutional Neural Networks

Convolutional neural networks (CNNs) [18] are multi-layered networks that can autonomously detect targets based on the properties learned from previous interactions with the targets. Some of the neurons in their many layers do convolutions, while others exhibit non-linearity activated, others are pooling, and still others are fully linked. A neural network has layers that transform activation of neurons, translating the input to a one-dimensional feature vector. In contrast to traditional neural networks, CNNs employ convolution in their layers rather than matrix multiplication. Kernel size is reduced relative to the amount of the input, thanks to parameter sharing linked weights and sparse interactions at each image point. A lower need for memory and processing power as a result are the primary characteristics of CNN architecture.

Figure 10.9 demonstrates the overall CNN structure. CNN muddies the whole picture using different kernels in the convolutional layers to produce feature maps. Rectified linear unit layer (ReLU), a non-linear activation layer, makes feature maps more non-linear and speeds up training. When the input volume is pooled, its spatial dimensions are decreased. The input volume's width and height are the only dimensions that are changed by the pooling layer. Down-sampling, or subsampling, is the process of reducing the number of samples used. Since the size is reduced, overfitting is avoided, and subsequent layers have a simple computational burden. By flattening out 2D feature maps into a 1D feature vector, fully connected layers carry out the neural network's high-level reasoning. The resulting vectors could then be used as feature vectors in subsequent processing or processed and subsequently fed back into existing classification systems. There have been many variations on the CNN design throughout the years. Some examples are the LeNet, AlexNet, ResNet, GoogLeNet, VGGNet, MobileNet, SqueezeNet, and Capsule Networks.

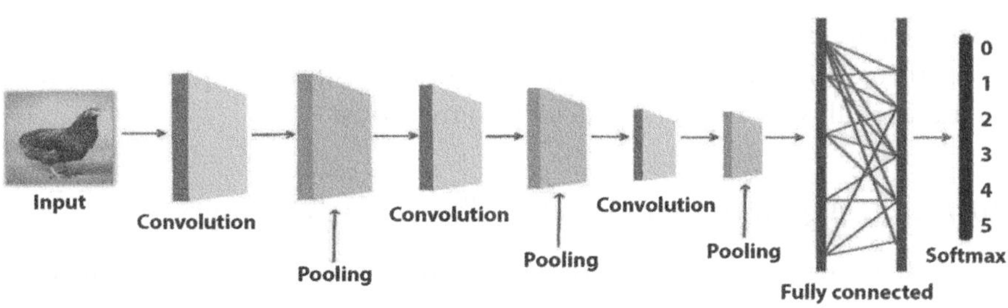

Figure 10.9 Architecture of CNN.

ResNets provided means of addressing increasingly difficult issues in CNNs (vanishing gradients). A ResNet is built up of many layers, each of which applies a different function to the input and from which the gradient signal can return to earlier layers via "shortcut" connections. ResNets are advantageous because they are highly modular, more precise, and need less weight in some circumstances. In addition, they can be engineered to estimate the potential depth of a network. The fundamental drawback of ResNets is that as the network depth increases, error detection becomes more challenging. In addition, the learning efficiency may be low if the network is too shallow. The CNNs are used in computer vision to interpret information about a visual image down to the pixel level, while deep learning recurrent neural networks are used to comprehend the relationships between pixels. The entry of data (picture) often obtained by sensors, cameras, or movies, and the subsequent processing of the data to change into the appropriate information is necessary for any visual inspection system.

The MobileNet can be thought of as a simplified, mobile-friendly variant of the Xception framework. The SqueezeNet is an effective deep learning architecture for low-bandwidth environments. It uses a CNN design like AlexNet, but with half as many parameters. The result is accuracy on ImageNet that is on par with AlexNet. An improved variant of conventional CNNs in terms of layering or internal structure is the multi-layer capsule system known as the CapsNet. Its resistance to geometric distortions and alterations makes it particularly useful for precise picture recognition applications. As a result, it may do extraordinarily well with rotations, translations, and other similar operations.

There are two main types of CNN-based algorithms: those that perform target detection in two steps Region based convolutional neural network (R-CNN), Fast R-CNN, Faster R-CNN, Mask R-CNN), and those that perform target detection in one step by having anchor boxes generated by a network of proposed regions called an R-CNN. The second type of algorithm is the one-stage target detection method, which includes Over Feat, SSD, YOLO, YOLO9000, and YOLO v3. Direct predictions of the target's position and category can be made by these algorithms. Therefore, they are used in real-time detection systems due to their superior speed compared to traditional two-stage target detection methods.

10.4.6.3 Algorithms for Detecting Targets in Two Stages

The first step involves passing the input image through layers of convolution, non-linear transformation, and pooling, all with the goal of extracting low-level information. In the second phase, the same three layers are used once more to extract high-level features.

10.4.6.4 Algorithms for Detecting Targets in a Single Stage

A one-stage target detector predicts all bounding boxes with a single run through the neural network. This is much more efficient and works well with mobile devices.

10.5 Computer Vision-Based Systems in Livestock Farming, Poultry Farming, and Fish Farming

10.5.1 Livestock Management

Precision livestock farming makes heavy use of computer vision techniques to guarantee the best possible outcomes for each animal. Real-time data from livestock monitoring devices help farmers make informed decisions. In recent years, non-invasive computer vision technology's application in recognizing animal behavior has been the subject of extensive study [13].

Figure 10.10 shows the structure of livestock farming.

10.5.1.1 Individual Cattle Tracing and Health Monitoring

To better track individual livestock, their well-being, and their productivity, a new Mask R-CNN instance segmentation approach is developed. Multiple research methodologies were used: The image is pre-enhanced with generalized color Fourier descriptors (GCFD). Multiscale semantic features are extracted from region proposals [19]. An existing fully connected layer of Mask R-CNN is integrated with a sub-network for improved

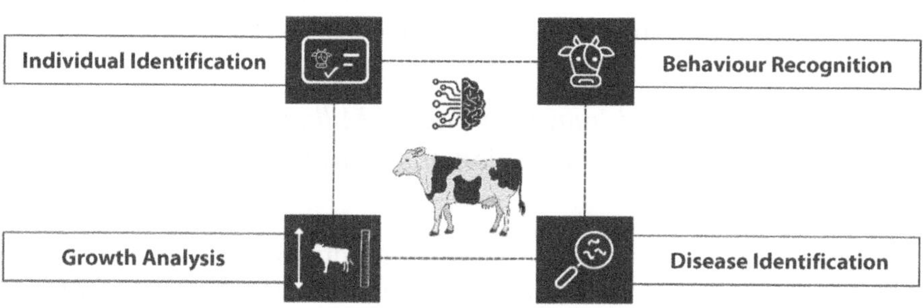

Figure 10.10 Livestock farming.

Figure 10.11 Cattle tracing and health monitoring.

segmentation, and the process culminates in a Grab cut. All these features are extracted using a filter size that is optimally smaller than ResNet101, the backbone of Mask R-CNN.

Figure 10.11 shows the structure of cattle tracing and health monitoring. Two components make up the improved Mask R-CNN cattle picture dissecting instances framework: Backbone component performs convolutional operations like feature extraction over the entire image, and the head component predicts category of objects, range of objects, and mask. The RPN computes region proposals from the GCFD-described feature map, which the region of interest (ROI) align layer matches with the feature map in order to send the fixed-size feature map to the fully connected network, it must first be processed. It is used to perform simultaneous tasks such as bounding box refinement, robust segmentation mask generation, and prediction of the cattle object class.

10.5.1.2 Automatic Prediction of Swine Stress Using Infrared Skin Temperature (Subscrofa)

Developing nations consume 5% more pork annually. Consumers want food safety and ethical meat production. Machine learning and paraconsistent logic were used to predict piglet stress using thermography of the skin infrared (IST). From the time they were born until they were about 52 days old, researchers monitored the infrared skin temperature of 72 piglets under various stress conditions [20]. The piglets' minimum IST, maximum IST, and sex were utilized to determine stress conditions (target). Data mining analyzed and classified the attributes. For contradictions and uncertainties in imaging, mathematical modeling is required [20].

10.5.1.3 CNN-Based Image Analysis of Milking Cows for Individual Identification and Tracking of Feeding Behavior

Precision livestock husbandry optimizes productivity and health monitoring by identifying and analyzing feeding behavior. The use of RFIDs, accelerometers, and other sensors feel intrusive, awkward, and stressful for dairy cows. Our non-invasive image analysis technique overcomes these constraints [21]. In a dairy cow farm, to track eating habits and distinguish between 17 Holstein dairy cows, researchers used an ROI centered on top of each animal's head and various classifiers based on the CNN model. With one in the feeding area, the CNN spots dairy cows. To tell if a dairy cow is standing or eating in front of the feeder, a second CNN analyzes its position. A third CNN checks the feeder for food and classifies it. The final CNN identifies dairy cows. We also investigate how a CNN connected to SVM and numerous CNNs affects person identification.

10.5.1.4 An Algorithm Based on Regularities for Deducing When Cattle are Grazing or Ruminating from their Auditory Signals

An algorithm for long-term foraging behavior study employs regularity to identify chewing the cud and thinking it over episodes. In an offline signal processing, acoustic signals are analyzed and categorized. To identify recurrent masticatory events and create foraging activity blocks, a whole recording is segmented [22]. This level defines no-foraging blocks (resting bouts). Sound envelope autocorrelation detects event regularity. To detect pauses and characterize regularity, sound signal energy is evaluated within a block. Rumination blocks halt, but grazing blocks do not. The algorithm performs well in segmentation and activity categorization. New multidimensional metrics examine both tasks.

10.5.1.5 Individual Pig Weights can be Automatically Estimated Using Image Processing Software

This study investigated if an automated approach to measure pig weight could cause diarrhea and reduce meal intake and weight gain. Shape recognition identified pigs by their unique painting patterns. On weight estimation, an elliptical fitting algorithm located pigs in the image. Then the ellipse's pig area was determined. Finally, dynamic modeling predicted pig weight. The projected weight was compared to manual twice-weekly pig weight measurements to validate the model. Pigs were marked on their backs and weight tracked using form recognition [23]. It calculated pig weights. On farms with stressed pigs, this method can replace weight evaluations.

10.5.2 Poultry Farming

10.5.2.1 Estimating Broiler Weight Using Computer Vision and Neural Networks

This study processed digital photos to test artificial neural networks at calculating live broiler chicken body weight. Grayscale and contrast enhancement brightened the photos. As part of their model, the Chen–Vese team cut off the chicken's head, and the Image Process Toolbox extracted feature characteristics such as surface area, boundary radius, convex area, major axis length, eccentricity, and minor axis length. Back propagation neural networks (BPNNs), called as multilayer perceptrons, are a popular type of neural network [24].

For live chicken weight estimate, four distinct three-layer BPNN training methods were developed and compared. The BPNN receives input vectors, passes them through neurons, and outputs value. In this investigation, the input layer comprised six neurons for area, perimeter, convex area, a combination of the major axis length, eccentricity, and minor axis length. The output layer had one chicken-weight neuron.

10.5.2.2 Deep Convolutional Neural Network-Based Poop Recognition and Classification in Broilers

Fine-grained pictures of broiler feces are analyzed by a deep CNN-based automatic broiler digestive disease detector to determine whether they are normal or abnormal. The R-CNN and YOLO-V3 are two deep CNNs built to compete with one another, but they were built much more quickly. The effectiveness of YOLO-V3 was improved through anchor box optimization. YOLO-V3 and Faster R-CNN were compared in this way. The extraction of the ROI is the first step in a faster R-CNN, followed by object/background recognition and boundary refinement. YOLO-V3 recognizes the complete image using the ROI [25].

10.5.2.3 Imaging Investigation of Flock Behavior at Multiple Feeders for Broiler Chickens

This study examined how three feeder kinds affected bird behavior. The F1, F2, and F3 feeders were examined. The factors were birds' successful eating, occupied area (OA), activity index (AI), and total birds (TB) presented in the area) birds effectively eating (EE). Computational image analysis [26] was used to determine OA and AI, whereas traditional methods were used to obtain TB and EE. Feeder type affected OA, TB, and EE, but not AI ($R2 = 0.01$). F2, having the most feed access, had 86.4% of birds consuming. It employs computational image analysis to evaluate the flock movements, floor occupied area, and eating in a commercial setting, the reaction of broiler chicks to three distinct types of feeders was observed.

10.5.2.4 Chicken Net, a Computer Vision-Based Method for Assessing Commercial Laying Hen Plumage

Here, we offer a method for evaluating the health of a bird's plumage that makes use of deep learning and computer vision. Chicken Net is an end-to-end convolutional neural network that can identify hens and provide accurate plumage predictions. Chicken Net is an enhanced CNN used for detection and plumage quality assessment, optimizing for both hen detection and plumage quality rating in a single model [27]. The system is based on machine learning; therefore, it should be robust against changes in bird appearance and environmental conditions. Chicken Net predicts each hen's plumage condition score based on the photo, as opposed to detecting injuries or naked areas.

10.5.2.5 Effects of Flooding on Litter-Directed Behavior, Anxiety, and the Pursuit of Novel Stimuli in Young Grill Chickens

Stimulating foraging activity in the first weeks of life by spreading food on the litter may avoid leg problems, reduce fear, and improve broiler welfare. Food enrichment research is inconclusive. The food resource may motivate animals to search the litter. This research looked at the effects of food enrichment on the activity levels, litter-directed behaviors, and fearfulness of broiler hens [28]. One hundred and twenty chickens (three batches of 40 animals) were randomly assigned to one of four enrichment groups, each receiving either mealworms (extremely attractive), whole wheat (less attractive), wood shavings (no nutritional benefit), or no enrichment at all. In 10-minute intervals, recordings were assessed for leg stretching, litter pecking and scratching, and physical activity. Tonic immobility and new object tests were given to all birds on day 16. Within 10 minutes of receiving a mealworm treatment, the number of times the treated birds scratched the litter box and pecked at the litter increased significantly (p 0.001). Activity, litter-directed behavior, and tonic immobility and new object test characteristics were unaffected. In conclusion, the broiler chickens' litter-directed actions were motivated by litter food. More research is needed to develop appealing food enrichments and complicated distribution techniques to boost activity time.

10.5.3 Fish Farming

Deep learning technology has been successful in aquaculture because of its rapid development. It opens new possibilities and problems for smart fish farming data processing.

10.5.3.1 Fish Behavior Analysis with Artificial Intelligence

In the proposed research, fish behavior is classified as either normal or starving using an efficient end-to-end CNN. To gauge CNN efficacy, we can adjust the number of fully connected (FC) layers and use either max-pooling or not. The detection algorithm is improved by 10%, thanks to the use of three FC layers and maximum pooling. The shallow CNN model that included a max-pooling algorithm and additional FC layers worked admirably, reaching an accuracy of 98%. This is a significant breakthrough in the development of an automated behavior recognition system for use in today's fish farming operations. Our study methodology makes use of a custom-built CNN [29]. The first model had two FC layers, whereas the second model had three. Initially, we tried CNN models without the max-pooling function. The second phase involved max-pooling both CNN models. The framework quickly recognized typical behaviors such as feeding and maturation in fish.

10.5.3.2 Strong School-Tracking with a CNN for Individual Fish Identification

The suggested method connects observations based on multiple frames projected place in motion and CNN Identifier Outcomes to track fish heads. Tracking and detection are performed on each frame. After sample verification, a determiner of the Hessian on a small scale (DoH) is used in order to disassemble the top part of every fish's body, and CNN input picture patches are created. CNN's identification and state prediction update each target's state [30]. Samples are kept in a pool during tracking. The Hessian as a function of scale (DoH) blob detector used efficiently obtains the DNA from each individual fish, blob-like head. It can identify similar-shaped fish heads by placing focus on the fish head, with the background disregarded. Fish might be detected using CNN fed with an image of a weighted head.

10.5.3.3 *Fish Detection and Classification in Aquatic Environments Using Deep Learning and Temporal Data*

Optical flow, Gaussian mixture models (GMM), and a YOLO deep neural network are used in this technique to identify and categorize species in unrestricted underwater video. The GMM segments the foreground against a model of the background obtained from dataset [31]. To train a backdrop model using only GMM to collect motion-inspired features, you must have images with no fish in them. To get beyond the constraints of GMM, we use optical flow to record fish motion in underwater videos. Blending the GMM and optical flow recovered blobs from the video frames used to create the input data, a hypothetical RGB image of a fish in the foreground is created. Potential ROI involving fish can be seen in the image's foreground. The generated ROIs are remapped onto the original RGB video frame, allowing for the extraction and accurate labeling of correspondingly colored areas as fish or non-fish. ResNet-50 CNN was trained using both "fish" (positive instances in training) and "non-fish" (negative examples in training) patches from the training set.

10.5.3.4 *Computational Object Classification in Marine Observation Using Active Learning System for Marine Observation Images (ALMI), a General-Purpose Active Learning System*

This method quickly collects and uses training materials for an AI-powered observatory picture interpretation. It uses deep learning feature representations and a specific active learning procedure to train efficiently. Active Learning System for Marine Observation Images (ALMI), a feature representation for deep learning that can be used in any situation active learning system for marine image object categorization, was proposed [32]. ALMI classifies image patches or sub-images, by content. Two concepts underpin ALMI: First, it selects the next sample for domain experts to classify and contribute to the training set using uncertainty and relevance sampling. Second, it uses deep learning characteristics rather than traditionally developed features such as "dominant" colors to increase versatility. Active learning workflow ALMI assigns semantic classes to a series of photos. A fully automatic initialization stage prepares the information needed for the largely automatic labeling procedure, which finds semantic classes and labels training samples of all classes for the training set. Each image is transformed into a 300-dimensional vector space feature vector in the initialization phase. Next, M clusters are formed out of the feature vectors. Furthermore, relevance scores are assigned to clusters. How well the cluster's parts help the classifier generalize to new classes not in the training set is measured by this metric.

10.5.3.5 Use of Machine Vision for Fish Weight Estimation

This work examines the relationship between fish weight and image processing-derived visual features and finds the best fit. The application collected dimensions including size, height, surface area, border, diameter, and major and minor axis measurements from the image [33]. A monochrome picture was used to compute the area, counting black and white pixels as one. The perimeter was calculated using grayscale image boundary pixels. Diameter of the circle used as a measure of equivalent diameter whose area matches intended shape. The oval encompassed by sample fish's major and minor axes were equal. Calculated area in square centimeter (cm^2) and other attributes in centimeter (cm). Linear and multiple regressions modeled fish weight and exterior factors. Grain elongation, as well as the major and minor axis lengths, equivalent diameters, and diameters were modeled. Linear, logarithmic, exponential, and power techniques assessed these factors. Seventy percent of data was utilized to model and 30% to evaluate equations; R^2 corrected. Best-fit model determined by R^2 and F test.

10.6 Computer Vision Systems for Intelligent Farming: Current Research Challenges

One of the most promising technologies that could help alleviate the world's growing food shortage is deep learning for computer vision, a relatively new field of study within the field of AI. Deep learning–computer vision is helping to overcome a number of previously intractable issues in the agricultural sector. However, there are always difficulties that come along with a highly innovative potential. Most deep learning applications are data-heavy, and the need for tremendous processing capacity is a fundamental obstacle to the study of seeing things on a computer employing deep learning. Infinite scalability, balanced loads, less time spent on upkeep, and maximum uptime are just some of the benefits of automatic scaling and are all capabilities offered by Cloud-based solutions that might be implemented to address this issue. The delay in accessing and retrieving data from the Cloud, however, makes it impossible for cloud solutions to be used for real-time processing. The rising price of processing such vast amounts of data, as well as privacy considerations, are other factors to consider. Modern edge devices equipped with accelerators can process live video feeds and return conclusions almost instantly. The latency constraints can be lowered by deploying machine vision technologies in equipment on the cutting edge. For many uses in agriculture, even the most advanced computer vision models fall short. Data quality analysis, code inspection, hyper-parameter tweaking, code versioning, developing the optimum deployment environment, thorough training, and re-training are just some of the things that need to be assessed thoroughly to make sure a promising model does not turn into an expensive liability.

Figure 10.12 shows the various challenges in computer vision-based systems implementation. Another major problem that must be addressed is the creation of effective data-driven solutions to ensure the quality of the data being used. The effectiveness of a deep learning–computer vision model is highly sensitive to the choices made for its hyper-parameters. Adjusting hyper-parameters like learning rate, batch size, momentum, and weight decay to optimize performance takes skill and a lot of experimentation. It is not easy to set the hyperparameter in a high-dimensional space. Some of the difficulties in practically implementing solutions to computer vision problems, such as object detection, include things like viewpoint movement, deformation, occlusion, shifting illumination conditions,

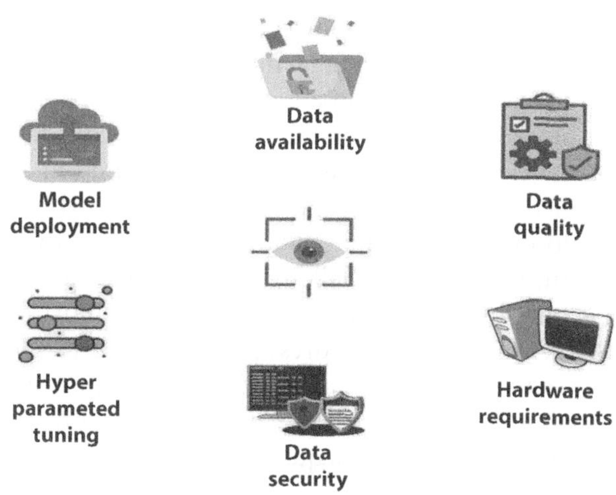

Figure 10.12 Implementing systems based on computer vision faces difficulties.

complicated backgrounds, and speed. The item's appearance might change depending on the observer's perspective, making viewpoint variation a prevalent feature of object detection and segmentation difficulties. As an example, the appearance of a crop can change depending on the viewpoint used to capture it. Apparently, the occlusion is to blame for the supplementary difficulty. In recent years, smart aquaculture's efficient, automated, and accurate expansion has helped the aquaculture business thrive. The use of AI in aquaculture is developing rapidly, however, there is more work to be done before completely automated systems are used. Although automation has made many aspects of fish farming possible, human observation, analysis of data, and judgment are still necessary, hence, the absence of human management presents a significant risk [13].

The field of precision agriculture has, thus far, been the subject of considerable study, and this trend is likely to continue all because of computer vision technologies. Central to both robots and AI, computer vision facilitates the efficient and autonomous completion of a wide range of activities. One major obstacle to the rapid design and evaluation of machine vision and machine learning algorithms' specific applications is the lack of publicly available image datasets. Despite these advancements, a comprehensive survey of such data sets remains unavailable. Determining whether the AI model can adapt to new environmental variables is the most difficult component of applying it in agriculture. Uncertainty is built into most agricultural tasks due to external elements such as rainfall, humidity, sunlight, temperature, and water supply. Therefore, a deep learning model's apparently ideal result could be suboptimal due to variations in input parameters. In order to produce accurate predictions, AI systems require large amounts of supervised data. However, the seasonal nature of agriculture makes it challenging to manage temporal data despite the ease with which geographical data may be obtained. For instance, harvesting a crop is the only time of year when data on that crop may be collected. Because of this constraint, developing reliable deep learning models in a timely manner is difficult. Further, there is no centralized system to which the millions of farms' agricultural data may be uploaded. Because of this, gathering a representative sample might be challenging. Taking advantage of satellite imaging data is one solution to this issue. As farmers must first give their permission, this will also have to address privacy concerns [34].

When it comes to farming, the consequences of a poor recommendation or forecast might be substantial. Contrast this with fields like customer service or IT, where the consequences of a wrong prediction are less severe. A year's worth of crop productivity may be lost if farmers followed poor advice, which would have serious consequences for their livelihoods and for global food security. A possible solution to this problem is to first have the farmers use it. Test out AI algorithms on a tiny plot of land first before implementing them throughout the entire farm. Probabilistic methods are another viable strategy for dealing with this issue. Predictions in a probabilistic model are shown in the form of a probability distribution, and each forecast is accompanied by a statistical confidence interval. The model considers its degree of certainty before making a forecast. If the prediction's confidence is low, it is not used, reducing the potential harm from being led astray by AI-prescribed adjustments. Since farmers typically lack technical expertise, The AI system's recommendations are meaningless unless someone else interprets and assesses them. Consequently, they are first hesitant to embrace the technology. Providing them with access to less complex technologies, like an online marketplace for agricultural goods, is one option. Offering AI features via a workshop will be feasible once students have mastered these elementary methods. The widespread availability of connected and sensor-equipped gadgets has liberated a myriad of user services while simultaneously making human life easier and more streamlined. Unfortunately, security and privacy concerns have been raised in response to the widespread use of Internet-connected devices and data-driven applications across many different fields, leaving these systems open to cyber-attacks [35].

The sale of livestock is a significant source of income for farmers. Cattle can be outfitted with sensors to track their health, allowing veterinarians to remotely administer medication or take preventative measures. Buyers can be given temporary access to the data of an animal they are interested in, just like in the case of livestock purchases, so that they can conduct an analysis before making a final decision. Production losses and illness outbreaks are possible consequences of remotely manipulating the temperature or creating otherwise unfavorable conditions in the barn, given that these animals are maintained in monitored surroundings.

There are several unresolved legal questions surrounding smart farming and precision agriculture. Some of the most pressing concerns in this area include the following.

10.6.1 Legal Documents

A smart farm requires cooperation between many entities, including farmers, Cloud service providers, and networking infrastructure providers. In order for contracts to be effective, it is necessary for these parties to discuss and reach agreements on a number of key points. Legal provisions protecting individuals' privacy, businesses' intellectual property, and networks' infrastructure are all incorporated into these agreements. A significant part of these agreements is dedicated to data security, which is crucial to the growth of smart farms. Such agreements between concerned parties are essential to safeguard the data's potential value and secrecy.

10.6.2 Privacy and Data Protection

Those in the agricultural industry who are considering the installation of smart infrastructure often worry that their data will be stolen by rivals or made public. As a result, including data security terms in contracts with technology vendors is essential. Even though most data processed by smart farming techniques is not personally identifiable, the potential for it to be linked to an individual raises severe privacy concerns. Information about a herd's livestock, for instance, may be traced back to its owner, or a farmer's crop yields could be tied to his or her contact information. Private information should be protected by including confidentiality clauses.

Hardware upkeep is a major issue for IoT solutions in farming because sensors are so widely deployed and easily damaged. Successful operation of smart agricultural apps is contingent upon first-rate internal infrastructure [36].

10.7 Conclusion and Future Scope

The preceding study and debate lead us to the conclusion that agriculture is an ideal setting for the implementation of computer vision technology, a developing technology. Technology based on visual intelligence generated by computers and fed by large datasets will be used in the future administration of agricultural output. The efficiency, scope, coordination, and strength of agricultural automation systems will all increase as a result of the incorporation of computer vision and AI algorithms. Deep learning, spectral analysis, and other cutting-edge technology will create intelligent agricultural automation systems. Future developments in computer vision technology will facilitate the extensive and swift growth of agricultural automation, which will boost the effectiveness and quality of agricultural production by giving farmers helpful advice and information to aid in making decisions and taking action [37].

Recent use of deep learning-based computer vision systems developed for livestock applications, as well as the strong predictive capacity displayed in published studies, demonstrate the potential of such an analytical approach for high-throughput phenotyping. Computer vision systems enable non-invasive, real-time predictions that can improve areas such as group and animal-level health, welfare, feeding, and reproduction. Research using deep learning as the primary foundation for image analysis in computer vision systems is still extremely small and novel in Animal Sciences, with 84% of articles dealing with livestock animals published in 2019 and 2020. However, it should be noted that the vast majority of the papers we looked at employed deep learning techniques that had only recently been proposed. Just two examples: 2016 saw the release of Xception, and 2017 saw the release of Mask R-CNN. So, it is reasonable to assume that there will be a lag between the development of deep learning algorithms and their use in industries like agriculture and livestock.

When applying computer vision to a whole new domain, it is anticipated that deep learning techniques will be widely adopted for picture classification tasks, as opposed to object identification and image segmentation. Since more complicated problems cannot be answered with simple picture classification, more nuanced image analysis is typically performed once object segmentation has been implemented. As an illustration, a deep learning algorithm that has just been taught to assert the presence or absence of a cow in an image will not be able to accurately predict where the cows will be, how far apart they will be, and how much space they will take up. Therefore, it is natural to progress from picture classification to instance segmentation as complexity increases in the issue of space. Labeling and annotating images are time-consuming and expensive compared to picture classification, hence, preventing the rapid proliferation of deep learning techniques for object segmentation [38].

Using cutting-edge technology has aided farmers in boosting animal output and well-being, and there is a wide range of machine vision techniques that have recently been published in academic journals and may soon be incorporated into widely available technological tools. The results of this study demonstrate the efficacy of machine vision and how it can be usefully applied to the monitoring of bovine and porcine activities such as lying, feeding, drinking, moving around, showing aggressiveness, and reproducing [39]. The present computer vision approaches perform admirably in a controlled environment with predetermined chicken motions. However, keeping tabs on chickens becomes more difficult in a genuine farm setting. There can be a shift in the morphological features of a flock of hens due to occlusion. Furthermore, sensor stability is greatly impacted by the presence or absence of shadows and the intensity of ambient light. To determine the economic viability and usefulness of these poultry monitoring devices, more study is required. To increase the overall efficacy and robustness of chicken monitoring in PLF [18], we need to solve these difficulties using a combination of methodologies from livestock science and engineering.

In order to attain technological adaptability and harmony in the years to come, it will be required to create a massive dataset. However, researchers should work to enhance hardware and software supporting computer vision systems' precision to ensure accuracy and robustness in a variety of difficult conditions seen in poultry farming. The challenges that researchers will face in the future are keeping the hardware active during detection for an extended period. Next, better ways of data processing and analysis must be discovered to lessen the effect of noise data. In the end, as computer vision technology continues to advance in the poultry farming industry, more fields will need to be integrated, and a greater number of qualified people will be needed [40].

The term "smart aquaculture" refers to a sustainable development trend in the aquaculture industry, which is characterized by an emphasis on automation and intelligence. Smart technology has helped various industries, including aquaculture, by decreasing the need for human labor, increasing aquaculture output, and minimizing environmental impact. To put it simply, machine learning is a branch of AI that uses model algorithms that have already been trained to recognize and learn features based on observational data. Identification by size, density, color, illness, and species are just some of the areas where machine learning has been put to use in smart aquaculture. Aquaculture has potential uses for AI, with a particular emphasis on the use of computer vision and machine learning. The use of algorithms and machine vision in smart aquaculture has been investigated by academics and farmers alike. Popular uses include measuring the growth and development of farmed fish, fish illness diagnosis, counting fish, identifying species, managing feed, and checking water quality. In the future, smart aquaculture will be more widely available in both onshore (in places like hatcheries and farms) and offshore (in places like reefs and other marine environments) aquaculture systems. In aquaculture, offshore, in confined spaces, machine learning and computer vision can be put to good use to monitor fish health, keep an eye on potential predators, and keep track of how much each fish weighs. The ability to identify sick fish or a damaged cage net is crucial. Farmers typically rely on manual processes for the entirety of the aquaculture manufacturing cycle. Diseased fish, for example, often swim to the surface because they are too weak to swim deeper, or else, someone would need to jump in the water and look for it. It is beneficial to use machine learning/vision coupled with a camera placed underwater. This technology has the potential to serve as a model for the future of offshore cage culture due to its ability to monitor and manage environmental factors such as temperature, humidity, and light intensity, as well as fish weight and size, in real time [41].

References

1. Vasileios, Moysiadis, Panagiotis, Sarigiannidis,VasileiosVitsas, *et al.*," Smart Farming in Europe" *Comput. Sci. Rev.*,Volume 39, 100345, February 2021.
2. Christian Zinke-Wehlmann & Karel Charvát *et al.*, "Introduction of Smart Agriculture", *Big Data in Bioeconomy*, pp 187–190, 14 August 2021.
3. António Monteiro ,Sérgio Santos and Pedro Gonçalves ,"Precision Agriculture for Crop and Livestock Farming-Brief Review" *Animals,* 2021, 11(8), 2345
4. ComputerVision in Agriculture, https://medium.com/swlh/computer-vision-in-agriculture-d84b69c6858e, Accessed 14.01.2023.
5. JakeAstill, Rozita A.Dara,Evan D.G.Fraser,BruceRoberts, *et al.,* "Smart poultry management: Smart sensors, big data, and the internet of things ", *Comput. Electron. Agric.*, Volume 170, 105291, March 2020.
6. Cong Wang, Zhen Li, Tan Wang, Xianbao Xu, Xiaoshuan Zhang, *et al.*, "Intelligent fish farm—the future of aquaculture" *Aquacult. Int.*, volume 29, 2681–2711 (2021).ages
7. Computer Vision in Agriculture: Introduction to Smart Farming, https://easyflow.tech/computer-vision-in-agriculture-introduction-to-smartfarming/, Accessed 15.01.2023.
8. Visoai,Top Applications of Computer Vision in Agriculture (2022) https://viso.ai/applications/computer-vision-in-agriculture/, Accessed on 14.01.2023
9. Nikesh Gondchawar, Prof. Dr. R. S. Kawitkar, "IoT based Smart Agriculture", *Int. J. Adv. Res. Comput. Commun. Eng.*, Vol. 5, Issue 6, June 2016.
10. K. A. Patil; N. R. Kale *et al.*, "A model for smart agriculture using IoT" *2016 International Conference on Global Trends in Signal Processing, Information Computing and Communication (ICGTSPICC)*, 22-24 December 2016.
11. Philip Thornton, Bruce M. Campbell, Tobias Baedeker *et al.,* "Climate-smart agriculture for food security", *Nat. Clim. Change*, 1068–1072, 2014
12. "*Computer Vision Techniques*", https://www.javatpoint.com/computer-vision-techniques, Accessed on 20.01.2023.
13. V.G.Dhanya, A.Subeesh, N.L.Kushwaha, Dinesh KumarVishwakarma *et al.*, "Deep learning based computer vision approaches for smart agricultural applications" *Artif. Intell. Agric.*, Volume 6, 2022, Pages 211-229, 30 September 2022,
14. Juliana Freitas Santos Gomes, Fabiana Rodrigues Leta, "Applications of computer vision techniques in the agriculture and food industry: a review", *Eur. Food Res. Technol.*, 235, 989–1000 (2012), 12 October 2012.

15. Nick Babich, "*What Is Computer Vision & How Does it Work? An Introduction*" https://xd.adobe.com/ideas/principles/emerging-technology/what-is-computer-vision-how-does-it-work/, Accessed on 15.01.2023.
16. "*The 5 Most Amazing Computer Vision Techniques Learn*", https://www.analyticsinsight.net/the-5-most-amazing-computer-vision-techniques-to-learn/, Accessed on 15.01.2023.
17. Xinting Yang, Song Zhang, Jintao Liu, Qinfeng Gao, *et al.*, "Deep learning for smart fish farming: applications, opportunities and challenges", *Rev. Aquacult.*, Volume 13, Issue 1 p. 66-90,29 June 2020.
18. Cedric Okinda, Innocent Nyalala, Tchalla Korohou, Celestine Okinda, *et al.*, "A review on computer vision systems in monitoring of poultry: A welfare perspective", *Artif. Intell. Agric.*, Volume 4, 2020, Pages 184-208, 6 September 2020.
19. Rotimi-Williams Bello, Ahmad Sufril Azlan Mohamed, And Abdullah Zawawi Talib, *et al.*, "Contour Extraction of Individual Cattle From an Image Using Enhanced Mask R-CNN Instance Segmentation Method", *IEEE Access* 2021.3072636, April 12, 2021.
20. Felipe Napolitanoda Fonseca, Jair MinoroAbe, Irenilzade Alencar Nääs, *et al.*, "Automatic prediction of stress in piglets (Sus Scrofa) using infrared skin temperature", Computers and Electronics in Agriculture, Volume 168, 105148, January 2020.
21. BrahimAchour, MalikaBelkadi *et al.*, "Image analysis for individual identification and feeding behaviour monitoring of dairy cows based on Convolutional Neural Networks (CNN)", *Biosyst. Eng.*, Volume 198, Pages 31-49, October 2020.
22. Sebastián R. Vanrella, José O. Chelottia Julio R. Gallib Santiago A. Utsumic, *et al.*, "A regularity-based algorithm for identifying grazing and rumination bouts from acoustic signals in grazing cattle", *Comput. Electron. Agric.*,Volume 151, Pages 392-402, August 2018.
23. Mohammadamin Kashiha, ClaudiaBahr, Sanne Ott, *et al.*, "Automatic weight estimation of individual pigs using image analysis", *Comput. Electron. Agric.*, Volume 107, Pages 38-44, September 2014.
24. S. Amraei, S. Abdanan Mehdizadeh & S. Salari. "Broiler weight estimation based on machine vision and artificial neural network", *Br. Poult. Sci.*, Volume 58, 2017 - Issue 2, 2017.
25. Jintao Wang, Mingxia Shen, Longshen Liu *et al.*, "Recognition and Classification of Broiler Droppings Based on Deep Convolutional Neural Network".
26. Diego PereiraNeves ,Saman AbdananMehdizadeh *et al.*, "Detection of flock movement and behaviour of broiler chickens at different feeders using image analysis", *Information Processing in Agriculture*, Volume 2, Issues 3–4, Pages 177-182, October–December 2015.
27. ChristianLamping, MarjoleinDerks, PeterGroot Koerkamp, GertKootstra *et al.*, " ChickenNet - an end-to-end approach for plumage condition assessment

of laying hens in commercial farms using computer vision" *Comput. Electron. Agric.*, Volume 194, 106695, March 2022.

28. Katarina Pichova, Janicke Nordgreen, Christine Leterrier, *et al.*, "The effects of food-related environmental complexity on litter directed behaviour, fear and exploration of novel stimuli in young broiler chickens", *Appl. Anim. Behav. Sci.*, Volume 174, Pages 83-89, January 2016.
29. Usama Iqbal, Daoliang Li and Muhammad Akhter, "Intelligent Diagnosis of Fish Behavior Using Deep Learning Method", *Fishes* 2022, 7(4), 2021.
30. Shuo Hong Wang, Jing Wen Zhao & Yan Qiu Chen, *et al.*, "Robust tracking of fish schools using CNN for head identification", *Multimedia Tools Appl.*, volume 76, 23679–23697 2017.
31. Ahsan Jalal, Ahmad Salman, Ajmal Mian, *et al.*, " Fish detection and species classification in underwater environments using deep learning with temporal information *Ecol. Inf.*,Volume 57, May 2020, 101088.
32. Torben Möller and Tim W. Nattkemper, *et al.*, "ALMI—A Generic Active Learning System for Computational Object Classification in Marine Observation Images".
33. Gerami M.H; Safiyari H; Jafari A; Mousavi-Nasab M "*Application of Machine-vision to assess weight of fish*" (Case study: Oncorhynchus mykiss), Iranian Journal of Fisheries Sciences, October 2015.
34. "Computer Vision and Deep Learning for Agriculture", https://pyimagesearch.com/2022/08/10/computer-vision-and-deep-learning-for-agriculture/, Accessed on 20.01.2023.
35. Mahmoud Abdelsalam; Sajad Khorsandroo; Sudip Mittal, "Security and Privacy in Smart Farming: Challenges and Opportunities" *IEEE Access*, vol. 8, pp. 34564-34584, 2020.
36. Mohd Javed, Abid Haleem, Ravi Pratap Singh *et al.*, "Enhancing smart farming through the applications of Agriculture 4.0 technologies", *Int. J. Intell. Networks*, Volume 3, Pages 150-164, 2022.
37. HongkunTian, TianhaiWang, YadongLiu, XiQiao *et al.*, "Computer vision technology in agricultural automation —A review, *Information Processing in Agriculture*" ,Volume 7, Issue 1, Pages 1-19, March 2020.
38. Dario Augusto, Borges Oliveira, Luiz Gustavo Ribeiro Pereira *et al.*, "A review of deep learning algorithms for computer vision systems in livestock", *Livest. Sci.*, Volume 253, 104700, November 2021.
39. Bozar Nasirahmadi, Sandra A. Edwards, Barbara Sturm, "Implementation of machine vision for detecting behaviour of cattle and pigs", *Livest. Sci.*, Volume 202, Pages 25-38, August 2017.
40. Balasubramaniam, S., & Kavitha, V., A survey on data encryption tecniques in cloud computing. *Asian J. Inf. Technol.*, 13, 9, 494–505, 2014.

41. Thi Thu Em Vo Hyeyoung Ko, Jun-Ho Huh , "Overview of Smart Aquaculture System: Focusing on Applications of Machine Learning and Computer Vision" *Electronics*, 2021, 10(22), 2882;
42. Balasubramaniam S and K Satheesh Kumar, "Optimal Ensemble learning model for COVID-19 detection using chest X-ray images", *Biomed. Signal Process. Control*, Volume 81, 2023 ,104392.
43. Balasubramaniam S, Mohammad Haider Syed, Nitin S. More, Vijayakumar Polepally, "Deep learning-based power prediction aware charge scheduling approach in cloud based electric vehicular network", *Eng. Appl. Artif. Intell.*, Volume 121, 2023, 105869.

11

Forestry Management with AI and Drone Technology – Digital Forestry

M. Shanthalakshmi*, M. Jeevasree, R. Kavitha, V. Madhumathi, S. Mythreye and A. Naafiah Yusra

Sri Venkateswara College of Engineering, Sriperumbudur, India

Abstract

Effective forest management protects the incredible variety found in nature. In order to preserve this invaluable asset, we need to employ cutting-edge forest management techniques. Remote sensing and drone technologies have emerged as the fundamental foundations of today's most well-known developments in technology due to the high-resolution satellite images employed in the data collecting and processing procedures. In the rapidly growing sector of "digital forestry," these technologies are indispensable. Using digital tools like remote sensing, big data, and artificial intelligence (AI), "digital forestry" measures, monitors, and manages forests to maximize their social, economic, and ecological advantages. Using AI and the Internet of Things (IoT) to analyze massive amounts of data might significantly improve our ability to predict, prevent, and deal with forest disturbances. In the end, this will allow for better forest management that can last for generations. With the use of technology, especially AI, public organizations, governments, and individuals may better respond to threats. By analyzing the data, looking for indicators of deforestation, and taking steps to prevent the process, conservationists are able to maintain track of the land's changes throughout time. Drones and IoT-connected sensors that upload data to the Cloud and relay the findings through satellite networks have become indispensable tools for forest managers. When applied to forest monitoring, for instance, satellites can provide early warning signals of fires or places at risk of fire before direct observations can be obtained. This expedites the time it takes to respond. Because of this, this article's major objective is to present the notion of "digital forestry," which works to reduce deforestation and boost sustainable development by employing tools that streamline forest evaluation processes.

**Corresponding author*: mshanthi.s@gmail.com

Rajesh Kumar Dhanaraj, Balamurugan Balusamy, Prithi Samuel, Malathy Sathyamoorthy and Ali Kashif Bashir (eds.) Computer Vision in Smart Agriculture and Crop Management, (259–282) © 2025 Scrivener Publishing LLC

***Keywords*:** Drone technology, remote sensing, IoT, artificial intelligence, sustainability

11.1 Introduction

The successful upkeep of forestry operations is critically important to the preservation of natural biodiversity. The different benefits that forestry may make to higher agricultural productivity include stabilizing the soil, increasing the rate at which groundwater is recharged, providing shade, regulating the temperature, and improving the cycling of nutrients. These are only some of the numerous ways that increased forest cover might aid agricultural output, but there are many more. For us to be able to keep this one-of-a-kind resource for future generations, we need forestry management that is cutting-edge.

Technology, such as intelligent sensors, is assisting in the improvement of forest management in Brazil; however, this is not the case in every country. Many of the most underlying difficulties in environmental protection may be partially overcome with the help of technological advancements not just in India but also everywhere else in the world.

Assessment of risks, digital asset registration systems, specialist forestry consulting, and research on forests that makes use of high-precision data are some of the services that are offered by Smart Forest as part of its suite of solutions. The program is presently awaiting clearance from two of the main forest firms in Brazil, where proof-of-concept testing has yielded good results. This approval is necessary for the initiative to move forward.

According to the first findings, which cover only a brief period of time, Smart Forest appears to have a higher rate of success in data collecting than traditional sources. One of the most important ways that the Internet of Things (IoT) is applied in nature conservation is through the gathering of statistics from internet-enabled devices and their subsequent dissemination in perfect sync. That is the apparatus at work in making the IoT a reality.

If India is going to take advantage of the IoT in forest management, and more specifically, the employment of green bots, the most significant impediment is the problem of poor web access in extremely wooded areas. Before green bots and IoT systems may be used in forests, there is a prerequisite that must be satisfied first: the increase of the evidence base. When it comes to planning in India's cities, it is vital to utilize the "bottom-up method" since each city is responsible for its own jurisdiction and the functional components of planning. This makes it such that planning must be done from the ground up.

An underlying source is necessary in order for these technologies to be employed in any way. When applied in combination with technologies that have been developed more recently, drone technology has the ability to perform a broad spectrum of functions. Some typical applications of these duties include but are not restricted to: monitoring climate change, carrying out search operations in the aftermath of natural catastrophes, photographing and recording and managing live stock.

One approach to addressing this issue is to employ urban municipal governments to facilitate participation. The United Local Governments were recommended to be strengthened and a financial structure to stimulate urban development was proposed in the 12th Five-Year Plan (2012–2017).

Every single forest in India has its own unique set of challenges, and those challenges can only be taken on and solved one at a time. In order to provide sufficient credible data and utilize it to make changes that are helpful, there needs to be a concerted effort that is both collaborative and coordinated on the part of persons working in a variety of sectors, such as forestry, land, and climate change. This effort is required in order to provide adequate credible data and utilize it to make changes that are helpful. In addition, the departments and units that are accountable for woodland maintenance need engagement in the method of deciding what to do so that the democratic administration that includes everyone may be assured. This can be accomplished by ensuring that they have access to relevant information. Partnerships between legislatures and commercial firms, conservationists, activists, and relevant academicians are needed to ensure the simple and widespread adoption of green bots. This will ensure that the full potential of green robots is utilized.

11.2 Drone Technology

Fixed-wing drones and multi-rotor drones are the two primary categories of drones. Two inactively lifting wings on a fixed-wing drone enable it to move at a set angle in the air while being controlled by two motorized wings. Multi-rotor drones move as a result of their motors, as opposed to multi-rotor drones, which move in reaction to the speed and direction of their motors. Multi-rotor aircraft can also be broken down into these subcategories as aircraft with single, four, six, and eight propellers as shown in the Figure 11.1. Because of the high energy requirements for lifting and maintaining altitude, multi-rotor drones go farther and fly for shorter distances than fixed-wing drones.

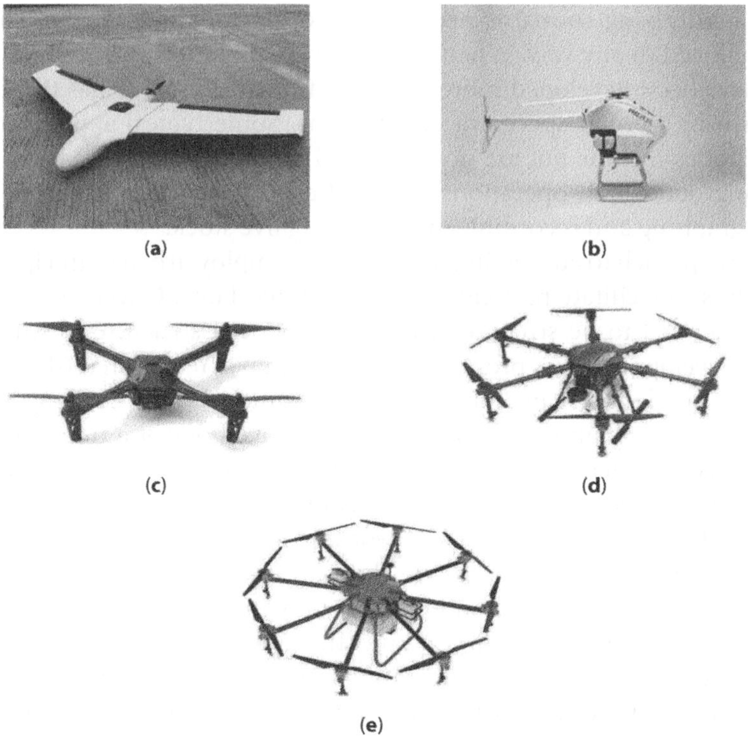

Figures 11.1 (a to e) Types of drones.

In the forestry industry, those who adopt the usage of drones stand to gain significantly from doing so. Good quality images may be captured with this fascinating and quickly developing technology as part of a larger data set, which might help businesses and scholars in the field quickly identify and track the welfare of forests and the wildlife that inhabit them.

Drones have the ability to monitor the health of trees, follow the positions of animals and their migratory trends, and could track down or communicate with rogue or mischievous individuals. When compared to the old techniques of data collecting, the technologies are both significantly more effective and economical than their predecessors. In addition, nimble drones are able to access areas that are either hazardous for humans to maneuver safely or are inaccessible altogether. We will explore the use of drones in the identification and prevention of tree diseases, the monitoring of animals, and human contact, to name just a few applications (security, as well as search and rescue).

11.3 Drones Employed for Disaster Management

The continued existence of humans depends on the natural surroundings of forests, woodlands, and other green areas. Regrettably, disaster management is already steadily raising the possibility that the circumstances of green areas will worsen and become destitute. The biggest and most ubiquitous hazard to European woods is from wildfires. They are to blame for climate changes, soil erosion, and biochemical alterations to the soil. A lot of interrelated factors, some of which are constantly changing over time, influence wildfires, making them a particularly complex event. People have always used slash-and-burn land management strategies all around the world, particularly to prepare areas for agricultural and residential enterprises. There has been a movement against the widespread abandonment of rural regions in the last few ages, which has led to the recolonization of numerous wooded and shrubby areas. The number of trees and shrubs in these places has increased as a result of this trend. These circumstances have produced a denser rural landscape, which has increased fire hazards, wildfire frequency, incidence, and volume. It also had detrimental consequences on inhabitants, infrastructural facilities, and the ecosystem. As a result of these actions, it is becoming more urgent to figure out how to put out forest fires.

Recently, there was a substantial increase in the usage of drones in forest fire management and response. As a way to provide real-time information, track the progression of a fire, and potentially reduce the need for ground-based firefighters to work in hazardous conditions, the implementation of drones that are upgraded with thermal cameras, putting out blazes systems, or other firefighting equipment has been proposed.

The Forest Service of the United States of America investigated drone usage in forest firefighting in research that was carried out earlier this year. In this particular investigation, drones that were fitted with thermal cameras were utilized to monitor the development of a fire and supply firefighters with information that was current at all times. According to the findings, the utilization of drones in the control of forest fires has the potential to significantly improve the situational awareness of fire controllers, which in turn leads to more effective decision-making on their behalf.

In another one of these case studies, drones that were armed with firefighting equipment were used to put out a blaze in a Spanish forest. The drones were utilized to put out fires in high-risk areas and stop the fire from spreading. According to the findings, employing drones in such a manner may considerably boost the effectiveness of firefighting operations, which in turn reduces the amount of time that firefighters on the ground need to spend working in hazardous situations.

There are potential benefits linked with the drone's usage in controlling fires in the forest; nevertheless, there are also obstacles related to their utilization. The requirement that drone pilots undergo appropriate training and obtain certification is one of the most significant obstacles. This includes training in the safe and effective operation of drones in forest fires, as well as training in the use of firefighting equipment such as thermal cameras and fire extinguishing systems. Additionally, this includes training in the safe and effective operation of drones in other types of fires. Also, the establishment of safety rules is needed to ensure the safe and effective use of drones in forest fires in order to prevent accidents and maximize their potential.

The requirement for reliable communication and coordination systems is still another obstacle to overcome. For drones to be used effectively in combating forest fires, ground-based firefighters and those operating the drones themselves need to maintain constant communication and cooperation. This involves the creation of well-defined protocols for the transfer of information between operators of drones and employees situated on the ground, as well as the utilization of appropriate radio or other communication networks.

In order for drones to be used effectively in fighting forest fires, there must also be regulatory concerns. There are stringent laws governing the use of drones in a number of nations. These regulations include guidelines for flying over populous areas as well as restrictions on the flight height. In addition, the utilization of drones in the battle against forest fires could call for the consent of regulatory entities like aviation authorities or firefighting organizations.

These unmanned aerial vehicles (UAVs) can consequently supplant batteries and refill quenching fluid, subsequently guaranteeing that the movement does not halt. The reason for this is to propose a conceptual system, with the assumption that in arranging a ramble swarm framework to be successful in combating fires, it has got to satisfy several characteristics, including the following:

(1) There cannot be any interruptions to the service for a very long period if it is going to ensure organizational consistency. (2) With the brief lifespan of modern strolls (about an hour), it must be capable of automatically replacing spent cells and incorporating themselves towards a recharging connection. (3) For accommodating a large number of users, replenishing the UAVs using a fitting quenching fluid set in holders docked to UAVs;

and it is fundamentally composed of a bolster unit that is able to effortlessly move and place itself in near vicinity to the front line of the fire battle. This unit is mindful of overseeing the swarm of UAVs.

In expansion, UAVs that will be flown in situations with high temperatures ought to be developed utilizing materials that are safe to fire. On the other hand, putting such a framework into hone is not reaching to be a straightforward assignment and will require a tall level of innovation. In expansion, the volume of water that can be conveyed by a drone is a bit lower when compared to how much fluid that can be carried using a fire-battling aircraft.

Thus, a huge proportion of UAVs could ensure both chronological advancements, as well as uniformity in dispersion. Both of which are outlandish to realize with an aircraft owing to the timing of the supply of the benefit. For forest fires to be appropriately quenched, intense dissipation is basic. However, a critical parcel of water that is dropped from airplanes in tall volumes does not contribute effectively to fire concealment since it does not have sufficient time to dissipate. Moreover, there is the plausibility that UAVs may be outfitted with a technology that seems to produce a stream of polydisperse beads of water. It has been illustrated that a water supply that is scattered all throughout the time and area can reduce the quantity of water that is necessary for putting out woodland blazes.

So, the ramble framework can create the so-called rain impact, which alludes to the release of modest volumes of fire-battling fluid by spraying it on the flames and on the field that is around it, instead of dumping the fluid in a concentrated way. This strategy has been demonstrated, both in hypothesis and in practice, to be especially valuable within the context of fire-battling systems in either private or commercial settings (fire sprinkler frameworks).

The reason for this is to decide whether an administration framework for a ramble swarm that fulfills the requirements outlined in [1–3] is competent in producing an adequate stream of quenching fluid to viably combat a timberland fire that is concentrated and incorporates a restricted zone of spread. We have a theory that each ramble can carry between 5 and 25 liters of water. This speculation is attainable when one considers the properties of variously produced UAVs, which are laid out in the Table 11.1.

Currently, the established design of drones in the hive cannot communicate with one another due to the design. Subsequently, this highlight is presently unavailable (example: no work arranged). The interaction between the stage with the UAVs uses progressive association architecture. This permits central insights, which is the stage or base station, to interact directly with

Table 11.1 Introduces several drones and the firms that produce them, outlining their features and uses.

Drone	Company	Type	Weight (Kg)	Dimensions (mm)	Folded dimensions (mm)	Propeller/rotor number	Speed (km/hr)	Payload (Kg)	Flight time (Min)
PD6B-Type 2	Prodone	Hexacopter	19.5(4 batteries Included)	L1874 W2060 H475	L1348 W600 H474	6	Max 60	No Payload ------------- 10 ------------- 20 (Practical use)	35 ----------- 15 ----------- 10
GD-40X	Gryphon	X8 octocopter	12 (dry weight) 40 (max takeoff weight)	D 1400	D1000 Detachable arms Reactable gear	8(4+4) Coaxial Propellers	Max 50 Avg 40	No Payload ------------- 22-25	50 ----------- 24
Vulcan D8	Vulcan	X8 Octocopter	16 (dry Weight) Max 55	L1400 W 1150 D1670	L1400 W400 HS00	8(4+4) Coaxial Propellers	Max 80 Avg 30/40	No Payload ------------- 10 ------------- 20	>30 ----------- 22 ----------- 14
Griff 135	Griff Aviation	X8 Octocopter	Max Takeoff weight 135	L2410 W2260 H470	L 1440 W770 H470	8(4+4) Coaxial Propellers	---------	No Payload ------------- 30 (max 50)	>30 ----------- 25-30

each person's ramble while it is refueling to grant them their claim flight plans. The stage can store information such as the current battery level, flying position, altitude, and speed, as well as transfer course plans to numerous UAVs. The ramble swarm can be viably requested and facilitated in this way, permitting synchronous flying along a course that is both directed and productively arranged. The platform has the potential to grant network to UAV's through the utilization of the 4G/5G remote organizing infrastructure. Those fast-moving remote design solutions built for slightly elevated mobile systems might overpower the stage to remotely work the UAVs and send and get information from them in real-time time. Also, these advances are created for broadband cellular systems. In ensuing articles, an examination of the specialized usage of the system and the control of ramble swarms at the firing front will take put. A broad investigation will be conducted to examine the method of creating control techniques, gaining experiences in the possibilities of organizing, and collaborating with drones in an arrangement to guarantee secure versatility, expand proficiency, and limit collisions.

In order to prepare for possible fire scenarios, the CA display has been used to mimic blaze's growth for drones to intervene and either extinguish the fire completely or as if it were a parcel of it. The full set could be a re-enactment, and there was no real-world point of reference for this situation to draw on. The locale covers a modest range with an entire surface region of approximately 40,000 square meters. It is almost completely level and is recognized by plant types that are practically equivalent to those of the Mediterranean scrub. A few frameworks are used to characterize the characteristics that are involved within the handle of a fire. These frameworks incorporate the following: wind speed and course, the thickness of vegetation, and the kind of vegetation (grass and bushes). The likelihood values that are related to the thickness and kind of vegetation have been experimentally decided. Extra components, such as the steady fire engendering likelihood, and wind coefficient have been figured out by running a few recreations with a non-linear optimization approach. These recreations have moreover been utilized to calculate. The entire cell lattice is covered with haphazardly produced frameworks that represent the vegetation thickness and the different types of vegetation.

In an arrangement to run the re-enactments, a platform that can manage UAVs that are each prepared with 1 liter of quenching fluid was chosen. A platform that can control UAVs that are each equipped with one liter of quenching fluid was selected in order to conduct the reenactments. In the unlikely event that the front develops diagonally, drones may disperse the material to a total, which is determined by calculating the orthogonal of a square by cutting along the battery's increasing span with a figure break equal with to. The number of cells where UAVs could spread fluid and become divided by the height of the battery with the height of the battery's side if the front develops horizontally.

After determining where the stage would be found (on the south side of the space within the recreation), we turned our attention to the burst and launched a frontal ambush on both the head and the flank of the blast. It has been decided that the status of the touching cells of the fire front that are closer to the stage ought to be equivalent, which implies that there ought to be a continuous stream of water that puts out the fire in these cells. Since the utilization of drones, the whole area affected by the fire might move in several diverse ways, depending on the conditions of the blaze. To be more particular, the inclusion of both types of vegetation within the show causes the beginning of a fire that has unassuming blazes but quickly spreads over the range. In expansion, the faster the breeze blows, the speedier the blaze stretches and noteworthy sum of water that is required to put it out, expecting that all other conditions stay the same.

The implications of the platform's effect on the progression of a wild arrival fire are portrayed within the figure that can be found underneath. In specific, the variation of the region in time without any quenching mediation and with the intercession of the fire battling procedure is shown in Figure 11.2a to portray the alter range promptly with no outside efforts to put out the fire. The UAVs have begun arriving 15 minutes after the fire began, and it was decided that both the grassland and shrub region had the same sum of dampness in their particular ranges. When there could be a wind speed of 10 kilometers per hour, a stage will be able to put out the fire. Compared to the spread that would have happened in the nonappearance of any intervention, the spread depicted in Figure 11.2b demonstrates how the range changed over time with a consistent wind speed. At other events, UAVs are the ones to start the intervention. In conclusion, Figures 11.2c and 11.2d show the progression of fire fronts without any extinction and with the association of UAV's (found along the blue line and compared to cells), individually. Both scenarios appeared within the same figure. In these conditions, a single stage may not be able to entirely extinguish the fire; despite this, it is still a viable strategy for holding back the progression of the fire. In this specific circumstance, the use of two partitioned stages makes it doable to accomplish the whole destruction of the issue.

The diagram that you can see above illustrates the effects that the suggested firefighting approach would have about the progression of a woodland blaze that was gathered in California. Figure 11.2a show changes in the size of a fire over the course during which no fires were put out and the efforts being made (non-dashed lines) with the application of a specific firefighting technique

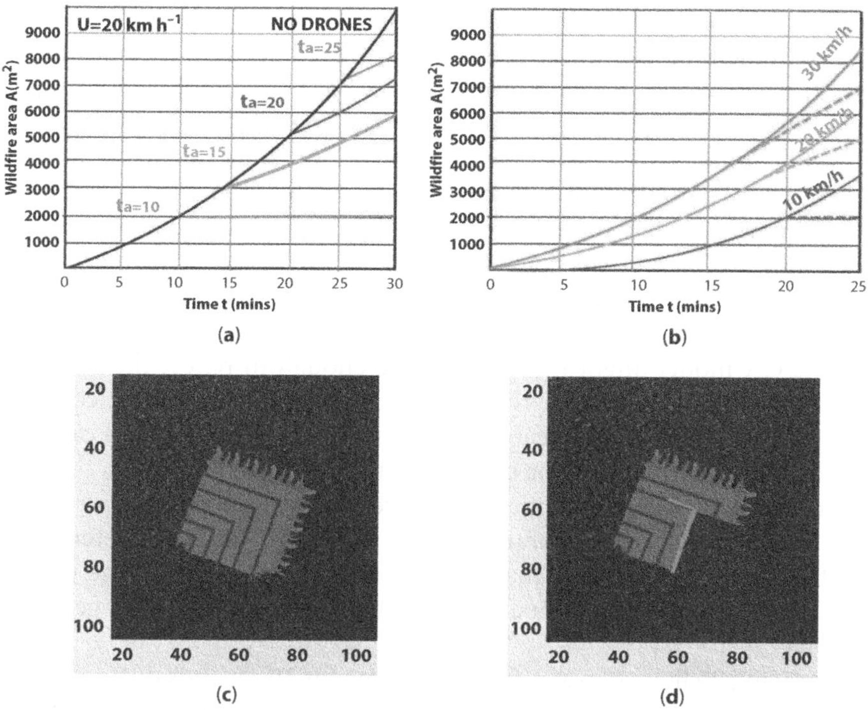

Figure 11.2 (a) Illustrates the spread of a forest fire over time without the use of drones; (b) Illustrates forest fire spread over time with constant wind and drone intervention (c) Depicts progression of fire front without the involvement of UAVs. (d) Depicts progression of fire front with the involvement of UAVs (found along the blue line).

(discontinuous lines) (dashed line). Fifteen minutes after the fire started, the first drones were spotted flying overhead. The progression of the fire is depicted by three separate curves here, each of which takes into account a different wind speed but maintains the same amount of moisture in the fuel (both for grassland and for shrub land). Figure 11.2b shows variations in the size among the affected area over time with a constant wind speed in comparison to the spread of the fire in the absence of any outside intervention (black line). At other occasions, drones are the ones to initiate the intervention. The blue line is patrolled by drones that are placed next to the line. The frontline of the growing blaze is represented by the cells that are red.

In the previous simulations, the interaction between the activity of drones and the firing front was taken into consideration; however, the simulations did not take into account any alternative methods that were intended to maximize the effect. As was said previously, the volume of water that can be conveyed by a drone is substantially less than the amount of liquid that can be held by firefighting UAV; nonetheless, UAVs are able to give consistency and stability throughout time of dispersion. The analysis of something like the connection involving UAV activity and wildfire frontline alteration needs thorough knowledge providing a brief liquid influence on the blaze's characteristics. The research has to account for how the liquid is dispersed, how much stress it is subjected to, along with ionized tiny particles. Even in comparison to the firefighting equipment that is now available, there is a paucity of published material on this subject, thus further study is going to be required to discover the mechanism of the continuous movement of the liquid (perhaps ionized into droplets) affecting the fire.

The CA model might not be adequate enough to reflect the influence of water on temperature, pyrolysis, and other processes, thus it could be essential to switch to physical fire models instead. These models can do a better job of representing the effect of water on these processes.

In this task, design specifications were not selected due to its reliance upon a large number of parameters the values of which are determined by the particular scenario that is being considered, whereas the goal of this duty is to investigate impacts caused by UAVs, which are not generally feasible.

Spraying 20 liters of water from the drone for 30 seconds during a flight above 10 meters could be one method used to create the illusion of rain falling from the sky. This results in an almost constant flow of water across the region where the drone is flying since it passes over it every 30 seconds. The utilization of a comparatively small number of drones might potentially supplement the conventional firefighting capabilities. For instance, they might be dispatched to either control flames in regions that are hazardous or are not readily available by firefighting teams, put an end to breakouts that would begin blazing overnight or hold the whole front until air support can arrive.

Noise pollution is a major problem with the current drone swarm method. In addition to this, drones are outfitted with imprecise sensors and communication devices that have a limited range. These difficulties need to be investigated in order to make the drone swarm system more reliable in the years to come. One of the challenges that designers have is coming up with a design for a system of clustering drones that is sustainable and can cover longer distances. Speech and text are the two modes of communication that may be used to convey information between drones that are on the ground and those that are in the air. When compared to technology based on text, technology based on speech is a more effective way. When dealing with time-sensitive applications or uses, text-based communication is seen as a challenging control solution. Technology is susceptible to being abused by adversaries in order to fulfil nefarious goals.

In conclusion, the utilization of drones in the control of forest fires has the potential to significantly improve the efforts that are made to put out flames. The case studies that were discussed earlier show proof of the benefits of using drones in forest fires, including greater situational awareness and increased efficiency of firefighting operations. Nevertheless, the implementation and regulatory components of this technology require careful attention. These factors include the necessity for adequate training and certification of drone operators, the establishment of safety regulations, and the necessity for effective communication and coordination systems.

11.4 Drones Equipped with Remote Sensing, GIS and LiDar for Geographical Dispersal Maintenance and Surveillance

The phrase "remote sensing" refers to a group of technologies that may convey information about an item without touching the thing itself. In present days, generating comprehensive 3D figures of forest ecosystem are used. It has also been used to learn more about mapping forest ecosystems. The variety of sensors available, in addition to their capabilities, has expanded alongside the development of technology. Conventional aerial photographs, which are identical to the image perceived by humans, to images that detect components that can be imperceptible to the human sight but are still detectable by the camera, such as the chemical composition of the object, span a wide spectrum of sorts. Conventional aerial photographs show the same thing. We will be able to precisely identify, characterize, locate, and describe the component portions of the Earth's surface once we have a view of the surface that has been acquired by remote sensing.

A sensor is a type of electrical circuit that is capable of recording the electromagnetic radiation that is incident onto it. A device known as a sensor is one that contains an optical component or system in addition to a detector that contains an electrical circuit. It is indeed capable of interpreting a change in the amount of energy being input in order to generate a corresponding change in either another kind of energy or identical energy. Each detector is utilized for telescopic stations on the planet's surface that rely on infrared energy for observational purposes to detect the different characteristics in the terrain. Several varieties of sensors are utilized for various tasks on the account of the necessities and objectives of those tasks.

The brightness of the objects being studied is measured by the sensors that are used in remote sensing at a certain wavelength. They are made up of a number of different parts, such as a system to the receiver, as well as a projector (the object), a diagnostic source and analyzer, an accelerator, along with a recorder. An imaging radiometer sensor, often known as a scanner, is fundamentally comprised of a radiometer that is supported by a system that acquires images one pixel at a time. This sensor is also commonly referred to as a scanner.

Multispectral scanning devices create photographs of the landscape in two dimensions for a swath that is located underneath the platform. A swath is the breadth of the strip that a scene takes up along the direction of the crosswise track. The surface of the planet is scanned using Across Track Scanners with a pattern of lines. The lines are aligned such that they point in a direction that is orthogonal to the movement of the sensor platform (i.e. across the swath). As a result, an optical-mechanical scanner, also known as a whiskbroom scanner, is utilized in the process of crosswise track scanning. This scanner moves from one side of the swath to the other. In the case of Landsat, examples of optical-mechanical scanners are the MSS and thematic mapper (TM), and in the case of NOAA, the AVHRR. These types of systems perform a scan of the environment by employing either a single detector or a collection of detectors for each band in order to capture the PAN or multi-spectral signals, respectively. The detector begins its examination by gazing at a neighboring element on the ground while using either a spinning or oscillating mirror. As a result, the information required to construct a strip or a scan line is gathered by the detector on a pixel-by-pixel basis. When the aeroplane travels ahead, successive scan lines are covered. This results in a sequence of continuous, or barely touching, thin strips of observation that create a two-dimensional picture of rows (scan lines) and columns.

Mounting sensors on platforms that are both adequate and sturdy are recommended for use in applications involving distant sensing. These platforms may have their origins on the ground, in the air, or even in space. As the platform is elevated to a higher altitude, both the spatial resolution and the amount of available observational area are enhanced as shown in the Figure 11.3 below. Hence, increasing the height at which the sensor is positioned results in a greater improvement in simultaneously increasing plans, strategies, and the synoptic view. The kind of sensor that will be attached to the platform as well as the intended use of the platform both have a role in determining the sorts of platforms and their properties. The platform can take the form of anything from a ladder to a satellite, depending on the mission. Moreover, sensors are sometimes installed on ground

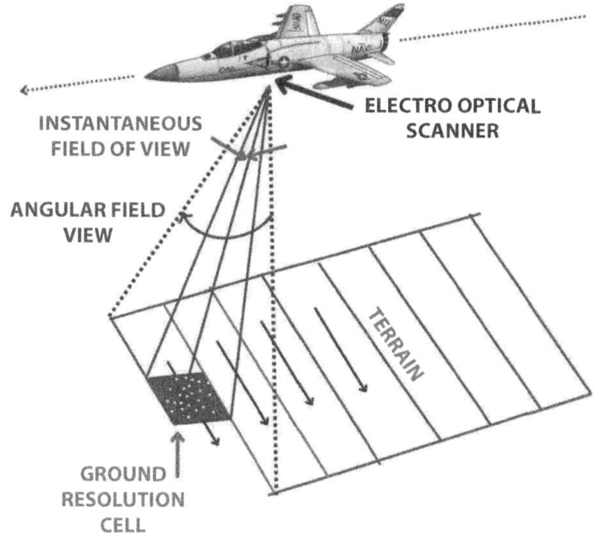

Figure 11.3 Illustrates the successive scan lines creating a two-dimensional picture of rows and columns.

platforms in order to complete certain tasks. Balloons and rockets are also utilized as platforms, despite the fact that aircraft and satellites are the most prevalent types used.

The recording of specific information on the objects or characteristics of the Earth's surface can be accomplished using ground observation platforms. They are being created with the purpose of enhancing scientific comprehension of the interactions between signals and objects, as well as between signals and sensors. A study in a laboratory and out in the field together make up what is known as ground observation. This method is utilized in the development of sensors, as well as the naming and categorizing of various land characteristics. Handheld platforms, cherry pickers, towers, portable masts, vehicles, and other types of structures can all be classified as ground observation platforms. In the laboratory and in the field, portable handheld photographic cameras and spectroradiometers are used extensively for the purpose of gathering reference data for ground truth verification.

The interpretation of images and other images through visual image interpretation is the process of extracting qualitative and quantitative information from a photograph or images using remote sensing data. It entails the identification of various items on the terrain, which might be natural or artificial and could consist of points, lines, or polygons. The terrain itself could be natural or artificial. It is dependent on the manner in which various characteristics reflect or emit the electromagnetic radiation that is incident, as well as the recording of these phenomena by a camera or sensor. From the beginning, when digital pictures and computerized categorization were not yet accessible, the aerial photographs were studied only via the use of visual interpretation. The training, the experience, the scale of the image, location where the research is conducted, the linked map, ground observation data, and other factors all contribute to the accuracy of the interpretation. Following the availability of satellite photos, the data were separated into two distinct processing methods: digital satellite images and analogue aerial photographs. Even if photos taken from aircraft can be analyzed by computers and images captured by satellites may be understood visually. Certain things in a picture or photograph could be quite easy to recognize, while others might be more difficult. It is dependent on a person's unique experiences and perspectives. The resolution of an image or photograph and the size of the photograph both have a role in the level of detail that can be examined in an image or photograph. It is not possible to do stereoscopic analysis on satellite photos because of their relatively smaller scale compared to aerial shots.

The technologies of remote sensing and geographic information systems (GIS) are the ones that are used to monitor and analyze the periodic shifts in the spatial distribution of a forest's diversity conditions. One of the tools that is used to integrate spatial data along with maps, aerial photographs, and satellite pictures with other quantitative, qualitative, and descriptive information that is recorded in databases is a GIS. The GIS is one of the tools that is utilized. A system that can capture, manage, store and retrieve the data, and presentation are all part of the work that is done for data synthesis. There is no need for any form of human interaction for remote sensing to create a portrait of the surface of the forest since it collects electromagnetic radiation that is back scattered from the characteristics that are present on the forest's surface. The purpose of the onboard sensor on the satellite with multi-spectral bands is to examine all of the resources on the forest surface based on the characteristics of their reflections. This is one of the fundamental tenets of remote sensing technology, and it refers to the fact that every surface feature of the forest has its own distinct electromagnetic spectrum reflection properties. Exploration of the region is carried out with the use of temporally obtained data from remote sensing and GIS in order to determine the geographical distribution of the phenomenon and how it varies over time.

This technical innovation has made it possible to conduct research on the detection and estimate of changes in biodiversity in a more densely forested area with a better degree of precision than was previously achievable. This research is only viable using the remote sensing approach. Hence, with the use of satellite imagery and GIS technologies, a forest supervisor may compile data on forested land, forest species present in a certain region, the extent to which people have invaded forested or guarded regions, the spread of wilderness environments, and so on. To ensure that proper laws are placed to control and regulate the way by which available assets can be used, this data is crucial for the development of strategies for managing forests and for the judgement call procedure. Moreover, data from satellites and multi-criteria analysis may be used to determine if a specific location or natural forest is right for a particular kind of natural wildlife.

In the past, the primary purpose of forest management inventories was timber management, and the primary focus was on recording area and volume by species. During the course of the last few decades, the ranges of duties associated with forest management have expanded. In the United Kingdom, foresters are up against a particularly difficult foe: an illness that impacts tree thickness instead of altitude (the latter being more detectable by traditional aerial surveillance). Experts had to dispatch out individuals walking with tape measures to document the thickness of every tree and construct physical datasets so they could monitor and report on the healthiness of the woodland, as aerial footage was unable to acquire enough information to identify the thickness of specific plants. Since advances in drone technology, a process that used to be highly time and arduous can now be completed in a fraction of the time, which is great news for everyone concerned.

One potential solution for this problem and others like it might be the deployment of a fleet of unmanned ground vehicles (UGV) armed with compact and agile UAVs that are capable of capturing detailed images and measurements. This would be one example of a solution. The UGV, which would be a solution that is only partially autonomous, would travel through the ground of the forest using wheels or "legs," and it would do so by following a path that was either periodically re or untethered. The UGV would be a solution that is only partially autonomous. At this point, an aerial UAV would be mounted atop the UGV, where it would be supported by a mobile power outlet. The mission would then continue. The aerial UAV would disentangle, start taking off, and utilize its secluded detection systems to perform a variety of operations after receiving instructions. These tasks would include tallying the trees in the vicinity, differentiating among different tree varieties, examining the altitude and/or circumference, and even monitoring hydrological regime in the region. This information that is acquired by the Drone and UGV is subsequently translated to understandable and quantifiable metrics, which would afterwards make it feasible for researchers and foresters to monitor changes with a precision that has not been attainable up to this point.

(a) (b)

Figure 11.4 (a) Satellites vs (b) drones.

The aforementioned LiDAR technology has reached such a state of technological advancement that it is now capable of navigating even above the canopy of forests, all the while counting the different trees down below, as well as gauging their altitude and circumference. This provides foresters and researchers with access to a resource that is of the utmost importance. Although it is feasible to use LiDAR to precisely gauge, number, and identify trees, the abilities for monitoring humans or fauna are not yet as advanced at this moment.

However, it is essential to keep in mind that the size restrictions that now apply to LiDAR devices are rather severe. An agricultural drone may be used in a number of ways to improve farming efficiency, such as more regular and precise aerial surveying and cautious chemical spraying. Unmanned aerial vehicles are utilized for mapping and imaging, especially when compared to other viable technologies for the purpose like satellite images. Figure 11.4 shows the pictorial representation of satellite vs drones and Table 11.2 is a comparison between the same for the data collection, it is important to keep in mind that drones can go far closer to crops than satellites can due to their location in space.

Table 11.2 Comparison of satellites and drones.

Functions	Satellites	Drones
Speed	Wait for satellite	Deploy on time
Spatial resolution	25 cm resolution	Centimeter level accuracy with RTK
Cost	High	Low
3D models and point clouds	No	Yes
Map area	Unlimited	$3km^2$ in one flight

11.5 Drones for Livestock Management

The use of drones for the purpose of keeping an eye on the livestock has captured the interest of the farmers. The photographs of the livestock that were taken by the drone may be processed to identify the quantity of cattle, as well as their behavior. This information is then transferred to monitor where the farmers can respond appropriately. The strategies are quite effective, particularly while monitoring livestock on a huge farming property, in addition finding human poachers down. This in turn causes the application to provide the wrong information to the farmers. A significant contribution is also made using lighting and background effects when portraying false information.

Research that is conducted in the use of drones inside the cattle farming industry has been the subject of more than one study. It has been shown that the utilization of a quadcopter type of drone in the monitoring of sheep livestock is permissible. Different flying controllers equipped with a machine learning algorithm to count the number of sheep and follow their moves and a 4G internet connection are the components, which make up the drone that was constructed. It was determined that processing data online gave a more desired outcome by means of accuracy.

According to studies, the minimum number of drones that should be deployed in order to track and monitor the maximum number of livestock can be determined. Each of the animals that were being monitored by the drone is outfitted with a GPS collar, which enables the receiver on the drone to receive information regarding the animals' location, as well as their health. The density-based clustering algorithm known as DBSCAN was utilized throughout the research project so that the minimum average distance between drones and animals could be determined.

Drones using thermal imaging cameras make it possible for a single operator to keep an eye on cattle. Because of this, farmers can maintain tabs on their cattle more frequently with less effort from their employees. It is easy for the drone pilot to check on the herd and observe if any animals are sick, wounded, or giving birth. The introduction of drones has made what was formerly a time-consuming and expensive process of monitoring the herd possible at all times. Some farmers will find thermal imaging particularly useful since it may detect predators of livestock. The revolutionary development of drone technology has had and will have far-reaching consequences for many aspects of modern life and industry. The agriculture sector appears to have enthusiastically embraced drone technology and is already making extensive use of these cutting-edge equipment to completely revolutionize the farming process. Farmers and the pilots who fly the high-tech drones can improve the efficiency of some parts of their operations.

Drones are incredibly helpful in various tasks, including agricultural visualization, cropping, food stock management, fertilizer spraying, and watering. Management of agricultural operations benefits from the use of drones because of the ease with which crop and livestock may be seen, measured, and acted upon using real-time data. It eliminates the element of chance from contemporary farming, empowering farmers to improve crop yields and manage more streamlined operations.

11.6 Conclusion

The majority of the research being done in this area focuses on farmers taking the role of drone pilots. While in reality, this can be done on its own, and farmers may direct their attention to other critically important duties. Farmers need expert understanding regarding the photographs acquired by UAVs in order to determine and establish appropriate methods, particularly in crop monitoring and mapping. Farmers must be aware that commercially available UAVs still have limited flight time. Drone manufacturing companies may soon prioritize developing farmland UAVs with improved range, image quality, and autonomy, and through research, as well as judgement alone without intervention of farmers is predicted to be the next frontier of research. Long range communication networks have also been more integrated with UAVs. Using farmland UAVs comes with a hefty price tag right now, but experts predict that will soon be shifting as a consequence of swift technical advancement.

There are several benefits to using drones in the forestry sector. This innovative and quickly developing medium can gather elevated photographs as component of a still greater aggregation of data which may relevant and effective and scientists in effectively comprehending as well as tracking the healthiness of woodlands and its populations.

The UAVs may be used to check on the well-being of trees, observe livestock migratory routines, and perhaps even search for and interact with missing or dangerous individuals. When contrasted with traditional means of gathering information, the emerging innovations provide significant advantages in both effectiveness and affordability. Furthermore, maneuverable UAVs could access areas that are too risky, even unattainable for people to get securely. They have many uses, some of them are pest control, animal observation, and the identification and avoidance of plant diseases including rescuing and providing security.

Things like the economy, illness, and weather are just a few of the many variables that cannot be manipulated in farming. Nevertheless, thanks to drone technology, farmers can now make informed judgements. Major technology companies have launched drone services to aid entrepreneurs and simplify many industries. Drone services are available to lend a hand to farms that may use some assistance using drone technology. Licensed drone pilots are available to assist farms in gathering the necessary information.

Bibliography

A.K., G.B. Wood, K.R. Skidmore and K.R. Shepherd, "Remotely sensed digital data in forestry: A Review," *Aust. For.*, 50(1), 1987, DOI: 10.1080/00049158.1987.10674493.

A Brief History of Forestry Policy and Forestry Management, 1960-1996. *Rural Probl. J.* https://doi.org/10.7310/arfe1965.32.57

Winkle, P. Digital forest management: the experience of Cantor. 67(6):630–635 in *For. Chron.*.

NIE, J., CHAO, X, WANG, Y., and LI, Y. (1 January 2022). A survey of the use of artificial intelligence and digital twins in sustainable agriculture and forestry. *Turk. J. Agric. For.*, 46(5), pages 642–661. https://doi.org/10.55730/1300011x.3033

Katsch, C and Kunneke, A. (2006, March). Forest inventory in the era of remote sensing technology. http://dx.doi.org/10.2989/10295920609505243 *South. Afr. For. J.*

Demir, N. Using unmanned aerial vehicles to detect trees from digital surface models. *J. For. Res.*, 29(3), pp. 813–821. https://doi.org/10.1007/s11676-017-0473-9

Kardoš, M. Žíhlavník, &Chudý, F. Applying digital photogrammetry to forest mapping. 222–230. *J. For. Sci.*, 53(5).

Hoffer, Richard M. Techniques of digital analysis for forest applications. *Remote Sens.*

R Zhou, L., Meng, R., Tan, Y., Lv, Z., Zhao, Y., Xu, B., & Zhao, F. Comparative analysis of UAV-based LiDAR and digital aerial photogrammetry for measuring crown-level canopy height in an urban setting. *Urban Green. Urban For.* https://doi.org/10.1016/j.ufug.2022.127489

Wang, X., Wang, Y., Zhou, C., Yin, L., and Feng, X. Monitoring of urban forests. 58, 126958 *Urban For Urban Green.* https://doi.org/10.1016/j.ufug.2020.126958

Koptev, S., & Skudneva, O. (2018, January 30). Concerning the Applicability of Unmanned Aerial Vehicles in Forestry Practise. Higher Education

Institutions Bulletin. Lesnoi Zhurnal, 1, 130–138 (*For. J.*). https://doi.org/10.17238/issn05361036.2018.1.130

Rai, S. (February 1, 2021). Drone/Unmanned Aerial Vehicle Forensics. *J. Digit. Forensic* (4n6). https://doi.org/10.46293/4n6/2021.03.01.02 13)TAN, Z.& OZET, E., KARAKOSE, M.(2022, August)

Drone-based drone tracking using deep learning. https://doi.org/10.24203/ijcit.v11i3.238

Nonami, K. (2016). Drone Technology, Innovative Drone Enterprises, and Future Prospects. 28(3), 262–272, *J. Robot. Mechatron.*. https://doi.org/10.20965/jrm.2016.p0262 15)Kale, A. A. (2021).

The Applications of Drone Technology. https://doi.org/10.2139/ssrn.3922787

Joglekar, M. N. U., Drone Camera Technology. 7(3), 1744–1747, *Int. J. Res. Appl. Sci. Eng.Technol.*. https://doi.org/10.22214/ijraset.2019.332

Niță, M. D. (2021). Testing a Forestry Digital Twinning Workflow Using a Mobile LiDAR Scanner and Artificial Intelligence Platform. Forests occurred in 1576. https://doi.org/10.3390/f12111576 18)Tang, L., & Shao, G. (2015).

Remote sensing by drones for forest research and management

Du, L., Zhou, Huang, K.,T., Zou, Z., Zhao, X., & Wu, H. (2014). Utilising Remote Sensing and the National Forest Inventory to Map Forest Biomass. MDPI. https://doi.org/10.3390/f5061267

Simons, N. K., Felipe-Lucia, M. R., Schall, P., Ammer, C., Bauhus, J., Blüthgen, N., F., Jung, K., Manning, P., Nauss, T., Boch, S., Buscot, F., Fischer, M., Goldmann, K., Gossner, M. M., Hansel, Oelmann, Y., Pena, R., Polle, A., Renner, S. C., Weisser, W. W. (2021, January 7). National Forest Inventories in Germany capture the multiple functions of managed forests. *The Roles of Aerial Photographs in Forestry Remote Sensing Image Analysis*. http://dx.doi.org/10.1186/s40663-021-00280-5

Hall, R. J.. The Roles of Aerial Photographs in Forestry Remote Sensing *Image Anal.*. https://doi.org/10.1007/978-1-4615-0306-4_3

Zhang, Y., Han, M., & Chen, W. (2018, November 20). Digital scenic area planning strategy from the perspective of protecting intangible cultural heritage doi.org/10.1186/s13640-018-0366-7

Krůček, M., Imbach, B., Král, K., Trochta, J., Vrška, T., & Zgraggen, C. ,Kellner, J. R.,Armston, J., Birrer, M., Cushman, K. C., Duncanson, L., Eck, C., Falleger, C., (2019). New Prospects for Forest Remote Sensing Utilising Ultra-High-Density Drone Lidar doi.org/10.1007/s10712-019-09529-9

Burghardt, F., Gradener, E., Elfgen, M., & Wohlgemuth, V. (2022, November). Comprehensive Drone System for Deployment in Disaster Scenarios, with an Emphasis on Fighting Forest Fires. Comprehensive Drone System for Deployment in Disaster Scenarios, with an Emphasis on Fighting Forest Fires | doi.org/10.1007/978-3-031-18311-9_9

Salam, A. (2019, December) Internet of Things for Sustainable Forest Management. *Sustainable Forestry and the Internet of Things* | SpringerLink. 7 March 2021. https://doi.org/10.1007/978-3-030-35291-2_5

Ausonio, E., Bagnerini, P., and Ghio, M. A Conceptual Framework for Drone Swarms in Fire Suppression Operations. July 2018 https://doi.org/10.3390/drones5010017

Staub, G. Case Studies on the Use of Remote Sensing to Detect and Monitor Trees in Diverse Environments in Chile. Chile Case Studies | IntechOpen. https://doi.org/10.5772/intechopen.72903

Zhang, Y., Han, M., & Chen, W. (2018, November). *EURASIP J. Image Video Process* - The strategy of digital scenic area planning from the perspective of intangible cultural heritage protection. SpringerOpen. https://doi.org/10.1186/s13640-018-0366-7

9 September 2021. Ochoa-Zezzatti, A., Ochoa-Ruiz, G., and Aguilar-Lobo, L. M. Using Hybrid Drone Data and Satellite Images for Geo-Referenced Fire Correlation in a Smart City Urban Forest. *Geo-Referenced Fire Correlation Utilising Hybrid Drone Data and Satellite Images* | https://doi.org/10.1007/978-3-030-68655-0_28

Perzl, F., & Teich, M. (2021). Geodata Requirements for Mapping the Protective Functions and Effects of Forests. *Effects of Forests* | IntechOpen. The application of a landscape lens to digital cultural heritage - Built Heritage.

The establishment and application of spatial decision support systems for digital forestry. (n.d.). DOI: 10.1186/s43238-020-00002-w. Establishing and Implementing Spatial Decision Support Systems for Digital Forestry | *IEEE Conference Publication* https://ieeexplore.ieee.org/document/5980806

A Digital Footprint from Multisource Data for Digital Technologies and Automation in Livestock Production Systems.

A Digital Footprint From Multisource Data: Digital Technologies and Automation in Livestock Production Systems | *IEEE Conference Publication* https://ieeexplore.ieee.org/document/9628544

Estimation of Forest Stand Height Using Ziyuan-3 Tri-Stereo Imagery and Lidar. Estimation of Forest Stand Height Using Ziyuan-3 Tri-Stereo Imagery and Lidar *IEEE Xplore*. https://ieeexplore.ieee.org/document/8897913

12

Drone Application and Use Cases in Smart Agriculture and Crop Surveillance: Future Research Directions

Nilotpal Das[1], Atin Kumar[1]* and Rohit Kumar[2]

[1]School of Agriculture, Uttaranchal University, Premnagar, Dehradun, Uttarakhand, India
[2]Faculty of Agricultural Sciences, GLA University, Mathura, Uttar Pradesh, India

Abstract
Drones (small unmanned aerial vehicles) present a variety of opportunities for the agricultural industry; common drone uses include pesticide application, soil sampling and fertilization, farm animal observation, real-time aerial imaging, and sensor data collecting. Drones are being extensively promoted for use in agricultural spraying by the Indian federal government. These "Kisan drones" are primarily designed to reduce time, improve resource usage in agricultural sprays (pesticides, etc.), and lessen the negative health impacts of manual pesticide spraying. In the future, airborne imaging, surveying, and transportation are all anticipated to benefit from the use of drones in agriculture. According to this analysis, the government is actively pursuing a liberalized strategy to encourage drones and provide institutions, individual farmers, and businesspeople with significant financial incentives to buy and utilize or manufacture drones. Drones reportedly use resources more effectively and save a significant amount of water. They also save time. In comparison to manual labor costs, the unit economics of drone spray cost per acre is coming up. For young people in rural areas, drone flying for agricultural purposes can be a good option. If farmers start to trust drones for spray-related tasks, there is a strong chance their use will be expanded to other cutting-edge use cases as well, which will be beneficial for Indian agriculture.

Keywords: Drone, agriculture, farmers, sensor, airborne imaging, surveying

**Corresponding author*: atinchaudhary0019@gmail.com

Rajesh Kumar Dhanaraj, Balamurugan Balusamy, Prithi Samuel, Malathy Sathyamoorthy and Ali Kashif Bashir (eds.) Computer Vision in Smart Agriculture and Crop Management, (283–302) © 2025 Scrivener Publishing LLC

12.1 Introduction

Drone technology is an amazing advancement that has the possibility to revolutionize the way agricultural regular manual tasks are carried out. Drone technology is being used more frequently by agricultural businesses worldwide to modernize farming. Remotely piloted aircraft systems (RPAS) are known as drones that are equipped with a propulsion system, programmable controller, the use of a satellite navigation system, computerized flight planning tools, and the ability to carry payload for completing a task, such as cameras and spraying systems. Unmanned aerial vehicle/systems (UAV/UAVs) and unmanned aircraft systems (UAS) [20] are two additional acronyms that are frequently used interchangeably; nonetheless, remotely piloted aircraft systems are the most official and global way to refer to such systems. Agriculture drone refers to a drone that is utilized for agricultural purposes. Drones are created to contain detectors that can transmit real-time data about the condition of crops or the movement of livestock, enabling efficient and accurate control of agricultural activities. The drone can either be programmed to follow a predetermined route using sophisticated operating either using inbuilt controller navigation algorithms or be remotely commanded using wireless connection. It can be refitted with various payload configurations of digital image sensors, such as multispectral systems and devices for high-resolution cameras used in field surveys, crop surveillance, scouting, spraying, and spreading, as well as monitoring of animals and fisheries. Farmers can precisely determine the size of their field, identify the type of crops and varieties, create soil maps for control of pests, and schedule the crop harvesting process using the data collected by drone-mounted cameras and data analytics. The operation's scope and productivity can be further increased by fully automating these drones.

Figure 12.1 Drone technology in agriculture [11].

Although there are many uses for drones in agriculture (Figure 12.1), these uses are restricted by national policies on their use.

Drones can be used to spray pesticides more effectively than labor-intensive, dangerous traditional methods, especially in challenging terrain like hills. Drone-taken, high-resolution photographs using the normalized difference vegetation index (NDVI) imaging technology can be combined with machine and artificial intelligence (AI) learning to predict crop output and better understand soil conditions. It is possible to find a stressed plant on its own and analyze it using image processing techniques. By taking preventive measures and using this result as a guide, producers can stop the transmission of illnesses that affect other crops. Action may be taken quickly to minimize the effects of climate change and unpredictable weather, optimize fertilization, rationalize irrigation, and biotic stressors like illnesses and pest insects should be avoided to minimize losses using insights derived, based on information gathered from remote sensing satellites and drones.

12.2 Definition of Drones

Unmanned aerial vehicles or remotely piloted aircraft or dynamic remotely operated navigation equipment system (DRONES) are aircrafts that can

Table 12.1 Specifications of drone types [14].

Item	Fixed wing drone	Quadcopter
Speed	11-17 m/s	8-10 m/s
Battery	Polymerized lithium	Polymerized lithium
Weight (kg)	2.5 – 6	1.5 – 5
Flight duration (minutes)	90	45
Flight altitude (meter)	150	90
Capturing photos capability	3 photos/second	3 photos/second
Area covered in single flight (km^2)	4 – 5	2 – 4
Type of terrain	Plain area	Hilly area

Table 12.2 Drone classification on weight basis [21].

Drone types	Weight	Remote Pilot License
Large	More than 150 kg	Required
Medium	More than 25 kg up to 150 kg	Required
Small	More than 2 kg & Up to 25 kg	Required
Micro	More than 250 g and up to 2 kg.	Not necessary for non-commercial use and necessary for commercial use
Nano	250 g	Not required

fly totally under a geographical positioning system (GPS) control without a pilot [18, 20]. Drones are another name for UAS. In other words, drones are presently aircrafts that can fly by loading a flight training or that can be sent using a remote control by a pilot on the ground to complete a mission. By 2022, the agriculture industry is expected to deploy drones in the second-largest global share, according to a Goldman Sachs analysis. Different specifications of a drone's type and their classification on weight basis is depicted in Tables 12.1 and 12.2.

12.3 Classification of Drones

12.3.1 Components of Agriculture Drones

The different components of the agriculture drones are listed below [15]:

- Frames (depicted in Figure 12.2)
- Controller systems (depicted in Figure 12.3)
- Propulsion systems (depicted in Figure 12.4)
- Camera systems (depicted in Figure 12.5)
- Navigation systems (depicted in Figure 12.6)
- Batteries
- Other Components (connectors, power systems, wires, carrycases, sprayers and sprinklers).

Figure 12.2 Drone frames [6].

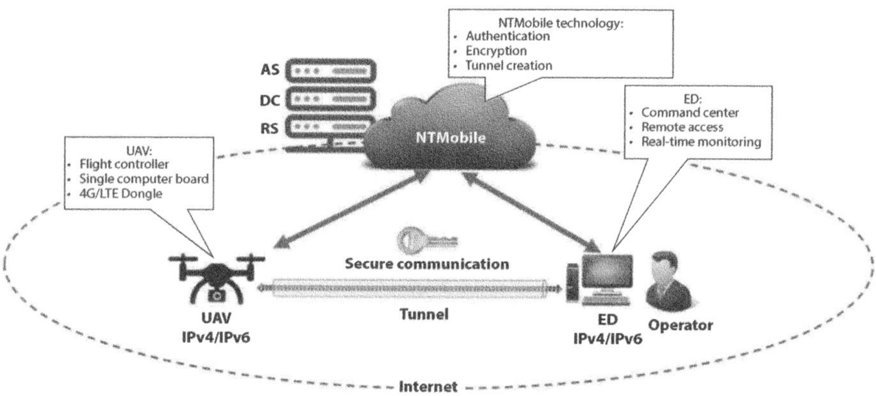

Figure 12.3 Controller systems [8].

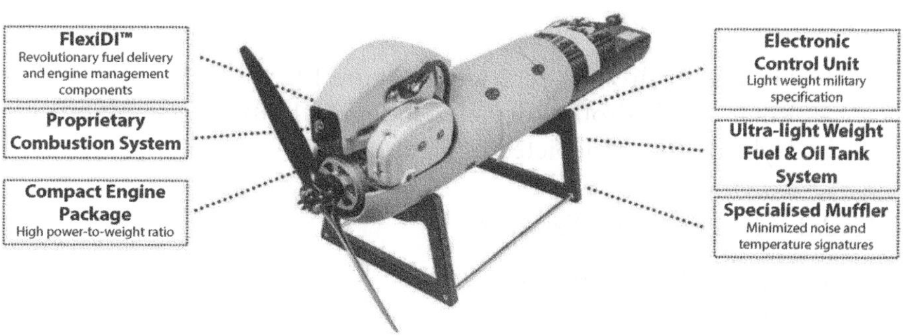

Figure 12.4 Propulsion systems [9].

Figure 12.5 Camera system [7].

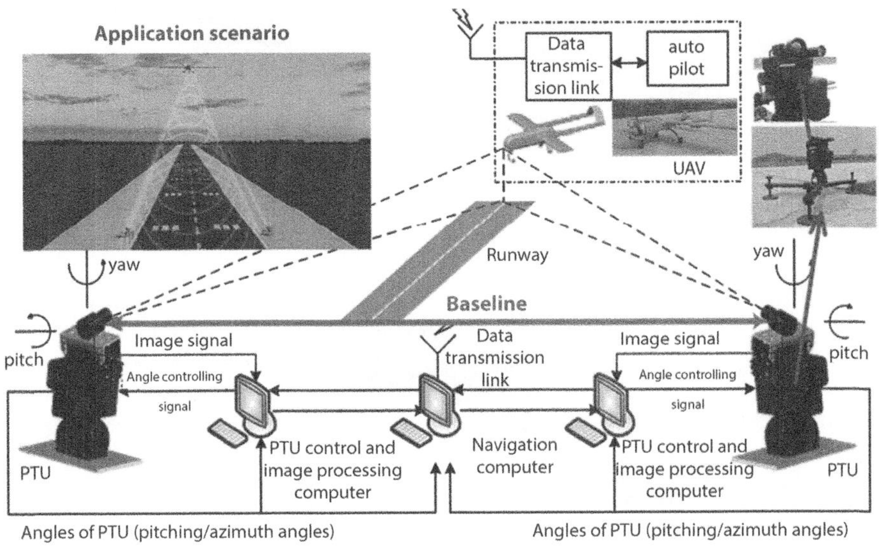

Figure 12.6 Navigation systems [12].

12.3.2 Drones for Agriculture by Offering

12.3.2.1 Hardware

- ➢ Fixed wing drones (depicted in Figure 12.7)
- ➢ Rotary blade drones (depicted in Figure 12.8)
- ➢ Hybrid drones (depicted in Figure 12.9)

12.3.2.2 Software and Services

- ➢ Data analytics software
- ➢ Imaging software
- ➢ Data management software
- ➢ Others

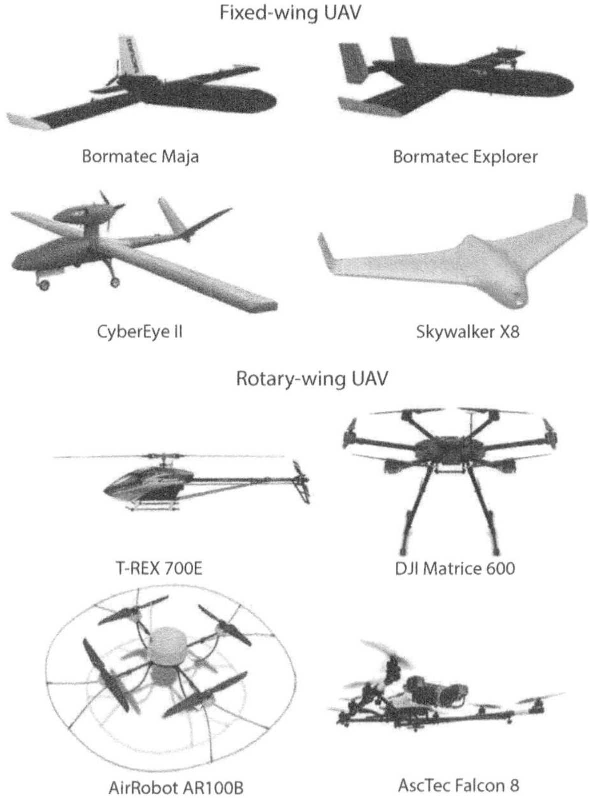

Figure 12.7 Fixed wing drones [13].

Figure 12.8 Rotary blade drones [10].

Figure 12.9 Hybrid drones [5].

The Government of India, in accordance with the regulations of the Directorate General of Civil Aviation, requires anyone who want to operate drones heavier than nano drones to possess a remote pilot license. Others call for operators to have remote pilot licenses.

12.4 Application of Drones in Agriculture

The following is a list of potential uses for drone technology:

- **Planning fields with soil analysis:**
  Drones can be used to monitor the nitrogen levels in the soil, the timing of planting, and the irrigation needs of fields and soil. Additionally, drones are useful for creating precise 3D maps that may be utilized for soil research to determine soil characteristics, content of moisture, and soil erosion.
- **Seed Pod Planting**
  Some businesses have developed an extra connection underneath the drone systems that can shoot pods holding soil that has previously been prepared with seed and plant nutrients, albeit it is not yet widely used. This lowers the cost of planting.

  A seed-dropping system that uses a drone can drop up to 60 seeds per minute and 28,800 seed balls in an eight-hour period. Seeds are disseminated once they reach the predetermined areas [4]. Start-ups have created drone planting systems that have a 75% adoption rate and an 85% reduction in planting costs [1].

- **Crop Monitoring:**
  Crop monitoring causes the most trouble for farmers, as well as many other parties involved in agricultural activities. This problem has gotten worse as unpredictable weather patterns have increased, increasing the danger of crop loss, and raising maintenance costs. By collecting multispectral geographic and temporal information at predefined scales relating to crop growth and health, drones can be utilized to plan its monitoring routes. Data analytics assist in gaining knowledge about crop health well before manual field scouting can reveal it.
  Drones with sensors can gather spectral data and produce maps that show changes in crop health. Along with lowering the cost, this technique improves crop management's consistency and effectiveness [17].
- **Crop Spraying:**
  For crop spraying on vast regions quickly, drones can transport appropriately sized reservoirs that can be filled with fertilizers, herbicides, or pesticides. Due to its self-contained and pre-programmed operation on predetermined schedules and routes, crop spraying is significantly safer and more economical. To ensure uniform and ideal spraying outcomes over diverse topography, additionally, drones have self-adjusting software employing time-of-flight (TOF) lasers, Global Navigation Satellite System signals (GNSS) signals, and ultrasonic echoes to determine altitude and speed. Drones are used by smart farms to spray agricultural products, reducing human exposure to toxic chemicals like pesticides and fertilizers. In terms of spotting treatments that are automated with stress sensing technologies that employ drones are unmatched in that they operate on sensors and cameras while maintaining the healthy portions. Spraying efficiency is approximately five times quicker with drones than with conventional apparatus.
- **Irrigation:**
  Multispectral indices can be used by drones equipped with thermal, multispectral, or hyperspectral sensors to locate the areas of the field that have moisture shortages. This aids in the precise and on-time arranging of irrigation for the designated locations.

- **Crop health assessment:**
  The amount of visible and near-infrared light that plants reflect changes according to their state of health and level of stress. Drones having sensors that really can scan crops in infrared and visible light can also be used to track crop overall health and evaluate how well corrective measures are working.
- **Crop surveillance:**
  In big fields, it is practically hard to gauge the general health of the crops. Farmers that use drones for agriculture mapping can identify which fields need attention and stay up to speed on the state of their plants. Drones use infrared cameras to scan the area and measure light absorption rates to gauge the health of the crops. However, the future possible use will be determined by the financial ramifications in relation to the chosen insurance type and the practical possibilities.
- **Controlling weed, insect, pest, and diseases:**
  Drones may identify and alert farmers to field regions affected by weeds, disease, and insect pests in addition to soil conditions. Based on this knowledge, farmers can reduce costs and improve field health by maximizing the amount of chemicals required to combat pests.
- **Scaring birds:**
  After planting the seeds of various crops, birds pose the biggest challenge. To keep the field protected, labor is required. Birds in fields can be scared off with a few drone flights.

12.5 Agriculture Using Drone Technology

Due to management techniques like precision agriculture, industry professionals have more alternatives than ever to improve the efficiency and accuracy of processes. For the development of SOP and algorithms, tasks related to plant growth such as the making of computerized field maps, monitoring crop conditions operationally, assessing germination, attempting to predict crop yields, assessing the effectiveness of soil preparation, keeping track of agricultural land monitoring tools, etc. are being researched [2]. The drone market for agricultural has grown significantly during the past few years. Depending on the brand and size of the spray tank, drones with spray accessories can cost anywhere between 3 and 7 lakhs. The companies currently registered in India to produce drones on

Table 12.3 Registered firms of remotely piloted aircraft systems (RPAS).

RPAS name	Producer's name	Optimum take-off weight (kg)	Max. height obtainable (ft)
Ninja	Idea Forge Technology Pvt. Ltd. (Research and product development in Navi Mumbai, Maharashtra)	1.98	400
Agribot UAV (AGUAV)	Io Tech World Avigation Pvt. Ltd. (Gurgaon, India)	23.2	33

the online sky service of the Director General Civil Aviation (DGCA) are listed in Table 12.3.

Under the strategy, cloud services provided by the public and private sectors can support the infrastructure for large-scale digital soil mapping, hydrological, and weather spatial information.

12.5.1 Execution Strategy

12.5.1.1 Production of Drones

Drones are produced in India by less than twenty businesses, including suppliers and assemblers. The establishment of local drone manufacturing and assembly facilities can be supported.

12.5.1.2 Suppliers of Drone Services

Recently, 50 new service providers have emerged through incubation facilities. To cultivate 180 million hectares of land, thousands of such enterprises must be supported.

12.5.1.3 Trained Human Resources

Given that flying drones is a skill-based operation, capacity development is necessary. Establishing training facilities should be done in conjunction with businesses who have experience with drones.

12.5.1.4 Ease of Doing Business

With a single-window idea, an enabling ecosystem must be created for entrepreneurs.

12.5.1.5 Application of Drone Research

There are not enough studies on drone use cases in different types of crops, animals, fisheries, and forests. For every use case in agriculture, several operational parameters need to be researched and standardized.

12.5.1.6 Extension System

Field demonstrations of different drone applications for farming should be the focus of field extension officials.

12.5.1.7 Custom Hiring Centers

By partnering with start-ups that provide drone services, tailored hiring centers should create a system to provide a one-stop shop for specialized drone application services. This mechanism should be similar to the one that allows for the provision of various agriculture machinery available for hire. The supply and demand networks for drones used in agriculture will be efficiently bridged thanks to this use of the already existing farmer–custom hiring network.

12.5.2 Scope for Public-Private Partnership (PPP) Mode Operations

The following factors should be considered when using drones in agriculture in public-private partnership (PPP) mode:

1) In accordance with the Directorate of Civil Aviation regulations, drone operators must register on the DGCA's Digital

Sky platform. The application programming interface for the digital sky are still in beta.

2) Provisional 'no permission – no take-off' compliance has been made available to several Indian original equipment manufacturers (OEMs).
3) The June 2, 2020, publication of the Unmanned Aircraft Systems [12] Rules part VII includes rules ports for drones and lanes (i.e., dedicated drone airspace).
4) A non-profit, sector organization called the Drone Federation of India (DFI) supports and aims to create a scalable and secure unmanned aircraft market in India. The DFI works in conjunction in order to share information and create a vibrant UAV market in India, experts, thought leaders, visionaries, and decision-makers will be invited.
5) Operating drones used in agriculture while adhering to the UAS regulations presents a potential opportunity for young entrepreneurs. The present labor migration brought on by the COVID-19 (2019 Corona Virus Disease) shutdown has created a special opportunity that is thought to have a beneficial impact on drone use in agriculture. This technology is expected to involve the next in agriculture and potentially buck the general trend of people moving away from farms. Suppliers of agricultural supplies like Bayer are actively promoting such technology.
6) A few private companies have already tested using drones to spray insecticides on farm fields and have made attempts to make the technologies more widely available. For instance, the Grape Growers Association used a drone to monitor vineyards in Maharashtra's Nashik district for disease.
7) The following would be a possible method of implementing drone usage in PPP mode:
   - The promotion of drone services for agricultural purposes by authorized companies that have registered with the DGCA, have pilots who have undergone the requisite training, and are in compliance with DGCA regulations.
   - Create block-level centers for drone-related agricultural operations that are spaced apart.
   - Decide which drone service provider each hub will use.
   - The Agriculture Department will stock up on necessary pesticides at the block level.

- Agricultural extension center Krishi Vigyan Kendra and the Agricultural Department must determine demand and coordinate other drone-related activities using the farmer database already in place and mobile devices.

12.6 Drone Use Rules and Regulations in India

The Indian government established a drone policy in December 2018 that permits the use of drones, moreover, as an illustration, deploying them in infrastructure projects. The DGCA and Government of India laws tacitly approve of the use of RPAS (drones, UAVs), except for pesticide spraying until specifically authorized.

The DGCA RPAS [3] Guidance Manual 17 outlines the procedures for issuing unmanned aircraft operator permits (UAOP), unique identification numbers (UIN), and associated activities.

Below is a list of the drone's general rules:

1) You should stay away from crowded places or densely populated regions.
2) Do not invade people's privacy.
3) Flying UAVs are prohibited five kilometers from airports and in locations where aircraft are flying.
4) Only travel during the day and in fair weather.
5) It is against the law to use camera drones in sensitive locations, such as buildings used by the military or by the government.
6) The drone user must have completed drone pilot training and must be at least 18 years old.
7) The drone must have a license plate displaying the owner's details and a phone number on it.
8) When employing RPAS [30], visual line of sight must be maintained.
9) No more than one UAV may be flown by a single person at once.
10) Drone flying is restricted inside 50 kilometers of the nation's border.
11) It is prohibited to fly drones more than 500 meters offshore.
12) In a similar vein, drone use is prohibited in Delhi within 5 kilometers of Vijay Chowk.

13) It is also prohibited to fly over national parks or animal refuges.
14) To cover any liability that may result from an accident, every drone must have a third-party insurance coverage that is in effect.
15) When operating a drone that weighs more than 250 grams, basic drone regulations must be fulfilled.

12.7 Policy Need

The privacy of individuals, public safety, international airspace, and civil liberties are the four main policy concerns posed by drones. Constructing a symbiosis combination of law, tax, and civil freedoms is necessary for the success of drone use in agriculture. Currently, regulations do not allow the use of drones for agriculture beyond visible line-of-sight activities, which presents difficulties when trying to cover an entire village and use PPP for such tasks.

The following legislative changes would promote drone use in agriculture:

1) Priority registration and regulatory training should be given to business owners that offer drone services.
2) The DGCA should reopen and expedite any drone registrations that may have been closed due to the COVID-19 lockdown-related incident.
3) Young entrepreneurs should have easy access to training for drone operators for agricultural purposes that is also reasonably priced.
4) The formation of higher education institutions for drone operations, such the Indian Institute of Drones, is due to the advancement of drone technology as a contemporary solution for difficult tasks. The modern agricultural graduate study must include this type of education and it must follow DGCA guidelines.
5) Apart from that, the drone-based pesticide administration paradigm can incorporate the current pesticide formulations with label claims. The new Pesticides Management Bill, 2017, also includes provisions governing aerial drone use spraying for consistent pesticide canopy coverage in important crops.
6) Liability and damages insurance should be a part of the 2017 Pesticides Management Bill when using drones to spray pesticides.

7) Creating Bureau of Indian Standards (BIS) standards for testing and evaluating recognized agricultural drone applications, as well as benchmarking drone specifications for use in agriculture.
8) Increasing the capacity of farm machinery testing centers to test and assess drone applications in agriculture for conformity with BIS formulation criteria.
9) It is important to plan drone corridors using unmanned traffic management systems, or drone-specific airspace.
10) Flights from the drones that operates completely by AI might be permitted, subject to certain limitations, for night operations and beyond visual line of sight the 400-foot limit. Thus, existing drone inventories will be able to maintain favorable benefit-cost ratios while operating for longer stretches of time at cheaper operational expenses.

12.8 Another Benefits of Drones in Agriculture

The different applications of drones in agriculture are mentioned below [20]:

1) Security: The drone pilots for the agricultural sprayers operate them from a distance. With the help of this process, farmers and farm workers are no longer exposed to hazardous chemicals or working conditions.
2) Excellent field usefulness and capability: Drones have relatively short delays in field operations such as turnaround time. The drones can spray between 50 and 100 acres each day, more than 30 times than the conventional knapsack sprayer.
3) Minimization of waste: 30% of the pesticide is conserved while spraying because of the high level of atomization. All levels of the crop can be sprayed with pesticides in the form of chemical fog.
4) Lower cost: The cost of using a drone to spray is 97% less expensive than using traditional methods.
5) Easy to use and maintain: Drones for agriculture are meant to be tough. It requires less maintenance, has a long useful life, and its parts are easy to replace as needed for the business providing drone services.

12.9 Drawbacks of Drones in Agriculture

1) High start-up costs
2) Conventional farmers are unable to analyze drone photographs in the required detail. You will need some fundamental knowledge and abilities to operate an agriculture drone. The farmer will need to acquire the abilities and knowledge of imaging software in these situations.
3) Among the several severe adverse effects on the environment, the threat to wildlife is a major worry.
4) It is tough to fly them in extreme weather.
5) The ability of the devices to fly or record the required space can be hindered by rain, fog, or wind. Rain can harm a drone's electronic components. To take images, there must be a lot of sunlight.
6) To utilize it, you will need official authorization.
7) Due to the continually changing laws and rules governing drone flight, the increasing number of constraints governing air space may result in monetary or legal consequences.
8) Drone survey limitations include battery life. The drone can fly for fewer hours as a result.

12.10 Drone Agriculture Cost

Typically, a drone used for agriculture with fixed wings can cost up to $25,000 (Precision Hawk's Lancaster), depending on the features and sensors required to fulfil their intended function. Certain drones have higher prices because they come with more expensive image sensors, software, hardware, and other parts [16].

12.11 Future Research Direction

The use of drones in agricultural production is expected to increase as they become more intelligent and capable of acting as mechanical pollinators. Drones are a promising and reasonably priced technology that can help address the problem of rising food insecurity.

The general public's opinion on drones is still mixed compared to other recent developments. This might be due to the absence of a framework that codifies drone behaviors, giving the impression that drones are an underdeveloped technology that needs further regulation.

More information on drone regulations at the global level is provided in a 2020 study [1]. The study also suggests a new strategy for better involving stakeholders and regulators in the implementation and enforcement of developed regulations, which will ultimately improve public opinion.

Despite conflicting opinions, a study [19] predicts that drones may be bringing about a new agricultural revolution and will eventually reach a market value of billions of dollars. The authors of the study speculate that this might be due to the world's present, swiftly changing environmental and demographic conditions.

By improving crop health and productivity, drones, for instance, can directly help address the issues of global climate change and rising food poverty. A study on "UAVs and agriculture," edited by Gerard Sylvester, was released by the International Telecommunication Union and the UN Food and Agriculture Organization. Drones are predicted to increase productivity across the board in agriculture, even as farmers seek to address other challenges like climate change.

Drone precision, duration, and applicability are anticipated to increase in a time of rapid technological advancement. Software integration offers very promising outcomes that will also increase usability and prices, making stakeholders better able to leverage technology.

Drone AG, a firm that develops drones and has incorporated several software applications, had this as its goal. In addition to crop-spraying drones and multispectral mapping drones, their models now feature a mobile-friendly crop scouting technology.

Such advancements will increase the effectiveness of drones and give farmers more tools to preserve food security in a world that is changing quickly.

12.12 Summary

1) The use of drone technology holds enormous promise for completing a variety of agricultural jobs quickly and effectively. According to the organization for UVS International, by the year 2020, over 900 companies would be deploying over 2900 UAVs.
2) The most challenging obstacles to overcome in order to make it work and be farmer-friendly are high start-up costs and governmental modifications. Despite limitations, the use of these technology and tools in agriculture to spread information about farming is growing swiftly.

References

1. Ahirwar S, Swarnkar R, Bhukya S, Namwade G. (2019) Application of drone in agriculture. *Int. J. Curr. Microbiol. Appl. Sci.*, 8(1): 2500-2505
2. Daponte P, De Vito L, Glielmo L, Iannelli L, Liuzza D, Picariello F, Silano G (2019) A review on the use of drones for precision agriculture. In *IOP conference series: earth and environmental science*, 275(1): 012022.
3. DGCA RPAS Guidance Manual, Revision 2 - 2020, Directorate General of Civil Aviation, Govt. of India, approved 11th March, 2020. https://dgca.gov.in/digigov-portal/jsp/dgca/InventoryList/headerblock/drones/DGCA RPAS Guidance Manual.pdf 13th August, 2020
4. Ghazali MHM, Azmin A, Rahiman W. (2022) Drone Implementation in Precision Agriculture–A Survey. *Int. J. Emerging Technol. Adv. Eng.* 12(4): 67-77
5. https://5.imimg.com/data5/SELLER/Default/2022/2/AE/NN/SR/2497808/agricultural-hybrid-drone-sprayer-500x500.jpeg
6. https://circuitdigest.com/sites/default/files/field/image/Drone-Frames.jpg
7. https://img.aeroexpo.online/images_ar/photo-g/175762-16301086.jpg
8. https://media.springernature.com/lw685/springer-static/image/art%3A10.1007%2Fs42452-020-2749-5/MediaObjects/42452_2020_2749_Fig2_HTML.png
9. https://orbitaluav.com/wp-content/uploads/2019/02/Propulsion-System-1024x414.png
10. https://www.alliedmarketresearch.com/assets/sampleimages/rotary-blade-uav-drones-1614055513.jpeg
11. https://www.equinoxsdrones.com/blog/importance-of-drone-technology-in-indian-agriculture-farming
12. https://www.researchgate.net/figure/Schematic-diagram-of-the-navigation-system-for-UAV-autonomous-take-off-and-landing-An_fig1_301506688
13. https://www.researchgate.net/figure/Several-examples-of-fixed-wing-and-rotary-wing-UAVs-Adapted-from-Reference-9-Creative_fig1_351037538
14. Kolhe P and Munde TN (2019) Use of Drone For Efficient Water Management - A Case Study. *3rd World Irrigation Forum (WIF3)* 1-7 September 2019, Bali, Indonesia.
15. M. Aleksandrova. "Why use Agriculture Drones?" [Online], https://dzone.com/articles/why use agriculture drones-main-benefits-and-best.
16. Pathak H, Kumar GAK, Mohapatra SD, Gaikwad BB, Rane J (2020) Use of Drones in Agriculture: Potentials, Problems and Policy Needs, Publication no. 300, ICAR-NIASM, pp 13+iv.
17. Press Information Bureau. (2021). *Ministry of Agriculture & Farmers Welfare* [Online] https://pib.gov.in/PressReleasePage.aspx?PRID=1783937.
18. Rani A, Chaudhary A, Sinha N, Mohanty M, Chaudhary R (2019) Drone: The green technology for future agriculture. *Harit Dhara* 2(1): 3-6

19. Ren Q, Zhang R, Cai W, Sun X, Cao L (2020) "Application and development of new drones in agriculture." In *IOP Conference Series: Earth and Environmental Science,* 440(5): pp. 052041.
20. The Unmanned Aircraft Systems Rules, 2020, Ministry of Civil Aviation, GOI, approved 2nd June, 2020. Available: https://www.civilaviation.gov.in/sites/default/files/Draft_UAS_Rules_2020.pdf
21. Varma A, Singh M, Parmar RP, Bhullar KS (2022) Feasibility study on hexacopter UAV based sprayer for application of environment-friendly biopesticide in guava orchard. *J. Environ. Biol.* 43(1): 97-104.

13

A Comprehensive Study on Machine Vision Techniques for an Automatic Weeding Strategy in Plantations

Manikandan J.[1*], Rhikshitha K.[2], Sathya Sudarsen G. S.[2] and Saran J. U.[2]

[1]*Department of Information Technology, St. Joseph's College of Engineering, OMR, Chennai, India*
[2]*Department of Computer Science and Engineering, Rajalakshmi Engineering College, Thandalam, Chennai, India*

Abstract
Agriculture is an essential occupation to the people of India. It is considered as the backbone of most of the Indian population. However, one of the biggest concerns of agriculture is the growth of weeds. These weeds have to be removed to get a fruitful harvest. This process of removing weeds is weeding, which must be done with utmost care without affecting the valuable crops. Using agricultural chemicals is one of the most popular ways to manage weeds. However, weed identification is one of the challenging parts of cultivation, as the use of chemicals throughout the plantation is harmful to the environment and the agricultural ecosystem. In addition, manually removing the weed is possible but not entirely practical, considering human error and labor charges that must be paid to them. This identification of weeds leads to the demand for alternatives to weed control and identification techniques. Therefore, industries continue to seek human-free automated mechanisms that are relatively inexpensive. In this regard, machine vision comes into action for agricultural automation. Machine vision technology uses cameras rather than the naked eye to identify. In recent years, machine vision technologies have rapidly developed, and the progress achieved is remarkable. Machine vision technology has been proven to help build automation in agriculture resulting in cost-effective, highly efficient, and high-precision solutions. In addition, the increased computational power of the hardware, decreased costs, and advancements in the accuracy

**Corresponding author*: jmanekandan@gmail.com

Rajesh Kumar Dhanaraj, Balamurugan Balusamy, Prithi Samuel, Malathy Sathyamoorthy and Ali Kashif Bashir (eds.) Computer Vision in Smart Agriculture and Crop Management, (303–322) © 2025 Scrivener Publishing LLC

and efficiency of the algorithms have made it possible to construct feasible and practical automatic weeding strategies. This chapter focuses on the exploration of numerous machine vision strategies involved in automated weeding and their applications, use cases, and research challenges.

Keywords: Machine vision technology, computation, algorithms

13.1 Introduction

Agriculture is an essential part of the human ecosystem that plays a major role in boosting the economy of the nation. Agriculture is highly demanding and requires a lot of resources in order to produce a high-yielding harvest. Agriculture contributes to about 23% of India's GDP. Agriculture employed over 59% of the total workforce in the year 2016 [5]. Though agriculture is widely practiced, it also comes with a variety of challenges that need to be handled carefully. Weeding is considered as the most demanding challenge that needs innovative and reliable solutions. Any crop that is not desired but grows along with the desired crop and oozes out the nutrition is known to be a weed. These weeds must be removed, and this is time-consuming if done manually. These weeds, if not removed, complete with the desired crop for water and nutrition resulting in a decrease in productivity. By controlling these weeds, crop production can be increased drastically. Weeding is nothing but removing these unwanted crops, usually handpicked, to improve yield. Weeding is a labor-intensive task, identifying and using chemical sprays to remove the existing ones and to prevent the growth of new ones. Hence, there is a huge demand for weed control strategies [1]. Traditional practices followed in agriculture should be replaced with innovative technologies in order to improve production [2]. In order to remove these weeds, precise identification of the weeds should be done. The identification of weed in real-time is a major concern, which needs to be addressed. If not precisely identified, the chemicals to be sprayed can get in contact with the desired crops, which will have harmful effects on the original useful crops. Automated weeding is becoming a booming area of agricultural research that focuses on reducing the dependence of human labor for weeding.

Machine vision is a rapidly growing technology for automating the challenges faced in agriculture. In the context of weed identification, image segmentation, object detection, deep learning, hyperspectral imaging, LiDAR, thermal imaging, image filtering, convolutional neural networks (CNN) and deep learning techniques are used in machine vision. All these techniques have their own advantages and limitations. These machine vision strategies can be used separately or in aggregation. These features are helpful in differentiating weeds from useful crops.

In the object detection approach, the objects are located within an image to segregate them as desired plant or weeds and identify their species. These algorithms are trained to identify objects based on their color, type, and species, precisely. Deep learning is the computer science that utilizes neural networks made up of computers utilized to learn from data and discover the underlying patterns. Deep learning is also known as "supervised learning." These algorithms are to be trained using large datasets to give accurate results. These algorithms come under a classification that needs the labeled datasets of the weeds to the desired plants to be recognized for weed identification. Hyperspectral imaging is a technique that can capture an image from the plantation field at multiple spectral wavelengths or bands. These spectral images captured are then analyzed to differentiate the objects based on the spectral signature. Hyperspectral imaging can be used to analyze the spectral signature of the weeds and identify the weeds based on their spectral properties. Once identified, weeds can be easily distinguished between crops in the plantation.

LiDAR is a 3D mapping technique through remote sensing, which uses laser pulses to create a 3D map of the plantation field. This can be used for identifying weeds based on their shape and dimensions. LiDAR can be used to estimate the yield, so it can be used to develop site-based management strategies. Thermal imaging is also a technique that captures temperature signatures into images. These images of crop fields can be used for differentiating between weeds from crops based on thermal signatures. Thermal imaging can also be used for detecting weeds at night or low light through which a real-time weed control system can be developed. A summary of using deep learning techniques to complete agricultural tasks has been given [3], as have several studies on the usage of machine vision in agriculture [4]. Machine vision technologies are proven to be as effective as the human eye because they work on image processing and filtering techniques. With sensors, cameras, and high-speed processors, it would be possible to identify the weeds. Due to advancements in technology, it would be possible to detect and devise strategies for weeding in a fraction of the time compared to traditional methods of removing them manually. This chapter focuses on the summarization of applications, use cases, and research challenges involved in the machine vision systems of automated weeding strategies. This chapter explores the future advancements that can be made to the current approach. The entire workflow of the present work has been shown pictorially in Figure 13.1.

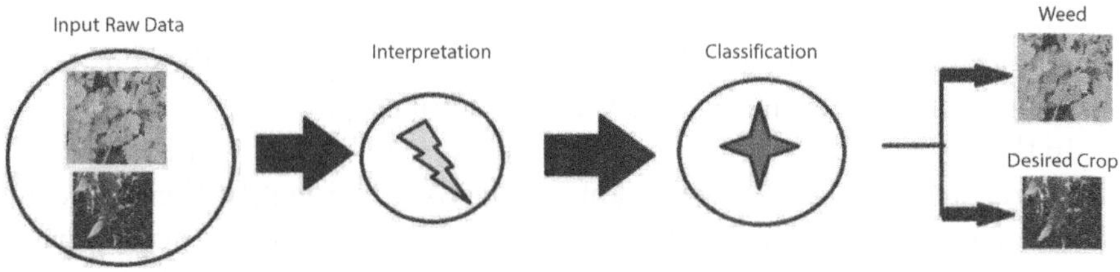

Figure 13.1 Overall workflow.

13.2 Related Study

Many other works related to machine vision in agricultural automation have been put forth for weed identification. Some of the summaries of the work are mentioned below.

Kargar *et al.* (2020) [6] proposed work on machine vision-based weed identification and control techniques that include traditional image processing and deep learning. The work focuses on various aspects of weed detection and control, such as image acquisition, feature extraction, classification, and control strategies. The authors examine both strengths and weaknesses of machine vision-based methods for weed detection and management, along with future directions of research. Their system utilizes image segments, feature extraction, and classification algorithms to accurately identify weeds. Naik *et al.* (2019) [7] have created an automated weed classification and detection system utilizing deep learning algorithms for robotic precision agriculture weed management. This paper utilizes CNN to classify images, demonstrating its accuracy at detecting and categorizing weeds. They propose that their system could be integrated with robot platforms for automated weed-control purposes. Peerbhay and Bansal (2019) [8] provided an in-depth examination of vision-based approaches for weed detection and control, including machine learning, robotic systems, and sensor technologies. Furthermore, they examined various aspects of weed control such as image capture, feature extraction, and classification strategies, as well as assess advantages/limitations of various techniques. Peerbhay *et al.* (2019) [9] and Rehman *et al.* (2019) [10] proposed that machine vision-based weed detection and control techniques could potentially help reduce herbicide usage in lettuce crops. They offer an automated weed control system using computer vision and robotics for lettuce crops as a potential solution to their problem. These authors utilize image segmentation, feature extraction, and control algorithms to detect and manage weeds in lettuce fields. Their system successfully reduces weed populations while simultaneously improving lettuce yield; additionally, they suggest it could also be applied to other crops.

Sun and Cheng (2019) [11] proposed an effective technique for postemergence herbicide application in row crops using image processing techniques, including color-based segmentation and morphological operations to detect weeds in real time. Tan *et al.* (2019) [12] presented an effective system to decrease herbicide usage and enhance crop yield. Their robotic weed control system utilizes machine vision and reinforcement learning technology. They used a CNN to identify crops and weeds in images, and a reinforcement learning algorithm to control a robotic arm for weed removal. The authors demonstrate the efficiency of their system for detecting and eliminating weeds under various conditions, suggesting it can serve as an automated approach to smart agriculture management. Wang *et al.* [13] proposed an automated weed recognition and mapping system using deep CNNs. They utilized CNNs to distinguish crops from weeds while clustering algorithms map infestations in the field. The authors demonstrated the efficacy of their system in accurately identifying and mapping weed infestations, as well as suggesting its application in precision herbicide applications or site-specific weed management programs. Zhang *et al.* (2019) [14] reviewed various machine vision-based weed detection and mapping techniques, including color-based segmentation, texture analysis, and machine learning algorithms. The authors suggested that machine vision-based weed detection and mapping techniques have the potential.

13.3 Methodology

First, we need to configure the software and hardware setting based on the requirements. Initially, the dataset of the images that needs to be identified is to be collected. The collected dataset is then labeled, that is all the images are classified. Labeling is nothing but tagging all the images of the dataset with the information they present. Once the labeling is done, the images are then had to be pre-processed. Pre-processing techniques include filtering, smoothing, and edge detection. Next is image segmentation whereby images are divided into regions based on color, texture, and feature to identify objects in an image. After segmenting has taken place, we must perform feature extraction – these extracted features help classify segments either as weeds or crops, which is then the final step of this workflow process – classification uses these extracted features to classify each segment accordingly. Classification involves the usage of several machine learning algorithms on the labeled dataset to learn to differentiate different types of

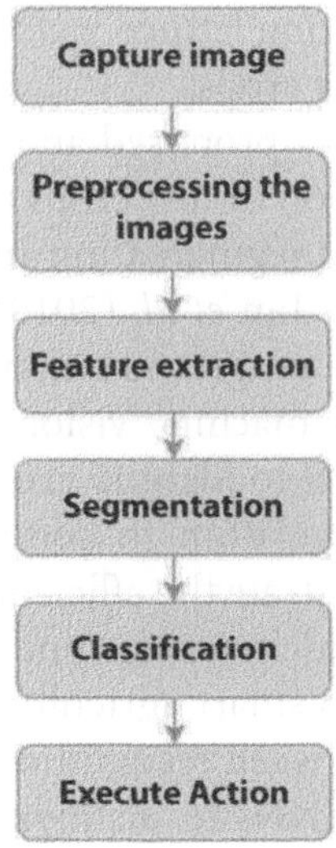

Figure 13.2 Proposed workflow.

objects based on their features. Once classification is done, results are used to find the location of weeds or crops in the field. Based on the location of weeds, appropriate actions such as spraying agricultural chemicals can be done. In the methodology section of this chapter, all the methods summarized in the proposed system will be discussed in detail with the appropriate diagrams and equations.

The algorithm template of the proposed system has been elucidated in Figure 13.2.

13.3.1 Collection of Datasets of Images of the Undesired Weed

Harvesting is a critical stage in farming as it decides the result of the time and effort that was put in during the cultivation. To make the harvesting fruitful, weed control must be highly efficient. Crops such as maize highly depend on weed control to grow as expected. The weeds must be removed in the early stage, say six to eight weeks since they highly compete for food and nutrition during this stage only. Identification and removal of weeds in the later stage may result in a decreased crop yield. Weed infestations are one of the major reasons for low crop yield. In general, a 10% to 100% decrease in crop yield can occur as a result of weed infestation. This elucidates why weed identification is important. Hence, the prerequisite is the identification of weeds. This can be done by collecting large datasets of potential weeds. There may be different species or types of weeds based on different features. The collected images may be grouped into classes based on their similarity or based on those that represent the common species. The images may be of variable size and in any of the common format say PNG or JPG. The collected dataset should be proper and appropriate in order to have precise outcomes. The dataset should not have inappropriate images such as the image of the desired crops, which can have dangerous results like classifying the desired crop as weed.

13.3.2 Data Labeling

The collected dataset will have images of a variety of weeds as mentioned. Weeds in general, can be classified as competitive or useful based on their behavior towards the useful crops. The collected dataset must be labeled appropriately in order to differentiate the intensify of danger that the weed threats to the desired crops. This is being done in order to make the data appropriate input to the training model. The labeling of crops and their respective description of the labels have been shown in Table 13.1 [15].

Table 13.1 Labeling of weeds and their description.

Class	Label	Description of the label
P	Highly Dangerous and Competitive	• Highly competes with the desired crop for water and nutrition. • Poses a serious threat to the desired crop. • Hinders the growth of the desired crop.
Q	Competitive	• Poses a threat to the desired crop. • Must be eradicated.
R	Not very dangerous and competitive	• Not as dangerous as P and Q but still need to be carefully monitored.
S	Useful	• Not competitive and dangerous. • Need to be conserved and not eradicated. • Useful for desired crops.

13.3.3 Pre-Processing of Dataset

Raw data that is collected will always be dirty. Hence, it needs to be cleaned in order to have efficient results. The raw data may be noisy and inconsistent. Pre-processing is the process of cleaning up these data. Pre-processing is required to make the images suitable for further processing [16, 18].

i. **Removal of noise:** The noise that is undesired randomness must be removed so that the images are clear and proficient.
ii. **Motion-blur removal:** Some areas in image may be blurry due to motion. This blur must be removed to get a clean and clear image.
iii. **Hazard removal:** Hazard is nothing but the disturbance in the image. This disturbance is removed to get a clean image.
iv. **Removal of rain drop:** There may be raindrops in the images since it is a more common phenomenon that can happen while raining or through irrigation. This is removed to avoid any ambiguity.

13.3.4 Transformation into Binary or Grayscale Image

To transform into a binary image, each pixel of the image is multiplied by 0.9 to enhance the contrast. The image is then added with Gaussian noise. Then a Gaussian filter is used to remove the noise. Based on the color of the weed, a respective color filter is applied to extract the colored part of the weed. This is crucial in differentiating the weed from the soil and other objects [22].

The formula for this are given as:

$$T = argmax(t)sigma_{B^{2(t)}} \tag{13.1}$$

Where T is the threshold value, t is the intensity value of the pixel, $sigma_{B^{2(t)}}$ is the between-class variance at threshold t. $sigma_{B^{2(t)}}$ is defined as

$$sigma_{B^{2(t)}} = w_{1(t)} * w_{2(t)} * (m_1(t) - m_2(t))^2 \tag{13.2}$$

Algorithm.1.

---

*1: **START***
2: Compute the histogram of the image.
3: Compute w_1(t) and w2(t) for all possible threshold values of t.
4: Compute m_1(t) and m_2(t) for all possible threshold values of t.
5: Compute sigma_B^2(t) for all possible threshold values t.
6: Select the value for T such that it maximizes $sigma_{B^{2(t)}}$.
7: END

13.3.5 Image Segmentation

Image segmentation is the next step. Image segmentation is a process that divides an image into homogenous segments. Image segmentation is used to distinguish the target object against the background of the other objects. Image segmentation can be done in several ways. This chapter discusses the watershed algorithm. The watershed algorithm is applied on grayscale images. The implementation of the watershed approach is that the flooding process can be applied on a grey tone image. This technique detects the lines that are formed by ridges and basins. Each segment of the image is marked with unique labels by the algorithm. The segments that are marked are stored in a CSV file, which is then fed as input the CNN. The CNN takes the input and classifies them as weed or crop. Figure 13.3 shows the workflow of transforming an RGB image to a binary image.

Algorithm.2.

---

*1: **START***
2: Convert the image to grayscale, if necessary.
3: Apply a Gaussian filter to smoothen the image.
4: Calculate the gradient magnitude of the image.
- ***a.** Calculate the gradient in the x and y directions using Sobel operators.*
- ***b.** Calculate the magnitude of the gradient.*

5: Calculate the watershed transform of the gradient magnitude.
- ***a.** Find the local minima of the gradient magnitude.*
- ***b.** Create a marker image with the local minima as markers.*
- ***c.** Calculate the distance transform of the gradient magnitude.*
- ***d.** Calculate the watershed transform of the distance transform using the markers.*

6: Apply a threshold to the watershed transform to obtain the segmented image.

a. Threshold the watershed transform based on the minimum and maximum intensity values of the image.

b. Set all pixels below the threshold to 0 (background) and all pixels above the threshold to 1 (foreground).

7: Post-process the segmented image to remove small regions, if necessary.

a. Remove small regions based on their area.

8: STOP

13.3.6 Feature Extraction

Feature extraction involves analyzing the segments that are provided and identifying the unique characteristics or attributes of the weed and crop. These features are then used as input for the classification algorithm. There are multiple feature extraction techniques that can be used are:

i) **Texture based feature extraction:** This feature extraction technique extracts the texture features of the weed and crop from the images. Methods like local binary patterns and Gabor filters can be used to extract texture features.

ii) **Shape-based feature extraction:** This feature extraction technique extracts the shape feature of the crops and weeds from the images. Edge detection and contour detection techniques can be used to extract the shape features.

iii) **Deep learning-based feature extraction:** This technique involves using pre-trained CNNs like VGG, ResNet, or Inception to extract high-level features from the images. These features can be extracted from the layers of CNN.

iv) **Color-based feature extraction:** In this technique, color features of the weeds and crops are extracted from the images. Color features are extracted using color spaces like RGB, HSV or LAB.

13.3.7 Classification

Classification is a type of machine learning model that classifies or categorizes the input data or images into a predefined set of classes. The main goal

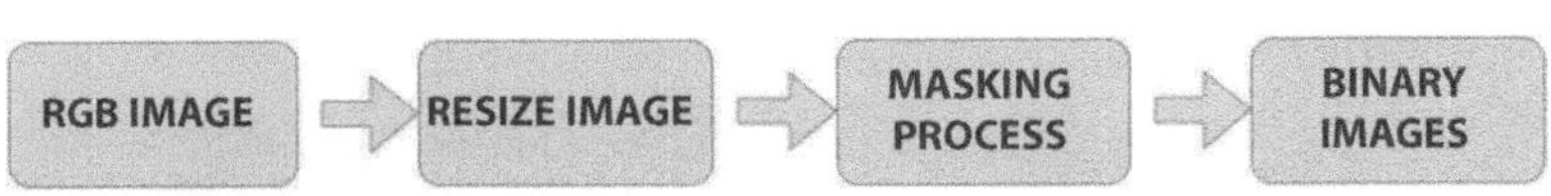

Figure 13.3 Transformation of RGB to binary.

of the classification model is to map the features to their corresponding output or labels. Classification models can be binary or multi-class. These classification models can be trained by using a variety of machine learning models such as support vector machines (SVMs), deep learning-based models like CNNs, decision trees, and logistic regression. The performance of these classification models can be measured based on the accuracy, F1 score, recall, and precision. These metrics measure the ability of the model to correctly classify the data.

Few of the models which can be used for automatic weeding strategies are:

Convolutional Neural Networks (CNNs)

The CNNs are a method based on deep learning that can automatically extract useful characteristics from the input photos. In a wide range of computer vision tasks, such as object identification and classification, they have attained state-of-the-art performance.

K-Nearest Neighbor (KNN)

The K-Nearest Neighbor (KNN) has been successfully utilized for weed detection in maize crops. One study utilized color, shape, and texture features extracted from images for classification into either "weed" or non-weed categories using KNN classifier achieving accuracy of over 95% [19, 26, 29].

Deep Convolutional Neural Networks (CNNs)

The CNNs have become widely utilized tools for automatic weed detection and classification across different crops. One study employed CNNs to classify both crop and weed plants in soybean fields with an accuracy rate exceeding 94% [17, 27, 28].

Support Vector Machines (SVMs)

The SVMs have been used for the detection of weeds in cotton fields using color, shape and texture features extracted from images to use an SVM for classification as either weed-bearing or non-weed images with an accuracy rate exceeding 95% [18, 23].

Random Forests

Random forests have been effectively employed for sugar beet weed detection using color, shape, and texture features extracted from images; then using Random Forest classifiers with over 93% accuracy [20, 24].

Naïve Bayes Classifier

Naïve Bayes classifiers have been successfully employed for weed detection in rice fields. One study extracted color, shape, and texture features from images taken of rice fields before classifying each image either weed-related or not-weed related using this classification scheme, with over 93% accuracy [21–25].

13.3.8 Execute Action

From the classification models, the images are classified into weeds or crops based on which the actions need to be taken. After the weeds are classified, they can be in the plantation fields, and the weeds can be marked, and different strategies can be implemented to remove the weeds automatically. Automated machines can be used for real time removal of weeds, thus ensuring proper nutrients for crops that are being cultivated.

13.4 Experimentation and Analysis

13.4.1 Dataset

The dataset references used in this chapter are aggregation of several datasets. The links for the dataset that are used from Kaggle are: https://www.kaggle.com/datasets/ravirajsinh45/crop-and-weed-detection-data-with-bounding-boxes. The above data set has been used widely for annotation of weed images and crop images. It contains 1300 images. Each of them is a 512 x 515 color image https://www.kaggle.com/datasets/vinayakshanawad/weedcrop-image-dataset. This dataset is hugely useful for labeling the images of the weeds. It includes 2822 images and is annotated by using the YOLOV5 algorithm. These datasets are aggregated to get the maximum number of various labeled weeds, which in turn increases the accuracy of the overall system. The images in the dataset are in JPG format,

Figure 13.4 Sample images from the dataset.

and these are commonly used for machine learning research. Figure 13.4 shows some of the images from the aggregated dataset.

The experimentation of this chapter was done in Python. OpenCV, TensorFlow and Keras were used to carry out the experiment. These packages play a major role in creating an efficient neural network.

13.4.2 Performance Evaluation Metrics

The measures that are used to evaluate performance serve as the basis for determining how effective the proposed system is. These metrics are utilized in order to assess how well the suggested model performs in comparison to other models. In research, the sorts of evaluation measures that are utilized the most often are accuracy, precision, recall, sensitivity, and specificity, as well as the f measure.

$$Accuracy = (\mathrm{TP} + \mathrm{TN}) / (\mathrm{TP} + \mathrm{TN} + \mathrm{FP} + \mathrm{FN}) \tag{13.3}$$

$$\mathrm{Precision} = \mathrm{TP} / (\mathrm{TP} + \mathrm{FP}) \tag{13.4}$$

$$\mathrm{Specificity} = \mathrm{TN} / (\mathrm{TN} + \mathrm{FP}) \tag{13.5}$$

$$\text{Sensitivity} = \text{TP} / \left(\text{TP} + \text{FN}\right) \tag{13.6}$$

$$\text{F1} - \text{measures} = 2^*(\text{Precision}^*\text{Sensitivity}) / (\text{Precision} + \text{Sensitivity}) \tag{13.7}$$

13.4.3 Results and Discussion

In this chapter, the usage of machine vision technology in automated weeding is reviewed. The efficiency of the proposed system depends on its accuracy. Thus, we run through all the techniques and select those with the utmost accuracy. The tables and the corresponding graphs of the results are given below.

Table 13.2 shows the results of the metrics without using the pre-processing techniques. This table shows that the performance of CNN is the best among the five with the accuracy of CNN being 95.31%. The recall and the specificity of CNN falls in the range of 93%. Next to CNN, KNN has an accuracy of 91.31%. But the lack of pre-processing would certainly have some disadvantages; hence, we would try the same after applying pre-processing techniques.

Table 13.3 shows the results after applying pre-processing techniques. This table shows that the performance of CNN is again the best among the five with the accuracy of it significantly improved. The accuracy of CNN is now improved to 98.11%. The recall and specificity also improved significantly to around 95%. Figure 13.5 and Figure 13.6 show the graphical representation of the Table 13.2 and Table 13.3.

Table 13.2 Results comparison of the algorithm without pre-processing.

Methods	Accuracy (%)	Precision (%)	Recall (%)	Specificity (%)	F1-Score (%)	Sensitivity (%)
Naïve Bayes	84.1	84.73	81.27	84.12	85.88	83.99
SVM	85.21	86.61	83.39	84.67	86.53	87.78
Random Forest	90.67	87.87	89.98	85.16	90.45	90.01
KNN	90.78	88.68	90.01	90.31	90.78	91.31
CNN	95.31	94.11	93.81	93.65	94.97	94.31

Table 13.3 Results comparison of the algorithm with pre-processing.

Methods	Accuracy (%)	Precision (%)	Recall (%)	Specificity (%)	F1-Score (%)	Sensitivity (%)
Naïve Bayes	85.1	85.32	82.54	85.12	86.18	84.89
SVM	86.21	87.19	83.98	85.78	87.53	88.37
Random Forest	91.78	88.21	90.67	86.14	91.56	90.99
KNN	92.78	89.43	91.23	91.76	92.22	92.45
CNN	98.11	95.24	94.56	94.59	95.67	94.56

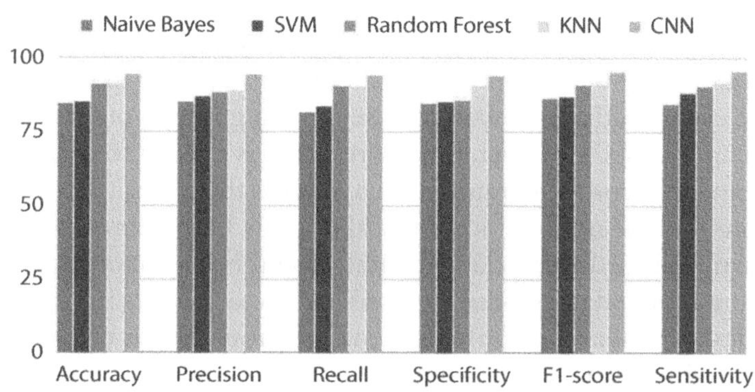

Figure 13.5 Graphical representation of result comparison without pre-processing.

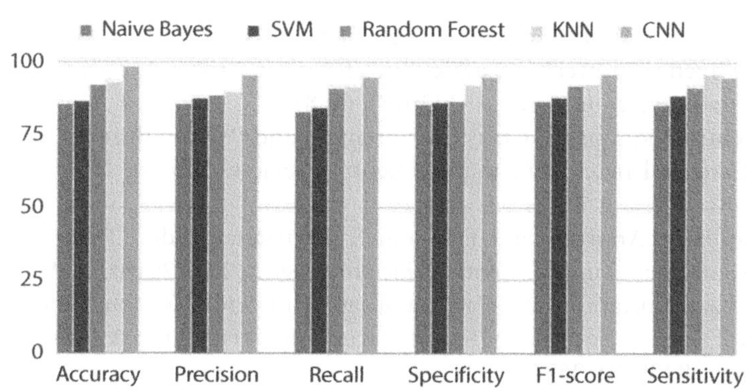

Figure 13.6 Graphical representation of the result comparison with pre-processing.

13.5 Conclusion and Future Enhancements

Machine vision when integrated with agriculture helps solve weeding problems more efficiently. Machine vision-based weed detection is faster, accurate and helps save human effort. A few promising machine vision techniques are CNN, SVM, KNN, Random Forest and Naïve Bayes algorithm. Machine vision-based weeding systems have great potential for reducing the influence of chemical herbicides in agriculture and reduce manual labor. Deep learning algorithms have greater potential than other classification models for plant identification and classification, including weed species. The algorithms can be trained on large datasets of labeled images to learn and recognize plant features and patterns and can be used to develop accurate and robust plant identification systems. It is concluded that the CNN is the most efficient and accurate model for the classification of weeds. The aim of our work is to analyze different machine vision techniques to find out the most efficient or accurate way to identify weeds from the crops so that our work can serve as a guide to other researchers and agriculturalists. However, further research is needed to optimize these systems and overcome their limitations, such as their sensitivity to lighting conditions and the need for accurate calibration. Our future work, focused deep learning-based weed classification, would be to include spectral and thermal images in data making it possible to detect weeds in real time, i.e., during day, as well as night. The current pre-processing is efficient but can be improved in turn by increasing the accuracy and overall performance of the model.

References

1. Patrício, Diego Inácio, and Rafael Rieder. "Computer vision and artificial intelligence in precision agriculture for grain crops: A systematic review." *Comput. Electron. Agric.*, 153 (2018): 69-81.
2. Bochtis, Dionysis D., Claus GC Sørensen, and Patrizia Busato. "Advances in agricultural machinery management: A review." *Biosyst. Eng.*, 126 (2014): 69–81.
3. Kamilaris, Andreas, and Francesc X. Prenafeta-Boldú. "Deep learning in agriculture: A survey." *Comput. Electron. Agric.*, 147 (2018): 70–90.
4. Liakos, Konstantinos G., Patrizia Busato, Dimitrios Moshou, Simon Pearson, and Dionysis Bochtis. "Machine learning in agriculture: A review." *Sensors*, 18, no. 8 (2018): 2674.

5. Parthasarathi, Prasannan. *Why Europe grew rich and Asia did not: Global economic divergence, 1600–1850.* Cambridge University Press, 2011.
6. Kargar, M., Moein, M., & Omid, M. "Machine vision-based weed identification and control in precision agriculture: A review." *Comput. Electron. Agric.*, 174 (2020), 105527.
7. Marchant, J. A., & Whalley, R. D. "Automated weed recognition and mapping using machine vision." *Precis. Agric.*, 16(4), (2015) 434–449.
8. Naik, H., Roy, A., & Pal, S. "Automated weed detection and classification using deep learning for robotic weed control in precision agriculture." *Comput. Electron. Agric.*, 162 (2019), 100–108.
9. Peerbhay, S., & Bansal, M. "Vision-based weed detection and control for precision agriculture: A review. " *Comput. Electron. Agric.*, 157 (2019), 205–219.
10. Rehman, S., Rodriguez-Leal, E., & Slaughter, D. C. "Automated weed control in lettuce using computer vision and robotics. " *Trans. ASABE*, 62(2), (2019) 387–397.
11. Sun, D. W., & Cheng, J. H. "Machine vision for weed detection during postemergence herbicide application in row crops. " *J. Agric. Food. Chem.*, 67(5), (2019) 1325–1333.
12. Tan, F., Li, X., Li, S., Li, W., & Liu, H. "Robotic weed control using machine vision and reinforcement learning. " *J. Field Rob.*, 36(4), (2019) 799–813.
13. Wang, C., Zhu, J., Zhang, M., & Li, W. "Automated weed recognition and mapping using deep convolutional neural networks. "*Comput. Electron. Agric.*, 160 (2019), 56-65.
14. Vijay, K., J. Manikandan, Babu Rajendiran, K. R. Sowmia, and Eugene I. Berna. "Deep Dive on Oversampling and Under Sampling Techniques in Machine Learning." In *Recent Trends in Computational Intelligence and Its Application: Proceedings of the 1st International Conference on Recent Trends in Information Technology and its Application (ICRTITA, 22)*, p. 423. CRC Press, 2023.
15. Ghazali, K.H., Mustafa, M.M. and Hussain, A., Machine vision system for automatic weeding strategy using image processing technique. *Am. Eurasian J. Agric. Environ. Sci.*, 3(3), (2008) 451–458.
16. Mustafa, M.M., Hussain, A., Ghazali, K.H. and Riyadi, S., Implementation of image processing technique in real time vision system for automatic weeding strategy, in: *2007 IEEE International Symposium on Signal Processing and Information Technology*, pp. 632–635, IEEE, 2007.
17. Pournima, S., C. Priyatharsini, G. Kirubasri and J. Manikandan. "Hybrid BILSTM Network for Improving Crop Pest Classification." In *2023 International Conference on Computer Communication and Informatics (ICCCI)*, pp. 1–5. IEEE, 2023.
18. Wu, Z., Chen, Y. ,Zhao, B., Kang, X. and Ding, Y., Review of weed detection methods based on computer vision. *Sensors* 21(11), (2021): 3647.

19. Mavridou, E., Vrochidou, E., Papakostas, G.A., Pachidis, T. and Kaburlasos, V.G., Machine vision systems in precision agriculture for crop farming. *J. Imaging* 5(12), (2019): 89.
20. Saleem, M.H., Potgieter, J. and Arif, K.M., Automation in agriculture by machine and deep learning techniques: A review of recent developments. *Precis. Agric.* 22(6) (2021): 2053–2091.
21. Adhinata, F.D., and Sumiharto, R., A comprehensive survey on weed and crop classification using machine learning and deep learning. *Artific. Intell. Agric.* (2024).
22. H. R. Pourreza *et al.*, "Automated weed detection in rice fields using a machine learning approach," *Comput. Electron. Agric.*, vol. 136, pp. 1–11, 2017. DOI: 10.1016/j.compag.2017.02.016
23. Mounashree, J. P., N. Sanjay, B. S. Sushmitha, B. G. Usha, and Mrs Anupama Shivamurthy. "AUTOMATED WEED REMOVAL SYSTEM USING CONVOLUTIONAL NEURAL NETWORK." *Int. J. Eng. Appl. Sci. Technol.*, 6 (2021): 231–235.
24. Sampurno, Rizky Mulya, Zifu Liu, RM Rasika D. Abeyrathna, and Tofael Ahamed. "Intrarow Uncut Weed Detection Using You-Only-Look-Once Instance Segmentation for Orchard Plantations." *Sensors*, 24, no. 3 (2024): 893.
25. Juwono, Filbert H., W. K. Wong, Seema Verma, Neha Shekhawat, Basil Andy Lease, and Catur Apriono. "Machine learning for weed–plant discrimination in agriculture 5.0: An in-depth review." *Artificial Intelligence in Agriculture* (2023).
26. López-Correa, Juan Manuel, Hugo Moreno, Diego Sebastian Pérez, Facundo Bromberg, and Dionisio Andújar. "Towards a true conservation zero tillage system: A proposed solution based on computer vision to herbicide resistance." *Computers and Electronics in Agriculture*, 217 (2024): 108576.
27. Gao, Wen-Tao, and Wen-Hao Su. "Weed Management Methods for Herbaceous Field Crops: A Review." *Agronomy*, 14, no. 3 (2024): 486.
28. Kakade, Subodh, Poonam Pawar, Devata Anekar, Yash Patil, Samadhan Rathod, and Fiyona Shah. "A Study of Weed Management in Agricultural Plants using IoT and Machine Learning." In *2024 International Conference on Emerging Smart Computing and Informatics (ESCI)*, pp. 1–6. IEEE, 2024.

29. Manikandan, J., Brahmadesam Viswanathan Krishna and K. Surendar. "Sign Language Recognition using Machine Learning." In *2022 International Conference on Innovative Computing, Intelligent Communication and Smart Electrical Systems (ICSES)*, pp. 1–5. IEEE, 2022.

14

An Effective Study on the Machine Vision-Based Automatic Control and Monitoring in Furrow Irrigation and Precision Irrigation

Manikandan J.[1*], Saran J. U.[2], Samitha S.[2] and Rhikshitha K.[2]

[1]Department of Information Technology, St. Joseph's College of Engineering, OMR, Chennai, India
[2]Department of Computer Science and Engineering, Rajalakshmi Engineering College, Thandalam, Chennai, India

Abstract

Water is an essential natural resource for sustainable living for all life forms on Earth. Water has become a scarce natural resource, and water usage in agriculture is a limited factor. In agriculture, without proper irrigation, farmers would be entirely reliant on natural rainfall, which may not be sufficient to meet the water needs of crops. Reduced yields, crop failure, and ultimately food instability can result from this. Irrigation has grown in significance due to climate change and unpredictable weather patterns for preserving agricultural output and guaranteeing food security. Population growth has led to increased crop demand, underscoring the importance of employing smart irrigation techniques that leverage machine vision. Smart irrigation combines hardware components, software apps, and various technologies; machine learning technology plays an essential role by harnessing data for optimizing irrigation systems to enhance their performance and enhance results. To overcome these challenges, we need to focus on efficient usage of water such as precision irrigation and furrow irrigation. Machine vision technology has proven itself as an efficient means for automating and monitoring furrow irrigation and precision irrigation systems, monitoring both water flow and soil levels with machine vision sensors. Furrow irrigation relies heavily on machine vision monitoring of its system for accurate irrigation water delivery and soil levels management. Precision irrigation

**Corresponding author*: jmanekandan@gmail.com

Rajesh Kumar Dhanaraj, Balamurugan Balusamy, Prithi Samuel, Malathy Sathyamoorthy and Ali Kashif Bashir (eds.) Computer Vision in Smart Agriculture and Crop Management, (323–342) © 2025 Scrivener Publishing LLC

systems rely on continual monitoring of soil conditions, plant health, and weather to deliver water and nutrients directly to crops at exactly the right moment. Machine vision-based control and monitoring systems comprise sensors, cameras, and algorithms, which use real-time data processing techniques to gather and interpret soil moisture measurements, water flow dynamics, and plant health metrics in real time for an informed irrigation management and fertilization decisions. Real-time analysis also allows informed decisions to be made concerning irrigation/nutrient application decisions – in this paper specifically, precision irrigation/furrow irrigation techniques will be described for automatic irrigation control purposes.

Keywords: Precision irrigation, machine vision, sensors, weather pattern, irrigation techniques, machine learning, furrow irrigation, accurate irrigation

14.1 Introduction

14.1.1 Introduction to Smart Irrigation and Machine Vision

Smart irrigation utilizes technology to utilize water resources while decreasing wasteful consumption and runoff effectively and efficiently. Sensors and data analytics are employed in agricultural fields to maximize water utilization efficiency. Furrow irrigation and precision watering are two artificial means used by agriculture for crop growth management and quality. Smart irrigation offers improved conservation compared with manual distribution systems – directly watering plants reduces runoff while sensors determine soil moisture levels more accurately than manual distribution does. It keeps track of watering schedules and the amount of water to make sure that plants receive the correct amount of water. By minimizing overwatering and ensuring that crops receive the right amount of water, smart irrigation systems can increase crop yields and reduce water waste. When we use a smart irrigation system, there will be no need for human intervention, which will save time. Smart irrigation systems can easily be controlled by smartphone computers or any such application; they will save you a huge amount of money that you need to spend on maintenance water bills and repairs costs. A smart irrigation system can reduce water usage and prevent water waste by delivering water only when it is needed, resulting in lower utility bills. It is more sustainable than manual irrigation systems. By conserving water, reducing energy usage, and minimizing the amount of maintenance required, smart irrigation helps to protect the environment and preserve natural resources [1].

Machine vision is a subfield of machine learning where certain techniques are utilized and illuminate the visual samples from images also videos. This algorithm, as the name vision suggests, makes the machine 'see' and recognize the world around them. Applying machine vision in irrigation helps enhance water usage and improve crop yield density. As part of smart irrigation solutions, machine vision sensors and cameras capture images in real-time of plants, soil, and weather conditions, allowing accurate regulation of irrigation systems. By pairing smart irrigation and machine vision together, farmers gain valuable insights into their crops for informed decision-making concerning water utilization. The integration of smart irrigation and machine vision can lead to more sustainable agricultural practices, with increased crop yields and reduced resource waste [2].

14.1.2 A Brief on Furrow Irrigation Water Management

To irrigate crops traditionally, furrow irrigation starts by creating trenches or furrows between plant rows and filling them with water. Furrow irrigation works in a way such that the water is applied to the crops in small channels that are constructed between each row of the plant. The water supply is delivered to the plants in small canals from the source of water to the destination. When the water reaches the plant, it gives moisture content to the plant [3]. Water seeps into the soil as it travels down the furrows, hydrating the roots of the plants. One of the earliest and most basic irrigation techniques, furrow irrigation is still utilized extensively throughout most of the world, especially in regions with scarce water supplies. Crops like corn, cotton, soybeans, and vegetables are frequently grown using this method. The regularity of the furrows, slope, and soil type are all variables that might affect how effective furrow irrigation [4].

14.1.3 A Brief on Precision Irrigation Water Management

Precision irrigation refers to watering crops accurately and precisely using sensors or other technologies, with precise soil moisture sensors measuring soil moisture content and determining how much water plants require for healthy growth. Machine vision technology, a cutting-edge advancement, can revolutionize precision irrigation by giving farmers access to more accurate and timely information regarding their crop's needs – helping improve both crop quality and yields. Precision irrigation ensures plants receive all the resources necessary for growth and success directly at their roots, maximizing water use efficiency while simultaneously decreasing waste and increasing crop yield and quality. Farmers using precision irrigation can grow more food using less water by using advanced sensors with data analytics and irrigation systems – leading to a more resilient and sustainable agriculture sector overall [5, 6]. Machine vision technology can provide valuable insights into precision irrigation water management, helping farmers optimize their irrigation practices, improve crop yields, and promote sustainable agriculture. By automating many of the monitoring and control tasks involved in precision irrigation, machine vision can also help reduce labor costs and improve efficiency [7].

14.1.4 Existing Irrigation Systems

Without adequate technological support, the existing irrigation system cannot accurately ascertain the exact water requirements of crops, leading to issues of under- or over-irrigation and inadequate irrigation levels that compromise growth or even kill off crops altogether. Inadequate irrigation may prevent crops from flourishing properly. Crop diseases and clogging issues are brought on by over-watering. An irrigation system that is both smart and feasible must be provided to counterattack their problems. Machine vision-based learning is used to collect images of such problems and counterattack them using precision and furrow irrigation [8]. Machine learning vision helps in the effective use of water, water conservation, and enhancement of both product quality and quantity.

14.2 Methodology

Initially, we must adjust the hardware and software settings in accordance with the specifications. The dataset contains photographs that must be identified and then must be assembled. The acquired dataset is then labeled, which classifies each image into clusters of its own. Labeling is nothing more than associating the data they contain with each of the dataset's images [9]. After labeling, the photos must be pre-processed in order to enhance image quality and remove noise or undesired items. Edge detection, smoothing, and filtering are examples of pre-processing methods. Image segmentation is the subsequent stage. To identify the items in the image, the image is then segmented or split into sections based on color, texture, feature, etc. We can separate the needed item from the background and other pictures using image segmentation [10]. Feature extraction must be carried out once the segments have been recognized. Each segment's attributes, such as form, size, color, and texture are retrieved. Segments are classified based on the plant's water needs with the use of these extracted characteristics. Classification is the process's last phase. Each segment is classified as either it requires furrow irrigation or precise irrigation.

Utilizing different machine learning techniques on our annotated dataset, our aim is to identify the model with the best performance based on desired outputs.

The methodology for implementing a machine vision-based automatic control and monitoring system in furrow irrigation and precision irrigation can be broken down into several key steps:

14.2.1 System Design

The main step of designing the system's main workflow and the best part to choose is done here. This may involve selecting appropriate sensors, cameras, and other hardware components, as well as designing software to control the system. Sensors may be used to keep track of a variety of environmental variables, including soil moisture content, temperature, humidity, and weather. The irrigation system may be informed by the sensor data and adjusted to change the water flow as necessary [11]. Images of the crops, soil, and other important environmental factors are taken using top-notch cameras.

The camera's pictures are analyzed by image processing algorithms in order to glean information about the surroundings. The irrigation may be adjusted in accordance with the data that the image processing algorithms produce and feed into the control system. Here, irrigation is done with precision and furrows. The visual data may be analyzed using machine learning techniques to forecast the requirement for irrigation.

The methodology for implementing a machine vision-based automatic control and monitoring system in furrow irrigation and precision irrigation can be broken down into several key steps:

14.2.2 Data Collection (as Images)

Images of the environment, including the crops, soil, and weather conditions, are taken using high-quality cameras. The cameras may be fixed on satellites, drones, or even on the ground. Since lighting affects image quality, cameras may come with extra lighting to guarantee that pictures are taken with enough contrast and color accuracy. Cameras must be calibrated to make sure they are correctly recording the surroundings before data gathering can start. This entails changing the camera's settings to take things like the lighting and lens distortion into consideration. The cameras' recorded pictures are kept in a database or data repository.

14.2.3 Image Processing

After the data has been gathered, techniques for image processing may be employed to examine the images and draw out pertinent data. Plants, crops, and soil are just a few examples of the kinds of items that object identification algorithms may discover and identify in a picture. To provide water to certain fields, a picture can be divided into several parts using segmentation algorithms based on color, texture, or other attributes. An image's crops and surrounding soil may be distinguished from one another using a segmentation method, for instance [12]. The kind of object in a picture, such as the

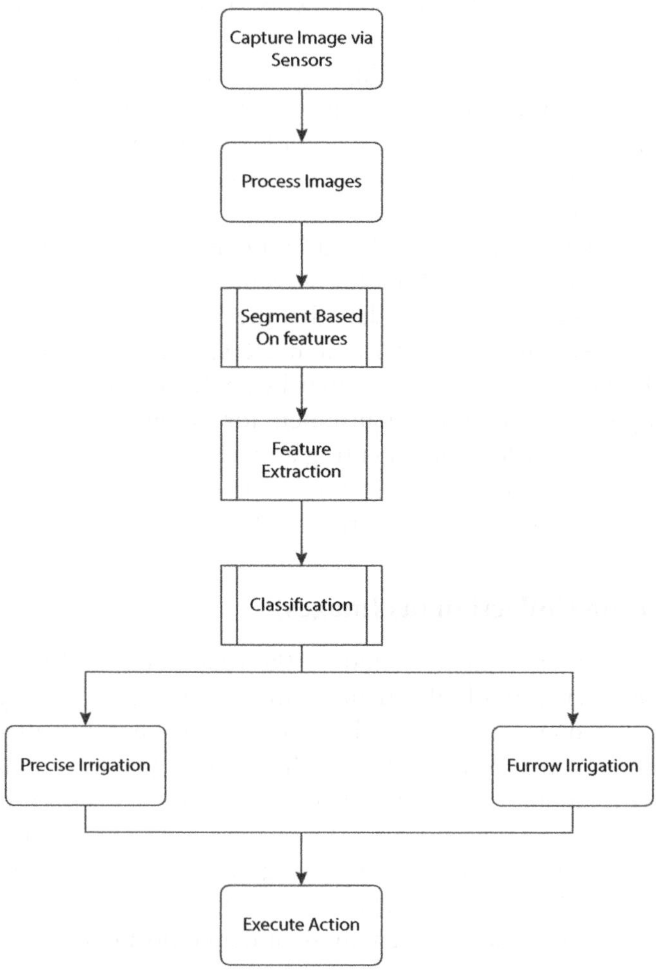

Figure 14.1 A flow of the proposed system.

species of a plant or the type of soil, may be determined through classification algorithms. Using this information, the irrigation system may be modified to provide each variety of plant with the proper amount of water and nutrients. Figure 14.1 schematically explores the flow of the proposed scheme.

The quality of the pictures that the cameras collect may be increased using image enhancement techniques. Enhancing the clarity of the photos may entail making changes to their brightness, contrast, and color balance. The form or texture of a plant, for example, can be recognized using feature extraction algorithms. The irrigation system may be modified in accordance with the development and health of the plants as a result of this information. Deep learning algorithms may be used to analyze a lot of visual data and discover patterns and connections between various environmental elements. This may result in more accurate estimations of the requirement for irrigation and more effective utilization of water resources [13].

The methodology for implementing a machine vision-based automatic control and monitoring system in furrow irrigation and precision irrigation can be broken down into several key steps:

14.2.4 Control Algorithms

Machine vision systems use data collected by sensors and cameras to autonomously regulate an irrigation system's functionality, including tasks like changing water flow rate, moving irrigation equipment around, activating or deactivating specific zones within an irrigation system and activating or deactivating specific irrigation zones. This is done by setting a few thresholds like:

i. **Threshold-based control:** Establishing a limit for soil moisture or other environmental variables, and then modifying the irrigation system accordingly. For instance, if the soil moisture drops below a predetermined level, the irrigation system will activate to raise the moisture level.
ii. **Feedback control:** This algorithm makes real-time adjustments to the irrigation system based on sensor and image processing feedback. For instance, the system can change the irrigation to supply more water to specific places if the image processing identifies that some parts of the field are drier than others.
iii. **Fuzzy logic control:** A more sophisticated algorithm that employs a set of rules and several environmental

considerations to regulate the irrigation system is known as fuzzy logic control. For instance, the system may make the best water delivery decision depending on soil moisture, temperature, and humidity levels using fuzzy logic.

iv. **Model-based control:** This method predicts how the system should be modified by using mathematical models of the surroundings. To estimate how much water to distribute at various points during the growing season, the system, for instance, may employ a model of crop growth rate [14].

14.2.5 Classification

Classification is a type of machine learning model that classifies or categorizes the input data or images into a predefined set of classes. Classification models aim to establish an association between input features and their respective labels or output, or labels. Classification models may either be binary or multi-class models; in this instance, the binary classification approach was taken. A variety of machine learning techniques such as the support vector machines (SVMs), deep learning methods like convolutional neural networks (CNNs), decision trees, and logistic regression were utilized during training [15] with evaluation measured against metrics like accuracy precision recall F1 score [15].

14.2.5.1 Convolutional Neural Networks (CNNs)

One of the most popular choices for irrigation systems is based on image classification. They may be used to examine pictures of crop health, soil moisture levels, and other environmental aspects and give input to the irrigation system so that water distribution can be adjusted as necessary. They consist of three types of layers, namely: convolutional layer, fully connected, and pooling layers, which work together to learn features from images and make predictions [10].

The basic formula for a convolutional layer in a CNN is:

$$\text{output} = \text{activation}(\text{convolution}(\text{input}, \text{weights}) + \text{bias}) \tag{14.1}$$

where input is the input image or feature map, weights are the learnable convolutional filters, bias is a learnable bias term, convolution is the convolution operation, activation is an activation function (e.g., ReLU), and output is the resulting feature map.

The step for the implementation of CNN is to:

1. Download the required dataset.
2. Read the label names.
3. Use the matplotlib function to depict the dataset via graphs.
4. Select the CNN model and apply it to the dataset.
5. Add the helper function.
6. Create the different three different layers that are required.
7. Create a fully connected layer and apply the loss function.
8. Create a variable to initialize the global variable.
9. Run the model.
10. Create a graph to depict the final results.

14.2.5.2 Support Vector Machine (SVM)

Classification tasks can benefit from machine learning techniques tailored specifically for this task, including crop classification using photos. These methods analyze crop photos and provide feedback to the irrigation system in order to adjust water distribution accordingly. During training, weight vector w and bias term b are learned via the SVM algorithm that attempts to identify the hyperplane that maximizes margin between classes [16]. Herein, the margin refers to the distance between hyperplane and the nearest data points of each class [15, 16]. To import this classifier, we use the sklearn model. The SVM algorithm steps include the following:

1. Import necessary libraries.
2. Fit the dataset into x and y.
3. Provide 80% of the dataset for training and the rest 20% for testing.
4. Import the SVM classifier model from sklearn.
5. Fit the SVM classifier into x train and y train.
6. Run the model and predict the function.
7. Evaluate performance using error and accuracy.

14.2.5.3 Random Forests

It is a form of machine learning, which embodies ensemble approach using bagging techniques, random samples are used to generate 'n' number of trees and are then voted based on soft voting probabilities:

$$y = mode\{f1(x), f2(x), ---- fn(x)\} \tag{14.2}$$

where y is the predicted class label for a new input x, fi(x) is the predicted class label by the i-th decision tree, and mode is the majority voting function that selects the most frequent prediction among the decision trees.

To implement the random forest model, we need to follow three basic steps.

1. Random Sampling of data.
2. Construction of binary trees based on these sampling data and then prediction.
3. With the help of voting, pick the most voted result from the predicted outputs.

14.2.5.4 Recurrent Neural Networks (RNNs)

These RNNs use time series data for their prediction. They can be utilized to analyses environmental sensor data over time and give input to the irrigation system, so that water distribution can be adjusted accordingly.

The formula for a simple RNN cell is:

$$h(t) = activation(w_x * x_t + w_h * h_{\{t-1\}} + b) \tag{14.3}$$

14.2.5.5 Deep Reinforcement Learning

It can be used to figure out how to automate and control irrigation systems based on input from the outside world. The agent's goal is to find the optimal policy by iteratively improving the Q-function using the Bellman equation and updating the policy based on the Q-function.

14.2.5.6 Artificial Neural Network (ANN)

The system can change irrigation frequency and volume to guarantee that the plants receive the right quantity of water by using artificial neural networks (ANNs) to analyze photos of crops and assess their health state. In order to optimize irrigation schedules, ANNs can also be employed to analyze data on soil moisture and other environmental parameters.

This model has to provide the highest accuracy and make sure there is a proper irrigation supply for the crops in a time based on ANN machine vision-based techniques, and hence farmers can increase crop production while using less water [17, 28, 29]. Additionally, by lowering the amount of labor necessary, this strategy can improve the effectiveness and cost-effectiveness of manual irrigation scheduling.

The formula for a simple feedforward ANN can be expressed as follows:

$$y = f(k_2 * f(k_1 * x + a_1) + a_2) \tag{14.4}$$

14.2.5.7 Feedback and Optimization

Closed-loop control: Closed-loop control algorithms make real-time adjustments to the irrigation system based on sensor data and image analysis; for instance, if the irrigation is turned on to raise the soil moisture level because the system has detected that it is too low. A constant environment can be maintained, and over- or under-watering may be avoided using closed-loop management.

Real-time optimization: Real-time optimization algorithms optimize for specified goals while using sensor feedback and image processing to adapt the irrigation system in real-time [18]. The system may be adjusted, for instance, to increase agricultural output while lowering water use. The system can adapt to shifting environmental conditions and accomplish specified goals with the aid of real-time optimization.

Model-based optimization: To determine how the system should be modified to accomplish certain objectives, model-based optimization algorithms make use of mathematical models of the surrounding environment [19]. To estimate how much water to distribute at various points during the growing season, the system, for instance, may employ a model of crop growth rate. The system may optimize for particular goals and make better judgments with the aid of model-based optimization.

14.3 Maintenance and Upgrades

Prioritize machine vision system maintenance and updates as necessary in order to sustain peak functionality over time, which may include replacing sensors or upgrading software components [20]. The primary aim of this study is integrating machine-based automatic control and monitoring into furrow irrigation and precision irrigation techniques using traditional neural networks, as well as ANNs as possible implementation methods [23, 24]. The resultant accuracy for both models should be found and can be compared for the best result. The accuracy of the cameras and sensors used in irrigation system machine vision can be impacted by regular cleaning and calibration since they can get filthy or misaligned over time [21]. Their functionality and accuracy may be maintained with regular cleaning and calibration.

Update software and firmware: To keep the machine vision system current and functional, it is essential to update the system's software and firmware when new features and enhancements are created. New sensors or cameras may need to be added to the system in order to give more information or capability as the system's needs alter [22].

Hardware update: As technology develops, it may be necessary to upgrade hardware elements like processors or memory to keep up with system needs [23].

14.4 Experimentation and Analysis

14.4.1 Dataset Summary

The dataset used here is from Kaggle, and it contains data of approximately 65 mb, the below link https://www.kaggle.com/datasets/aman2000jaiswal/agriculture-crop-images. The above dataset has been used to find certain metrics like humidity in the plants and their day-wise growth measure is

Figure 14.2 Sample images from dataset.

recorded. This dataset consists of around 210 images each of 520 x 520 color images. The above datasets are in JPEG format, and it is the most widely used format for machine learning via machine vision [26]. From the dataset, the sample images was given in Figure 14.2.

This experimentation was done in Python. OpenCV, Tensorflow and Keras was used to carry out the experiment. These packages play a major role in creating an efficient neural network.

14.4.2 Performance Evaluation Metrics

The performance evaluation measures serve as the foundation for the proposed system's efficiency. These metrics are used to evaluate how the proposed model performs in comparison to other models. The most often used performance evaluation measures in research include accuracy, precision, recall, sensitivity, specificity, and f measures [24–27].

14.4.2.1 Accuracy

The accuracy of the proposed system is given as the correct classification of the dataset on whether it requires furrow or precise irrigation methods.

The accuracy formula is given by:

$$Accuracy = (\mathrm{TP} + \mathrm{TN}) / (\mathrm{TP} + \mathrm{TN} + \mathrm{FP} + \mathrm{FN}) \tag{14.5}$$

14.4.2.2 Specificity

It is the measure of outliers present in the overall classified dataset.

The specificity formula is given by:

$$\mathrm{Specificity} = \mathrm{TN} / (\mathrm{TN} + \mathrm{FP}) \tag{14.6}$$

14.4.2.3 Precision

It is the measure of how accurate or how well the dataset has been classified.

The formula for precision is given by:

$$\mathrm{Precision} = \mathrm{TP} / (\mathrm{TP} + \mathrm{FP}) \tag{14.7}$$

14.4.2.4 Sensitivity/Recall

It is the measure of the proportion of actual ones that are correctly classified by the model.

The formula for sensitivity is given by:

$$\text{Sensitivity} = \text{TP} / (\text{TP} + \text{FN}) \tag{14.8}$$

14.4.2.5 F1 Measure/F1 Score

F1 measure is a harmonic mean of precision and sensitivity. It is calculated as:

$$\text{F1} - \text{measures} = 2 * (\text{Precision} * \text{Sensitivity}) / (\text{Precision} + \text{Sensitivity}) \tag{14.9}$$

14.4.3 Result and Discussion

In this chapter, we examine the application of machine vision technologies to precision irrigation and furrow irrigation systems for automatic control and monitoring. The precision of the suggested system determines how effective it is. As a result, we examine every technique and choose those that are the most precise. The results are shown below in the tables and related graphics.

Table 14.1 shows the results of the metrics without using the preprocessing techniques. This table shows that the performance of the ANN is the best among the six with the accuracy of the ANN being 94.66%. The recall and specificity of the ANN fall in the range of 95% and 96%, respectively. Next to ANN, deep reinforcement has an accuracy of 94.17%. But lack of preprocessing would certainly have some disadvantages; hence, we would try the same after applying preprocessing techniques.

Table 14.2 shows the results after applying preprocessing techniques. This table shows that the performance of the ANN is again the best among the six with its significantly improved accuracy. The accuracy of the CNN is now improved to 97.89%. The recall and specificity also improved significantly to around 96% and 97%, respectively.

Table 14.1 A results comparison of the algorithm without pre-processing.

Method	CNN	SVM	Random forest	RNN	Deep reinforcement	ANN
Precision	88.46	89.15	90.23	91.41	91.78	93.55
Recall	88.56	90.12	91.85	93.45	94.15	95.00
F-measure	88.19	90.26	91.84	92.58	93.94	94.87
Accuracy	91.56	91.87	92.84	93.78	94.17	94.66
Specificity	93.15	92.26	94.78	95.10	95.89	96.18
Sensitivity	91.56	91.14	91.89	92.74	93.62	94.55

Table 14.2 A results comparison of the algorithm with pre-processing.

Method	CNN	SVM	Random forest	RNN	Deep reinforcement	ANN
Precision	92.1	93.7	94.9	95.1	96.4	97.6
Recall	91.5	93.42	93.89	94	95.25	96.15
F-measure	90.46	91.00	91.56	92.47	94.17	95.35
Accuracy	92.33	92.65	93.44	93.58	95.11	97.89
Specificity	94.56	94.50	94.89	95.61	96.12	97.12
Sensitivity	93.69	93.88	94.51	94.90	95.82	96.54

14.5 Conclusion and Future Enhancement

Machine vision when integrated with agriculture helps solve weeding problems more efficiently. Machine vision-based weed detection is faster, accurate and helps save human effort. Few promising machine vision techniques are CNN, SVM, KNN, random forest, and Naïve Bayes algorithm. Machine vision-based weeding systems have great potential for reducing the influence of chemical herbicides in agriculture and reduce manual labor. In conclusion, machine vision-based automatic control and monitoring have the potential to revolutionize furrow irrigation and precision irrigation. Computer vision techniques enable accurate measurements of soil moisture, crop growth, and identification of weeds, pests, and diseases that can help monitor their distribution through irrigation systems

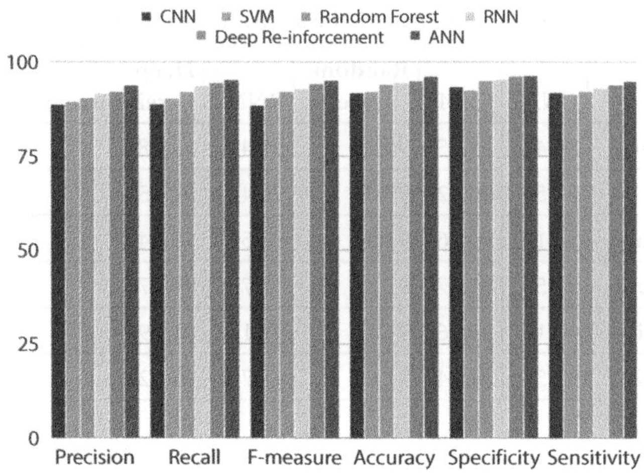

Figure 14.3 A graphical representation of a result comparison without pre-processing.

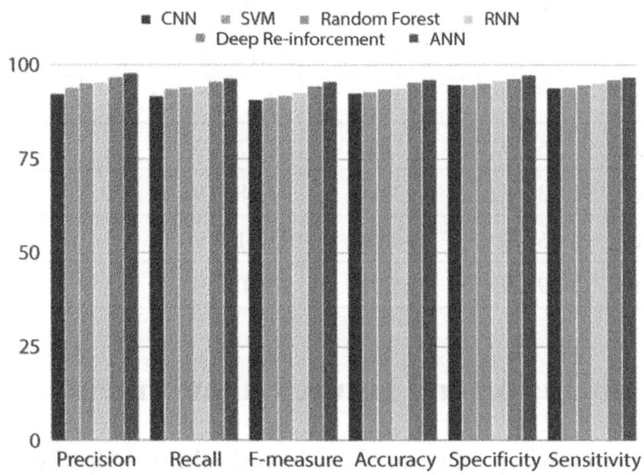

Figure 14.4 A graphical representation of a result comparison with pre-processing.

for maximum crop yield and lower consumption of water resources. With such valuable insights at our fingertips, improving crop yields while simultaneously decreasing consumption is possible. Figures 14.3 and 14.4 clearly shows the performance of the various algorithms with and without preprocessing technique.

Utilizing machine learning algorithms allows us to analyze data gathered from sensors such as cameras, humidity, and temperature to develop predictive models for irrigation management. These models help farmers make informed decisions regarding irrigation timing and volume to optimize crop growth while limiting water usage. Integrating machine vision-based automatic control and monitoring into furrow irrigation and precision irrigation systems holds great promise of increasing crop yields, decreasing water consumption, and optimizing overall irrigation management efficiency. As technology develops and becomes more readily accessible, automated agriculture systems may become more popular and promote efficient farming practices that support sustainable sustainability practices. Research in our field seeks to evaluate and select the most efficient vision techniques for classifying irrigation models accurately; whether precision- or furrow-based. However, further investigation must take place in order to improve these systems, address their limitations such as lighting sensitivity or precision calibration requirements, or just generally enhance our knowledge on this area of agriculture.

With smart irrigation and weather forecasting, it is possible to anticipate and adjust irrigation parameters based on predicted weather conditions by integrating machine vision-based systems with weather forecasting. As a result, crop growth can be optimized, and water can be conserved. To decrease manual work, automated maintenance systems can be adopted that develop machine learning algorithms that can predict equipment failure and schedule maintenance automatically. This will improve the efficiency of the system [27, 28]. It can prevent equipment downtime, reduce maintenance costs and ensure that the system is always running at its best. Energy can be optimized by using the following techniques for optimizing energy consumption. This can help prevent equipment downtime, reduce maintenance costs, and ensure that the system, while operating at peak performance, can still be improved.

References

1. D. Long, C. McCarthy and T. Jensen, "Row and water front detection from UAV thermal-infrared imagery for furrow irrigation monitoring," *2016 IEEE International Conference on Advanced Intelligent Mechatronics (AIM)*, Banff, AB, Canada, 2016, pp. 300-305.
2. Sugunedham, S.K., Rajeswari, M., Anandhi, S. and Manikandan, J., 2023, November. Fruits and Pest Diseases Detection using Deep Learning-Based Approach. In *2023 International Conference on System, Computation, Automation and Networking (ICSCAN)*, (pp. 1-5). IEEE.
3. Megersa, Girma, and Jemal Abdulahi. "Irrigation system in Israel: A review." *Int. J. Water Res. Environ. Eng.*, 7, no. 3 (2015): 29-37.

4. Talaviya, T.; Shah, D.; Patel, N.; Yagnik, H.; Shah, M. Implementation of artificial intelligence in agriculture for optimisation of irrigation and application of pesticides and herbicides. *Artif. Intell. Agric.*, 2020, 4, 58–73.
5. Wang M, Wang H, Xiao J, et al. A Review of Disentangled Representation Learning for Remote Sensing Data. *CAAI Artif. Intell. Res.*, 2022, 1(2): 172-190.
6. Yang, Niuzhuang, Gaoyang Gou, Xiaoli Lu, and Yue Hao. "Linear dichroism and polarization controllable persistent spin helix in two-dimensional ferroelectric ZrOI2 monolayer." *Nano Res.*, 15, no. 7 (2022): 6779-6789.
7. Wang, Yulin, Jiayi Guo, Shiji Song, and Gao Huang. "Meta-semi: A meta-learning approach for semi-supervised learning." *arXiv preprint* arXiv:2007.02394 (2020).
8. Majumdar, Parijata, and Sanjoy Mitra. "IoT and Machine Learning-Based Approaches for Real Time Environment Parameters Monitoring in Agriculture: An Empirical Review." *Agricultural Informatics: Automation Using the IoT and Machine Learning* (2021): 89-115.
9. Huuskonen, Janna, and Timo Oksanen. "Soil sampling with drones and augmented reality in precision agriculture." *Comput. Electron. Agric.*, 154 (2018): 25-35.
10. Gu, Jian, Guanghua Yin, Pengfei Huang, Jinlu Guo, and Lijun Chen. "An improved back propagation neural network prediction model for subsurface drip irrigation system." *Comput. Electr. Eng.*, 60 (2017): 58-65.
11. Tseng, David, David Wang, Carolyn Chen, Lauren Miller, William Song, Joshua Viers, Stavros Vougioukas, Stefano Carpin, Juan Aparicio Ojea, and Ken Goldberg. "Towards automating precision irrigation: Deep learning to infer local soil moisture conditions from synthetic aerial agricultural images." In *2018 IEEE 14th International Conference on Automation Science and Engineering (CASE)*, pp. 284-291. IEEE, 2018.
12. Goap, Amarendra, Deepak Sharma, A. Krishna Shukla, and C. Rama Krishna. "An IoT based smart irrigation management system using Machine learning and open-source technologies." *Comput. Electron. Agric.*, 155 (2018): 41-49.
13. Megersa, Girma, and Jemal Abdulahi. "Irrigation system in Israel: A review." *Int. J. Water Res. Environ. Eng.*, 7, no. 3 (2015): 29-37.
14. Chauhdary, Junaid Nawaz, Hong Li, Yue Jiang, Xuwei Pan, Zawar Hussain, Maria Javaid, and Muhammad Rizwan. "Advances in Sprinkler Irrigation: A Review in the Context of Precision Irrigation for Crop Production." *Agronomy*, 14, no. 1 (2023): 47.
15. Shi, Jiayou, Yuhao Bai, Zhihua Diao, Jun Zhou, Xingbo Yao, and Baohua Zhang. "Row detection BASED navigation and guidance for agricultural robots and autonomous vehicles in row-crop fields: methods and applications." *Agronomy*, 13, no. 7 (2023): 1780.
16. Chang, Chung-Liang, and Hung-Wen Chen. "Robust Guidance and Precise Spraying of a Four-wheeled Agricultural Robot based on Deep Learning Approach." (2023).

17. Pournima, S., C. Priyatharsini, G. Kirubasri, and J. Manikandan. "Hybrid BILSTM Network for Improving Crop Pest Classification." In *2023 International Conference on Computer Communication and Informatics (ICCCI)*, pp. 1-5. IEEE, 2023.
18. Wang, Huajia, Jinan Gu, and Mengni Wang. "A review on the application of computer vision and machine learning in the tea industry." *Front. Sustain. Food Syst.*, 7 (2023): 1172543.
19. Azghadi, Mostafa Rahimi, Alex Olsen, Jake Wood, Alzayat Saleh, Brendan Calvert, Terry Granshaw, Emilie Fillols, and Bronson Philippa. "Precise Robotic Weed Spot-Spraying for Reduced Herbicide Usage and Improved Environmental Outcomes–A Real-World Case Study." *arXiv preprint*, arXiv:2401.13931 (2024).
20. Yang, Qi, Licheng Liu, Junxiong Zhou, Mary Rogers, and Zhenong Jin. "Predicting the growth trajectory and yield of greenhouse strawberries based on knowledge-guided computer vision." *Computers and Electronics in Agriculture*, 220 (2024): 108911.
21. López-Correa, Juan Manuel, Hugo Moreno, Diego Sebastian Pérez, Facundo Bromberg, and Dionisio Andújar. "Towards a true conservation zero tillage system:"A proposed solution based on computer vision to herbicide resistance."" *Comput. Electron. Agric.*, 217 (2024): 108576.
22. Delveen Luqman Abd Al-Nabi, Shukri Shereen Ahmed. Survey on Classification Algorithms for Datamining (comparison and evaluation), *Comput. Eng. Intell. Syst.*, 4(8):18–24, 2013.
23. Yoga Pristyanto, Irfan Pratama, Anggit Ferdita Nugraha. Data Level Approach for Imbalanced Class Handling on Educational Data Mining Multi Class Classification. *IEEE Intel. Conf. on Information and Communications Technology (ICOIACT)*, 310–314, 2018.
24. Huang, Haocheng, Xiaohui Lei, Weihong Liao, Haichen Li, Chao Wang, and Hao Wang. "A Real-Time Detecting Method for Continuous Urban Flood Scenarios Based on Computer Vision on Block Scale." *Remote Sensing*, 15, no. 6 (2023): 1696.
25. Salcedo, E., Uchani, Y., Mamani, M. and Fernandez, M., 2024. Towards Continuous Floating Invasive Plant Removal Using Unmanned Surface Vehicles and Computer Vision. *IEEE Access*.
26. Yousaf, Arslan, Vahid Kayvanfar, Annamaria Mazzoni, and Adel Elomri. "Artificial intelligence-based decision support systems in smart agriculture: Bibliometric analysis for operational insights and future directions." *Front. Sustain. Food Syst.*, 6 (2023): 1053921.
27. Khadatkar, Abhijit, A. P. Pandirwar, and V. Paradkar. "Design, development and application of a compact robotic transplanter with automatic seedling picking mechanism for plug-type seedlings." *Scientific Reports*, 13, no. 1 (2023): 1883.

28. Khadatkar, Abhijit, A. P. Magar, C. P. Sawant, and R. U. Modi. "Development and testing of automatic seedling extractor in robotic transplanter using mechatronics for nursery seedlings." *Discover Applied Sciences*, 6, no. 2 (2024): 1-13.
29. Nithya, V., M. S. Josephine, and V. Jeyabalaraja. "IoT-Based Crop Yield Prediction System in Indian Sub-continent Using Machine Learning Techniques." *Remote Sensing in Earth Systems Sciences*, 6, no. 3 (2023): 156-166.

15

Applications in Agriculture for Assessing and Monitoring Soil Using Smart Sensing and Edge Computing

G. Padmapriya[1]*, V. Vennila[2], Prithi Samuel[3], Rajesh Kumar Dhanaraj[4], Balamurugan Balusamy[5] and Malathy Sathyamoorthy[6]

[1]Department of Computing Technologies, School of Computing, SRM Institute of Science and Technology, SRM Nagar, Kattankulathur, Chennai, India
[2]Department of Computer Science and Engineering, K.S.R College of Engineering, Tiruchengode, India
[3]Department of Computational Intelligence, School of Computing, SRM Institute of Science and Technology, Kattankulathur Campus, Chennai, India
[4]Symbiosis Institute of Computer Studies and Research (SICSR), Symbiosis International (Deemed University), Pune, India
[5]Shiv Nadar Institution of Eminence, Delhi-NCR, India
[6]Department of Information Technology, KPR Institute of Engineering and Technology, Coimbatore, India

Abstract
Agriculture continues to be the sector that gives the most to the nation. However, when we consider the technology that is currently being used in this field, people find that development is not extremely important. The significant technological advancements of today have an impact on a number of sectors, including medical care, the agricultural sector, and others. The main business sector in our country is agriculture. An innovative and effective method for relating things and people is to monitor the temperature and humidity at a specific location and make the information visible. The system controls and monitors environmental factors like temperature and relative humidity using sensors to transmit data to the system, which then plots the sensor data as graphical measurements. Using a rain gun, water is applied to the plant's roots drop by drop. When the amount of moisture

**Corresponding author*: padmaprg1@srmist.edu.in

Rajesh Kumar Dhanaraj, Balamurugan Balusamy, Prithi Samuel, Malathy Sathyamoorthy and Ali Kashif Bashir (eds.) Computer Vision in Smart Agriculture and Crop Management, (343–366) © 2025 Scrivener Publishing LLC

returns to normal, the sensor detects it and sends a signal to the microcontroller, which closes the value. When the moisture content of the soil drops, a moisture sensor attached to the Internet of Things (IoT) detects it and sends a signal to the microcontroller. The IoT facilitates gaining access to data and important decision-making by obtaining various values from sensors, such as moisture levels in the soil, detectors, and the quality of water. This method concentrates mainly on reducing the amount of labor required for irrigation on the land, as well as decreasing water waste in order to preserve the farmer's time, money, and energy.

Keywords: Smart sensing, edge computing, agricultural technology, precision agriculture, soil assessment, soil monitoring, environmental monitoring, humidity sensors

15.1 Introduction

One of the industries that benefits our nation's economy is agriculture. According to the government's motto, agriculture is existence for more than 85%, the government has increased funding for technology development to boost agricultural productivity. The most crucial challenge facing dry lands is water conservation. It is crucial for the plant's survival, as well, to avoid it from decaying or, in the most extreme case, dying, which determines how much water is in it. Considering this, different people drink different amounts of water. A cactus, for instance, can survive with only a little water.

In recent decades, the exponential growth of the population in comparison to its natural resources has made it imperative to create a completely digitized, highly intelligent, and smart agriculture system. To address this need, an Internet of Things (IoT)-based smart irrigation system have been developed to combat the continual hazard of water scarcity. The agricultural industry in Pakistan has two distinct types of plastic tunnel farming, namely low, high, and walk-in tunnels. The high tunnel offers greater convenience for sowing, spraying, and harvesting due to its larger size compared to the low and walk-in tunnels [6]. Conversely, traditional farming methods are unpredictable and result in more water wastage. The focus of this paper is the issue of smart irrigation, which is an application designed to increase the effectiveness and timeliness of the smart watering system. However, it is essential to note that simply relying on cloud technology for irrigation may not be sufficient, and further improvements are necessary to optimize the system.

15.1.1 IoT-Based Smart Agriculture

An enormous-scale IoT application might require more than just computing. To manage all of the different kinds of information generated by different sources, which includes sensors, an efficient and rapid utilization with an improved structure is required. A quick and efficient irrigation system's main objective is water conservation while ensuring that plants receive timely watering and protecting limited water supplies [1]. Several sensor-based automated irrigation systems are currently developed with their own mobile apps to tackle this issue [2, 12]. However, the reliability of these systems is questionable when dealing with growing data and increasing latency rates of IoT devices. Previous studies have used fuzzy logic to decide whether to water plants using input factors like relative humidity, temperature, moisture content of the soil, and the intensity of light [10]. Healthcare organizations that employ biosensors to track vital signs like humidity, oxygen saturation, blood pressure, and infections in wounds have also used fuzzy logic [4].

In recent years, the agricultural industry has faced significant challenges due to the exponential growth of the population in many countries, where natural resources are limited. The development of a completely automated, intelligent, and smart agricultural system has become a desirable solution to these challenges. In this context, IoT-based smart watering systems have been developed to address the constant threat of water scarcity. However, such systems need to be more efficient and timelier, and cloud computing alone may not be enough for large-scale IoT applications [3].

In order to manage different kinds of data obtained from various sources, like data sensors, a more effective and quick utilization by means of a better architecture is required. A quick and clever irrigation system's main goal is to conserve insufficient sweet water reservoirs and use water sparingly and promptly for the growth of plants.

Sensor-based automated irrigation systems with mobile applications have been developed to address this. Even so, there are still concerns about their dependability as data volumes increase and IoT device latency rates rise. In order to take this into account, this paper suggests an innovative approach that utilizes algorithms for machine learning and semantics, including input factors like the type of climate, the kind of crop, and the category of soil, along with the output parameters of the sensors—temperature, humidity, and soil moisture.

The suggested system addresses a number of drawbacks of current smart irrigation systems, including overloaded sensors, the absence of crucial parameters, and the loading of unwanted data onto the IoT server. The suggested method emphasizes machine learning and uses ontologies for information about the species of plants, soil kinds, and weather conditions to determine how much water a specific plant needs.

The system is designed for reliability and efficiency, where the value measured by a non-functional sensor is used to produce the result an hour before. The proposed approach offers a more efficient and advanced method of irrigation and contributes to the development of smart and intelligent agricultural systems.

15.1.2 Smart Sensing

Smart agriculture using smart sensing and edge computing involves enhancing the effectiveness, as well as the environmental impact of agricultural practices. Here are some ways in which smart agriculture is done using smart sensing and edge computing [8, 14, 20]:

15.1.2.1 Smart Sensing

Data on many different environmental factors, which include soil water content, temperature, humid conditions, brightness of light, and levels of nutrients, is collected using smart sensors. Then, this information is employed to enhance the cultivation of crops and decrease waste [6].

15.1.2.2 Edge Computing

Edge computing is the practice of dealing with and analyzing data locally, in this case, at the farm. This helps to reduce latency, improve real-time decision-making, and minimize the need for expensive cloud-based data storage and processing.

15.1.2.3 Internet of Things (IoT)

The collection of data on various facets of agricultural production, livestock farming, and management on farms is made possible by the integration of IoT devices like cameras, sensor technology, and drones. In order to gain knowledge about the health of the crops, the state of the soil, and other elements that influence farm productivity, this information is subsequently organized and investigated.

15.1.2.4 Predictive Analytics

Forecasting crop yields, spotting prospective disease and pest infestations, and optimizing schedules for irrigation are all possible with statistical analysis. This helps to prevent crop losses and improve farm profitability.

15.1.2.5 Precision Farming

Precision farming involves the use of data-driven insights to optimize farming practices such as seeding, fertilization, and irrigation. This helps to minimize resource waste and improve crop yields.

Overall, smart agriculture using smart sensing and edge computing helps farmers to make more informed decisions, reduce overheads, and recover the sustainability of their tasks.

15.1.3 Assessing and Monitoring Soil Using Smart Sensing

Around the world, traditional tunnel farms typically rely on rainwater harvesting or sprinkler systems irrigation techniques, which are thought to be more effective to customary flooding techniques. Both energy and water efficiency may differ depending on the type of irrigation method used. For instance, the water and energy utilization associated with surface and threshold irrigation techniques is low. Low-to-medium effectiveness is provided by sub-irrigation water supply, expenditure irrigation, and sprinkler-based irrigation methods. Both drip irrigation and sprinkler irrigation are thought to offer comparable energy efficiency. The use of sprinklers is less water-efficient than drip irrigation, on the other hand. Therefore, the choice of the irrigation method can have a significant impact on crop production, water usage, and energy consumption. Proper water management, including the use of technology and smart water management practices, is critical for increasing crop yields while reducing costs.

The effective oversight of water from the atmosphere is essential for increasing crop production and lowering costs. Farmers can now supply the exact amount of water needed for their crops thanks to the effective use of technologies, leading to a more profitable and environmentally friendly agricultural sector [15]. In order to guarantee ideal irrigation purposes with a preventative strategy on four pilots in Brazil and Europe, an IoT-based intelligent water management framework had been proposed as part of the SWAMP project in Europe.

The Europeans provided the SWAMP design, framework, and system deployed, which includes an evaluation of performance assessment of FIWARE components. In order to improve the project's effectiveness and productivity in managing water supplies, some of its parts will be reengineered to offer more flexibility while utilizing fewer computing resources.

By utilizing smart water management platforms like SWAMP, farmers can reduce water usage, improve crop yields, and optimize their use of resources. This proactive approach to water management is essential in combating the global water scarcity crisis, ensuring that we can continue to feed our growing population while preserving our precious freshwater resources.

Despite the increase in the quantity of terrestrial watered in the United States over the past decade, farmers are using significantly less water than before. This is largely due to advancements in irrigation technologies and management practices that have improved water use efficiency. In order to meet the needs of the population all year long, farmers have the ability to cultivate a large quantity of fruits, vegetables, nuts, and grains that are whole.

In the United States, two main conventional methods of irrigation are currently in use since 2013. The most common type is the gravitational system, which comprises 35–42% of the nation's irrigation systems. This system floods land-forming features like channels, rivers and streams, reservoirs, and creases to transport the water from the point of origin to the crop area. This category includes both regulated and unregulated flooded areas systems, including the furrow system.

The pressure system, which uses tubing or pipes to circulate water to the crop area, is the second major type of technology for irrigation used in the US. A sprinkler or drilled pipe is subsequently employed as an applicator to apply irrigation.

Throughout the past several decades, traditional techniques have been established and put into practice on regional scales in India to effectively satisfy the needs of the populace. The three main irrigation techniques used in the nation are diverting attention channels, small-scale businesses water bodies like rainwater tanks, and wells for gathering groundwater. Both small-scale and massive operations agricultural enterprises can use these techniques. India's population is expanding, but so has its water requirements for a variety of uses, including irrigation, personal consumption, hydroelectric power, industrial processes, mineral extraction, and restoration of the environment.

Despite having the world's biggest agricultural area in the entire globe, only 40% of the total agricultural land is covered by irrigation [13]. The widespread deployment of conventional irrigation techniques, that have an average water utilization rate of only 35–40% [11], is one of the main causes of this low amount of irrigated land.

When traditional irrigation techniques are used instead of the use of cloud computing and edge computing, the plant fertilizing system becomes unstable. Hence, the need for an efficient and well-planned watering system is of utmost importance. Over the past few decades, irrigation systems using sensor networks and various IoT approaches have been developed, providing solutions to the problem.

Research on the design and implementation of a mechanized irrigation system that is capable of utilizing water more efficiently [3]. They employed a wireless sensor network that enabled the management of water-saving measures. Their approach resulted in a significant reduction of 90% in water usage compared to conventional irrigation methods. In the same year, research studies on sensor-based irrigation systems were reported by Parameswaran and Sivaprasath [6], Kumar *et al.* [7], and Rawal [16]. Nelson created a system in 2015 that used sensor data, such as soil moisture and temperature, to remotely control the irrigation process. This system used wireless sensor and actuator networks to reduce consumption of water.

A smartphone-based irrigation system was evaluated in the field by Saab *et al.* [17], who discovered that the system reduced water consumption in Mediterranean settings by 25%. Saqib unveiled a wireless system for the HC-12 module in 2020 with the goal of enhancing communication capabilities and optimization efforts. These studies emphasize how crucial it is to effectively utilize irrigation systems and use less water by leveraging cutting-edge technologies like smartphone applications and wireless sensor networks. The implementation of such systems could lead to more sustainable and efficient agriculture practices in the future.

15.1.4 Smart Agriculture Using Edge Computing

Edge computing is gaining traction throughout a range of sectors in the age of digital transformation. By 2025, it is projected to be around 75%. Numerous sectors, including agriculture, are embracing edge computing. With smart farming, this machinery serves to establish the next generation of agriculture. Despite the fact that the cloud's architecture currently plays a significant role in agricultural development, edge computing takes the lead in the form of rapidity and effectiveness. Precision agriculture presents an opportunity when the use of edge computing is incorporated into smart farming technological innovations. Farmers rely on data when using edge computing technology to gain better control of the agricultural sector and improve the productivity of their operations, resulting in lower operational costs.

There are numerous use cases for smart agriculture and farming, ranging from monitoring crop/cattle ailments to greenhouse effect automation and IoT-enabled, end-to-end agricultural management remedies. In terms of edge computing applications in smart farming innovations, the key possibilities are referred to as "precision agriculture." Farmers use data to gain more oversight over their businesses, maximize the effectiveness of their operations, and, as a result, they minimize their operational costs.

15.1.4.1 Autonomous Tractors and Robotic Machinery

Through the aid of edge computing, autonomous agricultural equipment and robotic machines may operate automatically without the assistance of humans. The tractors are capable of communicating with nearby detectors to gather information about their surroundings. The robots are capable of assessing the most effective techniques for tackling the required area, taking into account the type of task carried out, the number of vehicles in the field, the size of the apparatus, and so on. Edge computing will allow agricultural robots to employ computer vision and pre-loaded growing information to generate interpret insights. Furthermore, if an obstacle, such as an animal or a human, is in the way, the automated tractors may redirect themselves automatically. Such intelligent implements can carry out a wide range of tasks.

15.1.4.2 Farm Automation

Using IoT edge computing, a greenhouse or even an entire farm may be activated as autopilot. This means that the closed ecological systems can take responsibility for itself rather than depending on a remote server to process data and determine choices about regular tasks such as watering plants, feeding livestock controlling the ambient temperature, sunlight, and moisture in the room. Edge computing, like agribots, will enable the farm or greenhouse function independently of the main computing system and make selections locally based on information acquired from local sensors. This has the prospective to growth reliability of processes and diminish waste, thereby making agriculture more sustainable. Agriculture IoT systems are capable of making informed choices about possible threats to the environment or catastrophic events by utilizing edge computing. Remote sensors, for example, are capable of gathering and analyzing data about shifts in the climate or the surroundings in order to forecast eventual catastrophes and, if specific danger warning signs are detected, alert every member of the command structure immediately. As a result, farmers will be able to take on time-preventative measures to safeguard the yields of crops, preferably partly, during the possibility of a wildfire.

15.2 Smart Agriculture Using Smart Sensing and Edge Computing

Smart agriculture has become the adoption of modern technology in agriculture that incorporates the IoT, detectors, analytics of large amounts of data, artificial intelligence (AI), and cutting-edge computing [18]. The goal is to improve farming methods, enhance the production of crops, and minimize the consumption of natural assets comprising fertilizer, water, and chemical pesticides.

Smart agriculture makes use of sensors to gather data on the moisture in the soil, the temperature, and various other factors related to the environment. This information is subsequently examined and processed using AI algorithms to supply farmers with perspectives and recommendations for improvement. Insights may involve when to supply agricultural products, when to apply nutrients, and precisely when to harvest.

Edge computing is extremely important in smart agriculture by processing data locally, close to the source of the data, rather than directing it to a centralized data center. This reduces latency and enables real-time decision-making.

Smart agriculture possesses the ability to revolutionize the agricultural sector through improving productivity, decreasing waste, and raising yields. It may also assist with tackling some of the agricultural industry's obstacles, such as water scarcity, global warming, and expanding populations [5].

15.2.1 Pest Control

Pest control is an essential aspect of smart agriculture. Pests have the power to seriously damage crops and reduce yields, which has an impact on farmers' bottom lines. Ecological imbalances can result from the use of chemical pesticides, which are traditional methods of controlling pests and have detrimental effects on the environment, as well as beneficial insects. The goal of smart agriculture is to apply technology to enhance pest management techniques, reduce the usage of toxic substances, and advance sustainable agricultural practices.

Using sensors and IoT devices to monitor and identify infestations of pests in crops is one of the key strategies in smart pest management. These sensors can detect a wide range of factors—light, humidity, and temperature—all of which are critical for the growth and development of pests. Farmers can identify infestations of insects promptly and take preventive action before the pests seriously harm the crops through analyzing this data.

Another strategy for intelligent pest management is the use of precision agriculture techniques, which include the variable rate application (VRA) of pesticides. The VRA determines the optimal pesticide dosage required in a given field area by combining device data with information from other sources. By ensuring that pesticides are only used where they are necessary, this lowers the overall consumption of chemical substances and lessens their negative effects on the environment.

The physiological control, which employs the utilization of natural creatures such as parasites and predators to control pest populations, is another potential pest control strategy that is supported by smart agriculture. Since no toxic substances are used in this process, it is ecologically sound and beneficial to the environment.

In conclusion, smart agriculture offers innovative solutions for pest management that promote sustainable agriculture, reduce the use of harmful chemicals, and increase crop yields. By utilizing technology such as sensors, IoT devices, and precision agriculture techniques, farmers can effectively manage pests and ensure the health and productivity of their crops.

15.2.2 Harvesting

Automation and technological advances are used during the harvest in smart agriculture to increase the process' precision and effectiveness. Machine vision, which uses recording devices and computational methods for image processing to recognize and categorize crops according to their stage of development and excellence, is one of the main innovations employed by smart harvesting.

Robotics, which may computerize the tedious task of grabbing and separating crops, is another crucial component of smart harvesting. Robots can pick strawberries or tomatoes, for instance, by recognizing them and using an automatic limb to carefully grab them.

Smart harvesting can also include the deployment of devices to track crop development and maturity together with machine vision and robotics. For instance, depending on the moisture level of the crops, sensors for moisture might be used to choose the best time for harvesting.

Overall, smart harvesting in agriculture can lead to increased efficiency, reduced labor costs, and improved quality and consistency of harvested crops.

15.2.3 Integrated Precision Agriculture Systems

In order to increase the efficacy as well as effectiveness of agricultural practices, integrated precision agriculture systems (IPAS) integrate various technologies, including sensors, unmanned aerial vehicles, robots, and GPS technology. In order to help producers, make decisions based on data that maximize resource use, production, and quantity, IPAS aims to give them access to real-time knowledge and insights about the current situation and requirements of their agricultural products, soil, and circumstances.

Integrated precision agriculture systems include various components and functionalities, such as:

1. **Sensing and monitoring:** To gather data regarding the soil, water, the environment, and crops, IPAS makes use of a variety of sensors. These sensors can be mounted on weather stations, affixed to drones, or buried in the ground. They collect data on various parameters such as temperature, moisture content, relative humidity, pH, levels of nutrients, and infestations of pests.
2. **Data analytics and decision support:** To process and evaluate the information gathered by the sensors, IPAS makes use of machine learning algorithms and data analytics. These algorithms can detect emerging trends, patterns, and abnormalities in the data, as well as offer insights regarding the well-being, potential yield, and growth of crops. Yields of crops and demand from consumers can all be predicted using this information, which may also be utilized to optimize pest control, fertilization, and irrigation techniques.
3. **Precision application:** IPAS applies resources like fertilizer, pesticides, and water primarily where and when they are required by using precision agriculture techniques. This lessens waste and the negative effects that agriculture has on the environment. Drones, for instance, can be deployed to spray pesticides on particular plants or regions, and technology with variable rates can be utilized for applying fertilizer in different quantities according to the nutrient levels of the soil.
4. **Robotics and automation:** To carry out operations like planting, harvesting, and pruning, IPAS uses automation and robots. As a result, agriculture needs less labor and can operate more accurately and efficiently.

All things considered, IPAS is a thorough method of smart agriculture that combines a number of technologies to give farmers access to real-time data and insights. This empowers farmers to make well-informed decisions and achieve their methods for sustainability and maximum yield.

15.3 IoT-Based Smart Agriculture

Using IoT-based smart agriculture, different aspects of crop cultivation are monitored and managed by utilizing IoT devices and technologies. These Internet of Things (IoT) devices are equipped with actuators and sensors to gather data on the moisture content of the soil, temperature and relative humidity of the air, light levels, and other environmental factors that affect plant growth which is illustrated in Figure 15.1. This data is subsequently transmitted to a central repository or a system that uses the cloud for analysis in order to make decisions about the process of fertilization, supply of water, and other aspects of crop management.

Smart agriculture powered by the IoT can help farmers optimize resource consumption—like water and nutrients—while reducing waste and environmental impact. One of its key benefits is this. By using real-time data and analytics, farmers are better equipped to determine when and how much irrigation or crop nutrition is needed. Higher crop yields, improved crop quality, and decreased crop costs could result from this.

Moreover, IoT devices can be used to automate fertilization and irrigation, two common agricultural management tasks. For instance, intelligent irrigation systems can be set up to water crops automatically when soil moisture levels drop below a predetermined point, and fertilizer can be applied precisely according to each crop's needs.

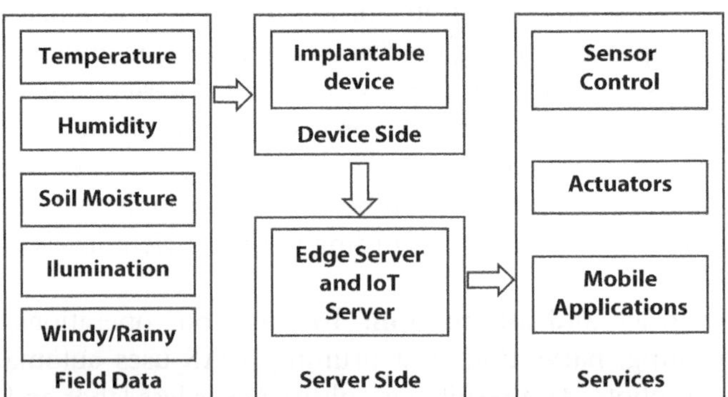

Figure 15.1 Concepts of smart agriculture.

The IoT-based smart agriculture, in general, has the potential to drastically change how crops are grown and managed by providing farmers with up-to-date data and knowledge that can help them make better decisions and boost their overall productivity and profitability.

It sounds like a highly intelligent irrigation system with a high level of sophistication that can make decisions based on a variety of environmental factors. Through the integration of an ontology and a machine learning model, the system can take into account a multitude of factors and generate decisions that are customized to meet the unique requirements of every crop.

The system can learn more about the traits of various crops and the particular soil and climate that support their growth with the aid of the ontology. By doing this, the system will be better able to decide when and how much water to give each crop. Figure 15.2 depicts the hardware based system which illustrates about the various hardwares that are used in smart agriculture.

The real-time data that the network of sensors can provide on temperature, humidity, and levels of soil moisture can be used to modify the irrigation schedule as necessary. By using a machine learning model, the system is able to learn from past data and forecast how different crops will respond to different environmental conditions. This technique can reduce water waste, have a negligible negative impact on the environment, and significantly improve crop yield and the effectiveness of irrigation.

The proposed design of IoT, comprised of four layers—application, processing, transport, and perception—offers a comprehensive and sophisticated way to implement an IoT system. The processing layer is an additional layer that offers a more sophisticated way of organizing and interpreting the data collected from the perception layer.

The perception layer of the architecture is equipped with sensors that measure temperature, relative humidity, and moisture content of the soil. This data is received by the transport layer after it has been sent over a range of networks, such as wireless, LAN, 3G, and 2G. Data must be moved from the layer that handles transport to the processing layer, in which databases, edge computing, and cloud computing will be used as tools for processing, storing, and analyzing the data.

Application-specific services, like an Android application that can assist in monitoring and controlling the IoT system, are supplied to end users by the application layer. Your architecture can offer a more complete and intelligent IoT system by integrating these layers. The suggested architecture uses cutting-edge tools and methods for data gathering, storing, and analysis to offer a more complex method of putting IoT systems into practice. In the end, this will benefit end users by enhancing the accuracy and efficiency of IoT systems.

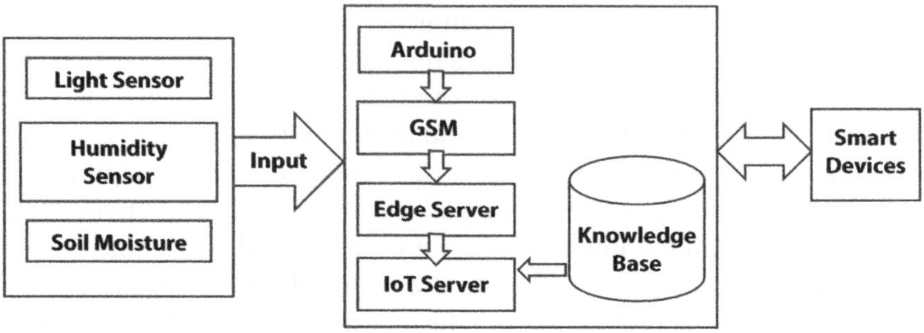

Figure 15.2 A hardware-based system.

15.3.1 Sensor Data

What has been described sounds like a standard workflow for an IoT system. The sensors, actuators, and microcontroller that make up the perception layer oversee gathering the temperature, humidity, and soil moisture data. After that, the data is sent to the transport layer, which may use a number of different communication protocols to send it to the processing layer.

In the processing layer, the data is stored and analyzed to derive meaningful insights. This layer may use a combination of cloud computing and edge computing technologies to process the data efficiently. With the data processed, the system can generate a schedule for watering crops, supervise the system, and provide suggestions for optimizing crop growth and yield.

Overall, the system you describe seems to be well-designed and capable of providing valuable insights for optimizing crop growth and yield. By leveraging IoT technologies, it can automate many of the processes involved in smart agriculture, reducing labor costs, and improving efficiency.

15.3.2 Semantic Knowledge Base

The system is made to gather sensor data from various angles within a sizable agricultural area. This can offer a more thorough perspective of the environmental factors influencing crop development and production. The sensor information you mentioned, including relative humidity, temperature, and brightness, is obtained at the yard level, whereas information on soil moisture is determined at the region's level.

To make sense of this data, you use a knowledge representation format called resource description framework (RDF). The RDF is a typical framework for describing and exchanging data on the web, and it can be used to represent data in a structured and standardized way. In addition to the sensor data, you also incorporate knowledge from an ontology that describes the crop species, climate types, and soil types in the area.

The control agent is responsible for processing this data and making decisions about watering the crops. It receives information about the plant requirements for water in specific soil textures, which can help it make more informed decisions about when and how much to water the crops. By combining all of this information, your system can provide more targeted and effective irrigation, which can improve crop yield and reduce water waste.

15.4 KNN-Based Smart IoT System

It appears that you are forecasting your crops' water needs using machine learning techniques. They specifically mention the supervised machine learning algorithm k-Nearest Neighbors (KNN), used for problems with classification.

In KNN, the algorithm uses the entire dataset to predict the class label of an unseen instance. It does this by finding the k closest data points to the instance and assigning the class label that is most common among those k neighbors. In your case, it sounds like you are using KNN with k=5, meaning the algorithm will look at the 5 closest data points to the instance to make a prediction.

Ideally, you suggest employing three different methods of feature selection (CBFS, FPRS, and KFRS) to choose the most crucial features from the data set in order to increase the precision of the predictions. The following may assist the information become less dimensional and boost the effectiveness of the machine learning model.

Overall, using machine learning to predict water requirements can help to optimize irrigation practices and reduce water waste. By accurately predicting when and how much to water your crops, you can improve crop yield and reduce the environmental impact of agricultural practices.

15.4.1 K-Nearest Neighbors Algorithm

Step 1: Determine the Euclidean distance that lies among each existing point Pn in the source dataset S and every new point of data X (4 features used for predicting the the consequence class, A, B, C, and D):

$$\text{Euclidean distance} = \sqrt{(XA-PA)^2+(XB-PB)^2+(XC-PC)^2+(XD-PD)^2}.$$

Step 2: Decide on "k," or the number of closest neighbors to the new data X. K=5.
Step 3: To determine the category of test data X, total up all "k" neighbors' votes.
Step 4: The results of the test X that received the most votes should be given to that class.

15.4.2 Smart IoT System Using KNN

The suggested smart IoT system makes use of sensors to collect environmental data and a module that supports GSM to send that data to an edge server. Specifically, using SIM808 GSM module with a data SIM card to facilitate real-time data transportation.

In terms of sensors, you mention using a hygrometer sensor for measuring soil moisture and an AM2302 DHT22 sensor depicted in Figure 15.3 for measuring moisture from the air, as well as temperature and humidity [19]. The electrically conductive properties of the soil, which is directly

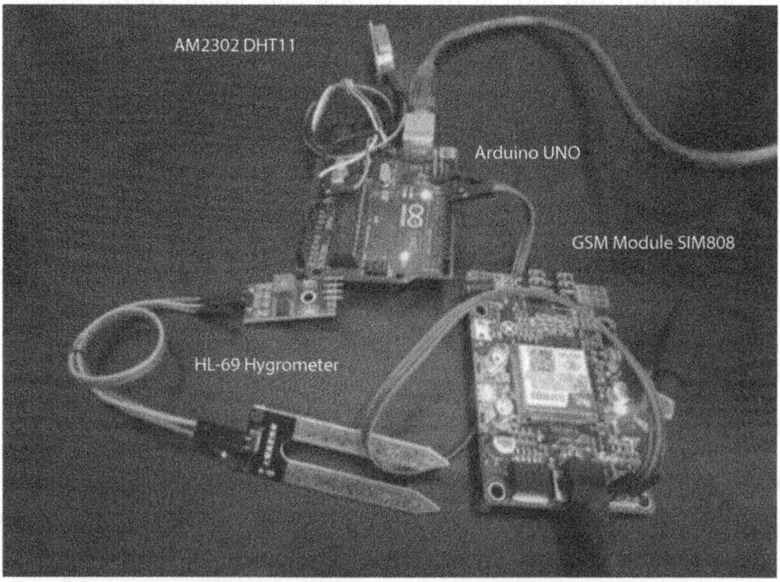

Figure 15.3 The AM2302 DHT22 sensor.

correlated to the soil's moisture, is measured by the hygrometer sensor in order to function. The AM2302 DHT22 sensor measures temperature with a thermistor and humidity levels with a conductive humidity sensor.

Using these sensors, you can collect data on soil moisture, temperature, and humidity, which can then be used to make informed decisions about when and how much to water your crops. Ideally, you can optimize water usage and make real-time irrigation decisions by utilizing a server at the edge for processing and analyzing the data [9].

15.4.2.1 HL-69 Soil Hygrometer Sensor

The HL-69 Soil Hygrometer Sensor is a type of soil humidity sensor used to gauge the water content of the soil as shown in Figure 15.4. The analog voltage that outputs corresponding to the soil's humidity content is provided by the sensor. Two metal probes are inserted into the soil, and the sensor measures the resistance between them. The amount of resistance between each of the probes decreases as the moisture content of the soil rises, leading to an increase in the output voltage. On the other hand, when the moisture content drops, the resistance that exists between the probes rises and lowers the output voltage.

The operating voltage range of the HL-69 Soil Hygrometer Sensor is 3.3V to 5V, and in order to read the output voltage, it needs an analog input pin. The sensor has a solid bolt hole on the digital board that makes installation simple. Two LED lights are also included; one for power indicator and the other for output indicator of digital switching. The sensor is widely used in applications where water content in the soil monitoring is essential

Figure 15.4 The HL-69 soil hygrometer sensor.

for the growth of plants and health, such as agriculture and greenhouse monitoring.

15.4.2.2 AM2302 DHT11 Sensor

The AM2302 DHT22 sensor and other digital sensors for temperature and humidity are widely used in IoT systems to monitor the environment. It uses a thermistor and a capacitive moisture sensor, respectively, to measure the temperature and relative humidity of the air surrounding it. The sensor's digital signal can be decoded by a microcontroller such as an Arduino.

The DHT22 sensor which is depicted in Figure 15.5 can function in a range of environments due to its temperature detection range of -40 to 125°C and the humidity metrics ranging from 0 to 100% RH. Because of its accuracy, which is within 2% RH and 0.5°C, this sensor can be trusted for applications requiring precise environmental monitoring. The sensor is a great option for IoT applications where electrical consumption is an issue because it runs on 3-5V DC and uses very little power.

15.4.2.3 BH1750 FVI Light Sensor

A digital light sensor called the BH1750FV shown in Figure 15.6 gauges the amount of ambient light present in a space. The detected light intensity is converted into a digital signal that can be read by a microcontroller using a 16-bit AD converter. The sensor can be used in a variety of lighting conditions due to its wide measurement range, which spans from 1 lux to 65535 lux.

Figure 15.5 The AM2302 DHT11 sensor.

BH1750 light sensor

Figure 15.6 The BH1750 light sensor.

With a straightforward I2C interface that facilitates communication with microcontrollers, the BH1750FVI sensor is user-friendly. Because of its small form factor and low power supply range of 3.3V to 5V, it can be used in small IoT devices. Many applications can make use of it, including backlight control for LCD screens, automatic lighting control, and outdoor light detecting for weather observatories.

15.5 Results and Discussion

The proposed tunnel farming watering system is so clever that it builds on and makes use of real machine learning decision-making support. The section before this one contains information on the system's architecture and hardware. To assess how the proposed system would respond, all sensors were placed in the actual field. The GSM module and an Android application are used to transfer data to the edge server, while a farmer can also view the results. The opening in the valve can then be opened or shut by a user acting in some way.

15.6 Performance Evaluation

This research study presents the results of comprehensive tests conducted on a randomly selected sample dataset, consisting of approximately 600 instances.

Table 15.1 Accuracy values of KNN and Naïve Bayes for number of instances = 300, 400, 500 and 600.

Sl. no	Number of instances	Accuracy rate in %	
		ANN	Naïve Bayes
1.	300	79.19	76.04
2.	400	79.72	74.90
3.	500	81.79	75.63
4.	600	84.63	76.89

Table 15.2 Analysis by descriptive statistics of mean, standard deviation, standard error mean values for KNN and Naïve Bayes with five optimal accuracy values.

Algorithm		N	Mean	Std. deviation	Std. error mean
Accuracy	KNN	10	84.63	4.92649	1.55789
	Naive Bayes	10	76.89	2.273447	0.86472

Table 15.3 The t-test for independent samples with a 95% confidence interval and a significance level of 0.047 (KNN significantly outperformed Naïve Bayes in terms of accuracy).

	Levene's test for equality of variances		t-test for equality of means						
								95% confidence interval of the difference	
Accuracy	F	Sig.	t	df	Sig. (2-tailed)	Mean difference	Standard error difference	Lower	Upper
Equal variances assumed	4.13	0.047	-10.8	18	0.00	-19.245	1.781	-22.98	-15.501
Equal variances not assumed	-	-	-10.8	14.06	0.00	-19.245	1.781	-23.06	-15.425

These two algorithms were executed with different number of datasets, accuracy was calculated. The accuracy values obtained for different number of datasets and depicted in Table 15.1.

This analysis provided descriptive statistical insights, including the mean, standard deviation, and standard error mean, which are presented in Table 15.2.

The difference in accuracy scores among KNN (84.63%) and Naïve Bayes (76.89%) is statistically significant.

A t-test via a sample that was independent was conducted to determine the t-test equivalent means, and the outcomes are displayed in Table 15.3.

15.7 Conclusion

The goal of the present study is to establish and open the door for cutting-edge computing and smart sensing in developing countries, where agriculture accounts for a large portion of their gross domestic product. Smart irrigation solution comprises three key modules. The first module encompasses a sensor network essential for monitoring various factors that impact the water requirements. To accomplish this, we have incorporated sensors like the DHT22 for temperature, the BH1750 for light, and the HL-69 hygrometer for soil moisture and air humidity. In the third module, we used an edge and fundamental IoT servers to enable smooth retrieval of information and transmit it by means of HTTP requests.

In the second module, we have employed the KNN algorithm on a representative dataset to train our model. This trained model plays a pivotal role in making informed decisions regarding water allocation. The system has been fully implemented within the Anaconda environment.

While our current system effectively utilizes KNN for decision-making, it is worth noting that alternative intelligent data-driven techniques can also be applied in the future. For instance, the presented irrigation system can be adapted to incorporate other decision-making methods such as random forests. Additionally, enhanced edge computing architecture by assigning the edge server the responsibility of processing data and generating results from the machine learning algorithm. In essence, deploying the trained KNN model on the edge server allows nearby devices to leverage its capabilities, significantly improving latency rates.

References

1. Cardenas-Lailhacar, M. D. Dukes and G. L. Miller, "Sensor-based automation of irrigation on Bermuda grass, during dry weather conditions," *J. Irrig. Drain. Eng.*, vol. 134, pp. 184–193, 2008.
2. H. Sattar, I. S. Bajwa, R. U. Amin *et al.*, "An IoT-based intelligent wound monitoring system," *IEEE Access*, vol. 7, pp. 144500–144515, 2019.
3. Joaquín Gutiérrez, Juan Francisco Villa-Medina, Alejandra Nieto-Garibay, and Miguel Ángel Porta-Gándara, "Automated Irrigation System Using a Wireless Sensor Network and GPRS Module", *IEEE Trans. Instrum. Meas.*, Vol. 63, No. 1, January 2014.
4. K. W. Migliaccio, B. Schaffer, J. H. Crane and F. S. Davies, "Plant response to evapotranspiration and soil water sensor irrigation scheduling methods for papaya production in south Florida," *Agricult. Water Manag.*, vol. 97, no. 10, pp. 1452–1460, Oct. 2010.
5. K. Xiao, D. Xiao and X. Luo, "Smart water-saving irrigation system in precision agriculture based on wireless sensor network," *Trans. Chin. Soc. Agric. Eng.*, vol. 26, pp. 170–175, 2010.
6. Khan, M.B. and J. Khan. 2020. An economic analysis of tunnel farming in enhancing productivity of off-season vegetables in district Peshawar. *Sarhad J. Agric.*, 36(1): 153-160.
7. Kumar, K. Kamal, M. O. Arshad, S. Mathavan and T. Vadamala, "Smart irrigation using low-cost moisture sensors and XBee-based communication," in *Proceedings of the Global Humanitarian Technology Conference (GHTC)*, San Jose, CA, USA, October 2014.
8. M. J. OGrady, D. Langton and G. M. P. O'Hare, "Edge computing: a tractable model for smart agriculture?" *Artif. Intell. Agric.*, vol. 3, pp. 42–51, 2019.
9. M. S. Munir, I. S. Bajwa and S. M. Cheema, "An intelligent and secure smart watering system using fuzzy logic and blockchain," *Comput. Electr. Eng.*, vol. 77, pp. 109–119, 2019.
10. M. Safdar Malik, I. Sarwar Bajwa and S. Munawar, "An intelligent and secure IoT based smart watering system using fuzzy logic and blockchain," *Comput. Electr. Eng.*, vol. 77, no. 1, pp. 109–119, 2018.
11. M. W. Rosegrant, X. Cai and S. A. Cline, "Global water outlook to 2025: averting an impending crisis," *Tech. Rep.*, International Food Policy Research Institute, Washington, DC, USA, 2002, 572-2016-39087.
12. N. Sales, O. Remédios and A. Arsenio, "Wireless sensor and actuator system for smart irrigation on the cloud," in *2015 IEEE 2nd World Forum on Internet of Things (WF-IoT)*, pp. 693–698, IEEE, Milan, Italy, December 2015.
13. Narayanamoorthy, "Economics of drip irrigation in sugarcane cultivation: case study of a farmer from Tamil Nadu," *Indian J. Agric. Econ.*, vol. 60, pp. 235–248, 2005.

14. Rai, P.K. An Eco-Sustainable Green Approach for Heavy Metals Management: Two Case Studies of Developing Industrial Region. *Environ. Monit. Assess.*, 2012, 184, 421–448.
15. Rai, P.K.; Kim, K.-H.; Lee, S.S.; Lee, J.-H. Molecular Mechanisms in Phytoremediation of Environmental Contaminants and Prospects of Engineered Transgenic Plants/Microbes. *Sci. Total Environ.*, 2020, 705, 135858.
16. Routray, S.K.; Javali, A.; Sharma, L.; Ghosh, A.D.; Sahoo, A. Internet of Things Based Precision Agriculture for Developing Countries. In *Proceedings of the 2019 International Conference on Smart Systems and Inventive Technology (ICSSIT)*, Tirunelveli, India, 27–29 November 2019; pp. 1064–1068.
17. S. Rawal, "IOT based smart irrigation system," *Int. J. Comput. Appl.*, vol. 159, no. 8, pp. 7–11, 2017.
18. Saab, M. Therese, I. Jomaa, S. Skaf, S. Fahed and M. Todorovic, "Assessment of a smartphone application for real-time irrigation scheduling in Mediterranean environments," *Water*, vol. 11, p. 252, 2019.
19. Tzounis, A.; Katsoulas, N.; Bartzanas, T.; Kittas, C. Internet of Things in Agriculture, Recent Advances and Future Challenges. *Biosyst. Eng.*, 2017, 164, 31–48.
20. Vermesan, O.; Friess, P. Internet of Things: Converging Technologies for Smart Environments and Integrated Ecosystems; River Publishers: Gistrup, Denmark, 2013.

Index

Also of Interest

Check out these other related titles from Scrivener Publishing

Ecosystem Management: Climate Change and Sustainability, Edited by Arnab Banerjee, Manoj Kumar Jhariya, Abhishek Raj, and Taher Mechergui, ISBN: 9781394231218. This book is essential for anyone who wants to understand the challenges of environmental degradation and learn about the sustainable solutions needed to address these critical issues.

Mathematical Modeling in Agriculture, Edited by Sabyasachi Pramanik, Niranjanamurthy M., Ankur Gupta, and Ahmed J. Obaid, ISBN: 9781394233717. The main goal of the book is to explore the idea behind data modeling in smart agriculture using information and communication technologies and tools to make agricultural practices more functional, fruitful and profitable.

Agroforestry, Edited by Abhishek Raj, Manoj Kumar Jhariya, Arnab Banerjee, Ramesh Kumar Jha, and Krishan Pal Singh, ISBN: 9781394231133. This book offers comprehensive insights into the management of agroforestry for livelihood security and sustainable development in the tropics, addressing ecological interactions, productivity, and the monetization of carbon credits, while also outlining a future roadmap and policy challenges.

Smart Agritech: Robotics, AI, and Internet of Things (IoT) in Agriculture, Edited by Santosh Kumar Srivastava, Durgesh Srivastava, Korhan Cengiz, and Pramod Gaur, ISBN:9781394302959. The main goal of Smart Agritech: Robotics, AI, and the Internet of Things (IoT) in Agriculture is to explore how emerging technologies such as robotics, artificial intelligence (AI), and IoT can be leveraged to improve efficiency, sustainability, and productivity in agriculture.

DIGITAL AGRICULTURAL ECOSYSTEM: Revolutionary Advancements in Agriculture, Edited by Kuldeep Singh and Prasanna Kolar, ISBN: 9781394242931. The book comprehensively explores the dynamic synergy between modern technology and agriculture, showcasing how advancements such as artificial intelligence, data analytics, and smart farming practices are reshaping the landscape to ensure food security in the era of climate change, as well as bridging the gap between cutting-edge research and practical implementation.

THE DIGITAL AGRICULTURAL REVOLUTION: Innovations and Challenges in Agriculture through Technology Disruptions, Edited by Roheet Bhatnagar, Nitin Kumar Tripathi, Nitu Bhatnagar and Chandan Kumar Panda, ISBN: 9781119823339. The book integrates computational intelligence, applied artificial intelligence, and modern agricultural practices and will appeal to scientists, agriculturists, and those in plant and crop science and management.

HANDBOOK OF AGRICULTURAL BIOTECHNOLOGY: Volume I Nanopesticides, Edited by Charles Oluwaseun Adetunji and Julius Kola Oloke, ISBN: 9781119836148. The book provides detailed information on the application of nanopesticides for the management of numerous agricultural pests and pathogens, as well as to foster an innovative pathway toward future sustainable agriculture and food systems.

HANDBOOK OF AGRICULTURAL BIOTECHNOLOGY: Volume II Nanobioherbicides, Edited by Charles Oluwaseun Adetunji, Julius Kola Oloke, ISBN: 9781119836155. The book provides a detailed examination of the application of nanobioherbicides that come from plants including information on the different metabolites derived from numerous plants that could become bioherbicides.

HANDBOOK OF AGRICULTURAL BIOTECHNOLOGY: Volume III Nanofungicides, Edited by Charles Oluwaseun Adetunji and Julius Kola Oloke, ISBN: 9781119836162. This book provides state-of-the-art information about recent advancements in the application of nanobiofungicides for effective management of post-harvest diseases and agricultural fungal diseases, including blights, mildews, molds, and rusts.

HANDBOOK OF AGRICULTURAL BIOTECHNOLOGY: Volume IV Nanoinsecticides, Edited by Charles Oluwaseun Adetunji and Julius Kola Oloke, ISBN: 9781119836179. The book provides detailed information about the application of repellent products that contain plant-based ingredients known as nanobioinsecticides, including the pesticide evaluation scheme guidelines for repellent testing, relevant information about the procedures to evaluate several repellent compounds, the development of new products that offer high repellency, and guidelines for consumer safety.

ECORESTORATION FOR SUSTAINABILITY, Edited by Arnab Banerjee, Manoj Kumar Jhariya, Surendra Singh Bargali, and Debnath Palit, ISBN: 9781119879718. With climate change and land degradation advancing at an alarming rate, the state of the global ecosystem and how it can be restored in our march toward sustainability is incredibly important, and this new, groundbreaking volume tackles these important issues head on.

FUNCTIONAL POLYMERS IN FOOD SCIENCE, Part 1: Food Packaging, Edited by Giuseppe Cirillo, Umile Gianfranco Spizzirri, and Francesca Iemma, ISBN: 9781118594896. Provides an extensive overview of the link between polymer properties and food quality. NOW AVAILABLE!

FUNCTIONAL POLYMERS IN FOOD SCIENCE, Part 2: Food Processing, Edited by Giuseppe Cirillo, Umile Gianfranco Spizzirri, and Francesca Iemma, ISBN: 9781118595183. Provides an extensive overview of the link between polymer properties and food quality. NOW AVAILABLE!

ADVANCES IN FOOD SCIENCE AND TECHNOLOGY Volume 1, Edited by Visakh, P.M., Sabu Thomas, Laura B. Iturriaga, and Pablo Daniel Ribotta, ISBN: 9781118121023. Written in a systematic and comprehensive manner the book reports recent advances in the development of food science and food technology areas. NOW AVAILABLE!

ADVANCES IN FOOD SCIENCE AND NUTRITION Volume 2, Edited by Visakh P.M., Laura B. Iturriaga, and Pablo Daniel Ribotta, ISBN: 9781118137093. This important book comprehensively reviews research on new developments in all areas of food chemistry/science and nutrition. NOW AVAILABLE!

Applied Water Science Volume 1: Fundamentals and Applications, Edited by Inamuddin, Mohd Imran Ahamed, Rajender Boddula and Tauseef Ahmad Rangreez, ISBN 9781119724766. Edited by one of the most well-respected and prolific engineers in the world and his team, this is the first volume in a two-volume set that is the most thorough, up-to-date, and comprehensive volume on applied water science available today. COMING IN SUMMER 2021

Sustainable Water Purification, by Safiur Rahman and M. R. Islam, ISBN 9781119650997. This is the only book that takes a zero-waste approach to propose 100% sustainable water purification techniques. NOW AVAILABLE!

Books in the series, "Bioprocessing in Food Science"

NUTRACEUTICALS FROM FRUIT AND VEGETABLE WASTE, Edited by Vidisha Tomer, Navnidhi Chhikara, Ashwani Kumar, and Anil Panghal, ISBN: 9781119803508. Nutraceuticals from Fruit and Vegetable Waste, the latest volume in the series, "Bioprocessing in Food Science," is an essential reference on the valorization of fruit and vegetable waste and the state of the art for fruit and vegetable processing.

NUTRITIONAL SCIENCE AND TECHNOLOGY: Concept to Application, Edited by Tejpal Dhewa, Anil Kumar Puniya, and Anil Panghal, ISBN: 97811198089. This new volume in the groundbreaking series, "Bioprocessing in Food Science," Nutritional Science and Technology, is a comprehensive resource covering topics such as human nutrition, food safety concerns, and emerging technologies like nanotechnology, and it also addresses the need for innovative food products, contamination, and regulatory challenges in the global food market.

HARVESTING FOOD FROM WEEDS, Edited by Prerna Gupta, Navnidhi Chhikara, and Anil Panghal, ISBN: 9781119791973. An exciting new volume in the groundbreaking new series, "Bioprocessing in Food Science," Harvesting Food from Weeds focuses on the latest processes, industrial applications, and leading research on this important, evolving new area, for engineers, scientists, students, and other industry professionals.

MICROBES IN THE FOOD INDUSTRY, Edited by Navnidhi Chhikara, Anil Panghal, and Gaurav Chaudhary, ISBN: 9781119775584. This newest volume in the groundbreaking new series, "Bioprocessing in Food Science," focuses on the latest processes, industrial applications, and leading research on microbes in the food industry, for engineers, scientists, students, and other industry professionals.

NOVEL TECHNOLOGIES IN FOOD SCIENCE, Edited by Navnidhi Chhikara, Anil Panghal, and Gaurav Chaudhary, ISBN: 9781119775584. The third book in the series, "Bioprocessing in Food Science," covers the latest and greatest new technologies in food science.

FUNCTIONAL FOODS, Edited by Navnidhi Chhikara, Anil Panghal, and Gaurav Chaudhary, ISBN: 9781119775560. Presenting cutting-edge information on new and emerging food engineering processes, Functional Foods, the second volume in the groundbreaking new series, "Bioprocessing in Food Science," is an essential reference on the modeling, quality, safety, and technologies associated with food processing operations today.

THERMAL FOOD ENGINEERING OPERATIONS, Edited by Nitin Kumar, Anil Panghal, and M. K. Garg, ISBN: 9781119775591. Presenting cutting-edge information on new and emerging food engineering processes, Thermal Food Engineering Operations, the first volume in the new series, "Bioprocessing in Food Science," is an essential reference on the modeling, quality, safety, and technologies associated with food processing operations today.

Printed and bound by CPI Group (UK) Ltd, Croydon, CR0 4YY

24/08/2026

14947831-0005

Printed and bound by CPI Group (UK) Ltd, Croydon, CR0 4YY
24/08/2026
14947831-0005